ABOUT THE AUTHOR

Mr. Amit Kumar is currently pursuing his Ph.D. in Horticulture (Fruit Science) from CoH, SVPUA&T, Meerut, Uttar Pradesh. He qualified for ICAR-ASRB NET in Fruit Science in 2021. He completed his B.Sc. in Horticulture (Hons.) in 2019 from CoH, BUA&T, Banda, Uttar Pradesh, and his M.Sc. (Ag) in Fruit Science and Horticulture Technology in 2021 from CoA, OUA&T, Bhubaneswar, Odisha. Mr. Kumar was selected for the award of the ICAR-NTS (PG) Scholarship. He was also awarded the Best Researcher Award (2021), Best Master Thesis Award (2022), Best Research Scholar Award (2022), and Young Horticulturist Award (2023) by various Professional Societies. He has published 11 Research Papers, 07 Review Papers, 30 Popular Articles, 09 Book Chapters, 01 Book, and More than 30 Abstracts as well during his academic years. He has actively participated in various National/International Conferences, Trainings, Webinars and Seminars.

Dr. Rahul Dev completed his graduation from CSJM University in Kanpur. He went on to earn postgraduate and doctoral degrees in Horticulture from GBPUAT Pant Nagar and IARI, New Delhi, respectively. Rahul Dev achieved the Junior Research Fellowship (JRF) in Horticulture and secured the 2nd rank in the Senior Research Fellowship (SRF) in Fruit Science. In 2014, he was appointed as an ARS-Scientist by ASRB, New Delhi. He has accumulated a total research experience of approximately one decade in the field of Horticulture and Underutilized Horticulture Crop Genetic Resources. For nearly six years, Rahul Dev served as a scientist at the ICAR-CAZRI Regional Centre in Bhuj, Gujarat. He has published around 80 articles, including two books, and has served as an editor for various journals and magazines. Additionally, he actively participated as a member in organizing several conferences, faculty and farmers' training programs. He also attended a training program on "Cactus Pear Evaluation & Best-Agronomic Practices" in Jordan and a virtual training on Agrobiodiversity in a changing climate conducted by FAO and Sapienza University, Rome. Currently, Rahul Dev holds the position of Scientist (Senior Scale) in the Division of Crop Improvement at ICAR-VPKAS in Almora.

Mr. Ajay Kumar is currently pursuing his Ph.D. in Horticulture (Fruit Science) at the ICAR-Indian Institute of Horticultural Research, Bengaluru (Karnataka), which is an outreach campus of IARI. He obtained his bachelor's degree in Horticulture (Hons.) from BUA&T, Banda, Uttar Pradesh, and his master's degree from Dr. RPCAU, Pusa, Bihar. Mr. Kumar has received the Junior Research Fellowship (JRF) in horticulture and has successfully cleared the ICAR-ASRB NET in Fruit Science. Throughout his ac-

A Textbook on Advances Production Technology of Temperate Fruits

ademic journey, he has authored approximately 14 articles, including research articles, review articles, and popular articles, along with 2 book chapters and approximately 10 abstracts. In addition, he has actively participated in various National/International Conferences, Seminars, and training programs.

Mr. Sajeel Ahamad has completed his B.Sc. (Hon's) Horticulture with Gold Medal in 2019 from CoH, BUA&T, Banda, Uttar Pradesh and M.Sc. (Ag.) Postharvest Technology in 2021 from ICAR-IARI New Delhi. Presently he is pursuing his PhD. in Horticulture (Postharvest Management) from ICAR-IARI New Delhi. He qualified ICAR-ASRB NET in Fruit Science and Vegetable Science in the year 2021,2023 respectively. He has secured AIR 8th rank in JRF and 1st rank in SRF conducted by ICAR AIEEA (PG), ICAR AICE-JRF/SRF (PhD) in 2019 and 2021 respectively. He was selected for award of ICAR-JRF and SRF fellowship. He was also awarded Best Master Thesis award, Best Research Scholar Award and Young Horticulturist Award by various professional societies. He has published few research papers, articles, book chapters and presented his work in various national and international conferences and seminar.

Mr. Maneesh Kumar is currently pursuing his Ph.D. in Horticulture (Fruit Science) at G.B. Pant University of Agriculture and Technology in Pantnagar, Uttarakhand. In 2019, he completed his B.Sc. in Horticulture (Hons.) from Banda University of Agriculture and Technology in Banda, Uttar Pradesh. He also obtained his M.Sc. in Horticulture (Fruit Science) from G.B. Pant University of Agriculture and Technology in Pantnagar, Uttarakhand in 2021. Mr. Kumar was selected for the award of the ICAR- JRF (PG) Fellowship. In 2023, he successfully qualified for the ICAR-ASRB-NET in Fruit Science. Additionally, he has been honored with the "Student Excellence Award-2022" by GBPUA&T Pantnagar. He has published few research papers, articles, book chapters and presented his work in various national and international conferences and seminar.

A Textbook on Advances Production Technology of Temperate Fruits

Editors

Mr. Amit Kumar
Ph.D. Research Scholar
Department of Fruit Science,
College of Horticulture
Sardar Vallabhbhai Patel University of Agriculture and Technology
Meerut-250110, Uttar Pradesh, India

Dr. Rahul Dev
ARS-Scientist, Crop Improvement Division,
ICAR-Vivekananda Parvatiya Krishi Anushandhan Sansthan
Almora- 263 601 (Uttarakhand), India

Mr. Ajay Kumar
Ph.D. Research Scholar
Division of Fruit Crops
ICAR-Indian Institute of Horticultural Research
Hessarghatta, Bengaluru-560089
Karnataka, India

Mr. Sajeel Ahamad
Ph.D. Research Scholar
Division of Food Science & Post-harvest Technology
Indian Agricultural Research Institute
New Delhi-110012, India

Mr. Maneesh Kumar
Ph.D. Research Scholar
Department of Horticulture
G.B. Pant University of Agriculture & Technology Pant Nagar
U.S. Nagar-263145, Uttarakhand, India.

PMW

PMW
FU 75, Pitam Pura
New Delhi 110034
publishmywork23@gmail.com
Mob. 91-8447075807

ISBN: 978-81-96458-81-2

ICAR- Central Institute of Temperate Horticulture

(Ministry of Agriculture and Farmer's Welfare, Government of India)
Old Air Field, Rangreth-191 132, Srinagar, J &K, India
Phone: 0194-2305044/ 2305045, E-mail: director.cith@icar.gov.in

FOREWORD

Temperate fruits are an important component of a balanced and healthy diet, which help to maintain good health and preventing various diseases. Important temperate fruits/nuts include apples, pears, peaches, plums, apricot, cherries, nuts, olive, kiwi, berries, and underutilized temperate fruits. The nutritional composition of temperate fruits varies depending on the type of fruit, but generally, they are low in calorific value with higher fibre, vitamins, mineral nutrients and anti-oxidant properties. Therefore, it is essential to promote the cultivation and consumption of temperate fruits for a healthier and prosperous life.

India is blessed with diverse agro-climatic conditions that provide ample opportunities to produce a wide range of fruits. India, the second-largest fruit producer after China, has continuously expanded its fruit production over the past few years. The current production of fruits in India is 102.48 million tons, and the area is 6.93 million hectares, and the demand for temperate fruits has also increased significantly. Hence, the need for a comprehensive information on the production technology of temperate fruits has become more important.

I am pleased to introduce this valuable book on **"A Textbook on Advances Production Technology of Temperate Fruits"**. This book covers a wide range of aspects of fruits related to their general information, importance, uses, nutritional composition, origin, distribution, taxonomy, botany, flowering, pollination, fruit set, varieties, rootstocks, soil, climatic, propagation, land preparation, planting, orchard establishment, training, pruning, growth regulator, tree nutrition, water management, intercultural operations, orchard floor management, insect pests, diseases, maturity indices, harvesting, post-harvest management, physiological disorders, yield, future prospects, and relevant references.

The authors of this book have done an excellent job of compiling all the necessary information related to temperate fruit crops. The book is written in a very meaningful language which is easy to understand for students, teachers, researchers, and innovative fruit growers.

This book will serve as a valuable resource for all the stakeholders who are involved in temperate fruit production and will be going to play a vital role in the upliftment of technology-based knowledge in India. I congratulate the authors for their excellent compilation of work with hard work and zeal.

Best regards

Place: Srinagar

Date : 23.06.2023

[M. K. Verma]

Director

Preface

The horticulture sector is widely acknowledged as the most vibrant and sustainable division within global agriculture. It encompasses the comprehensive management of a diverse range of crops, including fruits, vegetables, flowers, plantation crops, spices, medicinal plants, and aromatic crops, both during the pre and postharvest phases. Fruit, specifically, holds a crucial position within horticulture, playing a significant role in driving our economy forward. In recent years, the demand for temperate fruits has increased significantly, leading to the need for advancements in production technology to meet this demand. This has resulted in the development of innovative techniques for fruit cultivation. The advances in production technology have not only increased the quality and quantity of temperate fruits but also led to higher yields and improved economic returns for growers. Therefore, it is crucial to continue to invest in the development and implementation of advances in production technology to meet the increasing demand for fruits and ensure a sustainable and prosperous future for the fruit industry.

This comprehensive guide provides in-depth information on all aspects of temperate fruit production. This book satisfies the demand for a quality reference cum edited book among students. The authors have done an exceptional job of compiling all the necessary information related to temperate fruit production technology. They have presented it clearly and concisely, making it easy to understand and apply. Every effort has been taken to assure the accuracy of the data gathered from a variety of sources, including people, organizations, institutions, and publications. We expect that this book will be a helpful resource for educators, professionals, and learners seeking employment in ICAR institutions, State Agriculture University, State Public Services/ UPSC/ARS/JRF/SRF/ASRB-NET along with entrance examinations for B.Sc., M.Sc., and Ph.D. Degree in Fruit Science and also very useful for progressive farmers, and horticultural extension workers, and boost farmers' incomes.

We are thankful to our all authors, contributors, friends, seniors, and juniors for their excellent work, and thank them for sharing their knowledge with us and for their moral support. We are also thankful to Publication to bring the finest format of this book.

We are confident that readers will greatly benefit from the content of this book in their various pursuits. We wholeheartedly welcome any suggestions or ideas to enhance its educational value. We strive to make this book an even more enriching experience for readers by incorporating educational features that promote deeper understanding and practical application of the concepts discussed. Your input and contributions are highly appreciated in this endeavor.

27 May 2023 **Editors**

Contents

1

Advances Production Technology of Apple

Amit Kumar*[1], Ajay Kumar[2], Sajeel Ahamad[3], Maneesh Kumar[4] and Vishal Gangwar[1]

[1]*Ph.D. Research Scholar, Department of Fruit Science, College of Horticulture, Sardar Vallabhbhai Patel University of Agriculture and Technology, Meerut-250110, Uttar Pradesh*
[2]*Ph.D. Research Scholar, Division of Fruit Crops, ICAR-Indian Institute of Horticultural Research, Hessarghatta, Bengaluru-560089, Karnataka*
[3]*Ph.D. Research Scholar, Division of Food Science & Post-harvest Technology, Indian Agricultural Research Institute, New Delhi-110012, India*
[4]*Ph.D. Research Scholar, Department of Horticulture, G.B. Pant University of Agriculture & Technology Pant Nagar, U.S. Nagar-263145, Uttarakhand*
**Email: amitworld701@gmail.com*

Introduction

The apple (*Malus x domestica* Borkh.) is a significant temperate fruit growing in India and is widely cultivated in many regions around the world with cold winters, frost-free springs, and warm summers. It is a deciduous tree that belongs to the Rosaceae family. It is a widely cultivated fruit tree and is known by various names across different regions. In India, apples are commonly referred to as "seb" in Hindi, "sev" in Bengali, "sevva pazham" in Tamil, and "seema palam" in Telugu. These names reflect the cultural and linguistic diversity of the country (Indian Council of Agricultural Research, 2020). The apple is a round fruit that comes in a variety of colors including red, green, and yellow. The apple tree, like most plants, has a specific chromosome number that determines its genetic makeup and characteristics. The chromosome number of the cultivated apple, *Malus domestica*, is 2n = 34 (Janick *et al.*, 1996). This means that each cell of the apple tree contains 34 chromosomes, with 17 pairs in total. Most of cultivated apples (2n=34) are diploids, and a (2n=51) are triploids. Understanding the chromosome number is crucial for breeding programs and genetic research as it

Fig. 1.1: Apple Fruit

helps in identifying genetic variations and selecting desired traits. Delicious group apples are widely popular and cover 50–70% of the land. The apple, which is the most popular table fruit in the world, is regarded as the king of temperate fruits. Its positive qualities include a pleasing look and a longer shelf life at room temperature. Except for the Kashmir Valley and a small portion of the Kullu Valley, the majority of apple acreage is on hill slopes. Unlike other fruits, apples can be grown effectively in sod culture as long as the apple plants are given more nutrients than is recommended to prevent nutritional competition between grass and apple trees. They require a cool climate with cold winters and moderate summers for the fruit to develop properly.

Importance, uses and composition

One of the most popular fruits in the temperate region of India is the apple, a crisp fruit with bright colors. The proverb "An apple a day keeps the doctor away" is certainly familiar to you. Apples are rich in essential nutrients, vitamins, and dietary fiber, making them a popular choice among consumers seeking a healthy and nutritious diet. The demand for apples has been steadily increasing due to the rising health consciousness and changing dietary patterns of the population (Agrawal *et al.*, 2019). Apple production plays a crucial role in India's agricultural sector and economy. The cultivation of apples contributes significantly to the income of farmers, especially in regions with favorable climatic conditions such as Himachal Pradesh, Jammu and Kashmir, and Uttarakhand. Apples are not only consumed fresh but also processed into various value-added products like juices, jams, jellies, and apple-based desserts, which further enhance their economic importance (Singh *et al.*, 2018). Therefore, the advancement of production technology in the apple industry is crucial to meet the growing demand, enhance productivity, and ensure sustainable cultivation practices.

Apples are not only delicious but also packed with essential nutrients. They are a rich source of dietary fiber, vitamins (such as vitamin C, vitamin A, and various B vitamins), and minerals (including potassium and calcium). Additionally, apples contain phytochemicals like flavonoids and polyphenols, which have been associated with numerous health benefits, including antioxidant and anti-inflammatory properties (Boyer & Liu, 2004). The composition of apples varies depending on the variety, growing conditions, and maturity stage. The peel of apples is particularly rich in fiber, while the flesh contains more water content. The flavor and texture of apples are attributed to their sugar and acid content, with some varieties having a higher sugar-to-acid ratio than others (Singh *et al.*, 2018).

Apples have versatile uses, both in their fresh form and as processed products. They are widely consumed as a nutritious snack and are incorporated into various culinary preparations. Apple is used in the production of juices, ciders, sauces, pies, and baked goods, adding a distinctive flavor and aroma. Additionally, apples are used in the pharmaceutical and cosmetic industries for the extraction of natural compounds and their potential health benefits (Agrawal *et al.*, 2019).

Nutrient composition

According to the USDA, the nutritive value of an apple (100g of edible portion) is as follows:

Table 1. Nutritive value of apple fruit/100 g edible portion

Contents	Amount with Unit	Contents	Amount with Unit
Water	85.56 g	Vitamin	C 4.6 mg
Energy	52 kcal	Thiamin	0.017 mg
Protein	0.26g	Riboflavin	0.026 mg
Total lipid (fat)	0.17g	Niacin	0.091 mg
Carbohydrate	13.81g	Vitamin B-6	0.041 mg
Fiber, total dietary	2.4g	Folate, DFE	3 µg
Sugars	10.39g	Choline, total	3.4 mg
Calcium, Ca	6 mg	Vitamin A, RAE	3 µg
Iron, Fe	0.12 mg	Carotene, beta	27 µg
Magnesium, Mg	5 mg	Cryptoxanthin, beta	11 µg
Phosphorus, P	1 mg	Lutein + zeaxanthin	29 µg
Potassium, K	107mg	Alpha-tocopherol	0.18 mg
Sodium, Na	1 mg	Vitamin K	2.2 µg
Zinc, Zn	0.04 mg	Fatty acids, total saturated	0.028 g
Copper, Cu	0.027 mg		

Source: USDA Nutrient Database: g = grams, kcal = kilocalorie, µg = micrograms; mg milligrams

Origin and Distribution

Apples are believed to have originated in the area that comprises South Western Asia, the Caucasus, Soviet Central Asia, and the Hindu-Kush Himalayan region. Eastern Turkey, southwest Russia region of Asia Minor is the hub of diversity for the genus Malus. Over a span of thousands of years, early farmers presumably improved apples through selection. The shrunken apples that Alexander the Great is credited with bringing back to Greece from Asia Minor in 300 BC may have been the ancestors of dwarfing rootstocks. Apples were introduced to North America by colonists in the 1600s, and the earliest apple orchard on this continent is thought to have been located in or around Boston in 1625. Originally from New England, apples made their way west during the 1700s and 1800s with settlers, John Chapman (also known as Johnny Appleseed), and missionaries. Captain Lee introduced the apple to India in the Kullu Valley of Himachal Pradesh in 1865. Later, in 1917, American missionary Mr. Satya Nand Stokes introduced red-colored Delicious group varieties in Kotgarh in the Shimla district of Himachal Pradesh.

Malus sieversii is the wild parent of *Malus domestica* that evolved as a result of accidental hybridization between diverse wild species. Instead, it appears that a single species, which is still present in the Ili Valley on the northern slopes of the Tien Shan mountains at the border between northwest China and the former Soviet Republic of Kazakhstan, is the most likely ancestor of the modern apple. The DNA composition of leaves collected from trees in this region revealed that they all belonged to the species *Malus sieversii* and shared certain genetic sequences with *M. domestica*.

Malus baccata and *Malus sylvestris* are two other species that were formerly believed to have contributed to the genome of domestic apples, although there is no concrete proof for this in older apple cultivars.

Area and Production

Apple production is dominated by China, along with the USA, Turkey, Poland, India, and Italy as other major apple-producing nations. In India, Jammu and Kashmir, Himachal Pradesh, and Uttarakhand are the three states where apple is commercially grown. In North-Eastern states like Arunachal Pradesh, Sikkim, Nagaland, Meghalaya, Manipur, and the Nilgiri Hills of Tamil Nadu, it is also grown on a small scale. Apple trees cover 313 thousand hectares in India, producing 2276 thousand MT in total (Anon, 2020).

Taxonomy

The Rosaceae family of flowering plants includes the genus (*Malus*) "an apple," which contains about 30-35 species of smaller deciduous trees or shrubs. The name also refers to the fruit of these trees, particularly that of the domesticated orchard or table apple species *Malus domestica*. The other species are typically referred to as "crabs," "crab apples," "wild apples," or "crabs" because of their typically small, sour, and unpleasant fruit. The genus is indigenous to Europe, Asia, and North America in the temperate region of the Northern Hemisphere.

Domestic apples, a species of *Malus*, are capable of free hybridization. A globose pome, the fruit ranges in size from 1-4 cm in most wild species to 6 cm in *M. pumila*, 8 cm in *M. sieversii*, and much larger in cultivated orchard apples. One to two (rarely three) seeds are contained in each of the five star-shaped carpels that make up the fruit's central portion. One variety of *Malus trilobata*, native to southwest Asia, contains fruit with many structural variations and leaves with three to seven lobes, which superficially resemble maple leaves. It is frequently referred to as *Eriolobus trilobatus*. Here are some important apple species

Table 2. Apple species

Species	Traits
M. sieversi	The wild ancestor of domesticated apple
M. Floribunda	Small fruited spp.
M. sieversii	Sucker-free spp.
M. sieboldii, M. sikkimensis, M. toringoides & M. hupehensis	Apomictic spp.
M. baccata var. himalacia & M. sikkimensis	Indigenous to Himalaya
M. floribunda & M. atrosanguinea (Vf)	Resistant to scab
M. zumi	Resistant to PM and codling moth
M. hupehensis and MM series	Resistant to wooly aphids

Apple is classified into the following categories-

Table 3. Taxonomy of apple

Rank	Scientific Taxonomical Details
Kingdom	Plantae
Phylum	Magnoliophyta
Class	Magnoliopsida
Order	Rosales
Family	Rosaceae
Sub-family	Pomoideae
Genus	Malus
Species	domestica
Basic chromosome	17

Botany

Plant Growth Habit

Apple trees are deciduous, perennial plants that are small to medium-sized trees with a spreading canopy that can grow to 30 feet in the wild but is typically 6 to 15 feet tall when grown. Rootstock and training systems have a significant impact on tree size and shape. Ellipsoid, dark green leaves with serrated margins and mild pubescence on the underside. They are typically propagated by grafting.

Flowers

The "pink" bloom stage results from the fact that petals are initially white when open but have red-pink undersides. The inferior ovary, which is embedded in the floral cup or hypanthium, has five locules, typically housing two ovules per locule. The inflorescence is a cyme of 4-6 flowers, with the center flower opening first. Known as the "King bloom," the center flower can give rise to a fruit that is larger than those of the other flowers. On spurs and to a lesser extent on long stems, flowers are produced terminally from mixed buds. Spurs typically only grow a tenth of an inch every year and form on wood that is at least two years old. The most efficient pollinators are honey bees.

Fruits

The pome is a unique fruit category that is given to apples and related fruits. Instead of the ovary, the hypanthium or floral cup is where the majority of the fleshy edible component is derived from. Small, black, and very slightly toxic, seeds. Although some fruit may be produced in the second year, fruiting starts three to five years after budding. Depending on the dwarfing rootstocks more precocious and cultural techniques of excessive pruning delay flowering and fruiting.

Leaves

The leaves of the apple tree are simple, alternate, and lobed. They are dark green and have a glossy appearance.

Soil and climatic requirements

Apple grows most successfully on loamy soils that are abundant in organic materials. The ideal soil has a pH between 6.0 and 6.5, is well-drained, and is aerated. Deep, fertile soil free from hard substrates and wet situations is ideal.

The apple is a common fruit found in temperate climates and is typically farmed there. Its cultivation is thought to be best suited for a northern aspect. Although some low-chill apple cultivars only need 500–800 chilling hours, the majority of apple varieties need between 1000 and 1600 chilling hours to overcome their winter dormancy. During the active growing period, the average summer temperature should range from 21°C to 24^{0}C. Bee activity is restricted by low temperatures below 15°C, rain, and foggy weather during bloom; bee activity is fully suppressed below 50°C, which harms fruit set. Temperature18 to 22^{0}C is the ideal range for pollination, pollen germination, and fruit set. The best places for apple farming are those with a spring without frost and enough sunshine in the summer without significant temperature swings. The best rainfall for the season is regarded as 100–125 cm of evenly spaced rainfall. Long periods of drought during fruit development, as well as severe rain and cloudy weather as the fruit reaches maturity, reduce fruit size and quality.

Important Commercial Varieties

Apple cultivars are recommended in various Indian states.

Season and State	Early	Mid	Late
Jammu and Kashmir	Irish Peach, Benoni	American Mother, Razakwar, Jonathan(P) Cox" Orange Pippin(P), Queen"s Apple, Rome Beauty, Scarlet Siberian	King Pippin, American Apirouge, Kerry Pippin, Lal Ambri, Sunheri Chamure, Golden Delicious(P), Red Delicious, Ambri, Baldwin, Yellow Newton(P)
Himachal Pradesh	Tydeman"s Early Worcestor(P), Michael, Mollies Delicious, Schlomit, Starkrimson, Anna	Starking Delicious, Red Delicious, Rich-a-red, Vance Delicious, Top Red, Lord Lambourne(P) Red Chief, Oregon Spur, Red Spur, Red Gold(P), Silver Spur, Scarlet Gala, Well Spur, Super Chief, Scarlet Spur	Golden Delicious(P), Yellow Newton(P), Winter Banana, Granny Smith(P), Red Fuji, Coe Fuji
Uttarakhand	Early Shanburry(P), Fenny, Benoni, Chaubattia Princess	Red Delicious, Starking Delicious, Mclntosh(P), Cortland, Golden Delicious(P)	Rymer, Buckingham(P)

*Pollinizer (P)

Spur types and colour sports: Red Chief, Oregon Spur-II, Silver Spur, Well Spur, Red Spur, Super Chief, Starkrimson, Hardi Spur, Schelet Spur, Ace Spur

Standard colour mutants: Vance Delicious, Top Red, Skyline Supreme, Hardiman, Bright-N-Early

Standard varieties: Starking Delicious, Red Delicious, Rich-a-red

Low chilling varieties: Michal, Schlomit, Anna, Vered, Tamma, Tropical Beauty and Parlins Beauty.

Pollinizing varieties: Golden Delicious, Red Gold, Tydeman"s Early Worcester, Summer Queen, Golden Spur, Granny Smith, Winter Banana, Mclntosh, Scarlet Gala and flowering crabs like Manchurian, Snow Drift and *Malus floribunda*

Scab resistant varieties: Prima, Priscilla, Sir Prize, Florina, Flrdous, Shireen, Macfree, Coop-12, Coop-13, Liberty and Freedom.

Hybrid Varieties: Lal Ambri (Red Delicious x Ambri), Sunheri (Ambri x Golden Delicious) Chaubattia Princess and Chaubattia Anupam (Early Shanburry x Red Delicious), Ambred (Red Delicious x Ambri), Ambrich (Richared x Ambri), Amroyal (Starking Delicious x Ambri).

Intergeneric hybrid

Pamapple is a 1st Successful Pear x Apple Hybrid developed by Ellis Marks (1952) in John Innes Centre.

Important commercial varieties with their salient feature

Ambri: The Ambri Apple is the pride of Kashmir and is also known as Kashmiri Apple. North Indian reason widely consumes this variety due to its sweet aroma. It has a greenish and red exterior. It is medium to large, Ambri apples have a greenish-yellow with a reddish-pink blush and some minor striped patterns. The white flesh of the Ambri apple is creamy, crunchy, flavorful, and sweet

Macintosh: The Macintosh apple is a cultivar of the domesticated apple. The Macintosh apple is known for its distinctive sweet-tart flavor and crisp texture. It is typically red with a green background, but the color can vary depending on the growing conditions. The Macintosh apple is an early-ripening variety, typically harvested in late September to early October. It is a good choice for eating fresh, but it can also be used for cooking and baking, as well as for cider making. The fruit is medium to large in size and round in shape. Some of the features of this variety are its juicy and crisp texture, its sweet-tart flavor, and its early ripening.

Granny Smith: Granny Smith apples are a type of green apple known for their tart, crisp flavor, and firm texture. They are a popular variety for cooking and baking, as well as for eating fresh. They are also known for their long shelf life, making them a great choice for storing. Granny Smith apples are high in vitamin C and dietary fiber and are a good source of antioxidants. They are also relatively low in calories, making them a healthy snack option.

Honeycrisp: A newer variety that is known for its sweet taste, crisp texture, long storage capability and disease resistance.

Chaubattia Anupam: It is a hybrid variety found by the crossing of Red Delicious and Early Shanburry. Uttar Pradesh and Uttarakhand are the prime producers of this apple variety.

Golden Delicious: A yellow-skinned variety that is known for its sweet taste, crisp texture, and long storage capability.

Sunehari: Sunehari is a variety of hybrid apples, which comes from the crossing of Golden delicious and Ambri apples. It has crunchy and juicy flesh with a sweet-acetous taste.

Fuji: A Japanese variety that is known for its crisp texture, sweet taste, long storage capability, and disease resistance.

Gala: A hybrid variety that is known for its sweet taste, crisp texture, disease resistance and high yield.

Tydeman's Early: This variety comes from the hilly areas of J&K and Himachal Pradesh. It has a red colour with a yellowish-green base, making it more widespread. People use it in the form of desserts and fresh salads.

Red Delicious: Another popular variety, known for its bright red skin, sweet taste and long storage capability.

Fig. 1.2: Apple Varieties

Pink Lady: A pink-skinned variety that is known for its sweet taste, crisp texture, long storage capability, disease resistance and high yield.

Jonathan: A red-skinned variety that is known for its sweet taste, crisp texture, and disease resistance.

Royal Delicious: One of the most popular apple varieties in India, known for its sweet taste, crisp texture, and long storage capability.

Rome: A red-skinned variety that is known for its sweet taste, crisp texture, long storage capability and disease resistance.

CITH Lodh Apple-1

The Central Institute of Temperate Horticulture (CITH), Srinagar, identified the variety in 2012. It was created through clonal selection from the original Red Delicious plantation. It is a consistent-bearing bearing, a mid-season flowering cultivar that is fairly early. In terms of vigour and spreading, trees are average. Fruits are oblong-conical in shape and moderate to medium in size (185–190 g). The texture of the flesh is firm (8.5kg/cm2), crispy, and juicy with TSS 13.6 ^{0}B. The majority of fruits are born on thick, sturdy spurs as well as on shoots. Spur densities are average. Early to mid-season maturity is reached (117 days from full bloom to harvest). High yielding (28–32 t/ha at 10–12 years old). tolerant of serious infections and pests. Good shelf life. The variety is ideally suited for cultivation in the mid- to high-hills of the North Western Himalayan agro-climatic zones, where reduced fruit set on conventional high-chilling apple cultivars has been seen as a result of climate change.

HRMN-99

This variety can grow at a low elevation of 1800 feet above sea level and doesn't need chilling time. From a 7-year-old plant, the average yield is 1 quintal. Early June, ready for harvest (after three years of transplanting) and scab disease-tolerant.

Pusa Amartara Pride

Pusa Amartara Pride is an apple variety that is a result of crossing Royal Delicious and Prima apples. It is known for its larger fruit size compared to other varieties. The larger fruit size makes it visually appealing and can be advantageous for commercial purposes. This variety of apple also exhibits a higher TSS (Total Soluble Solids) content. TSS refers to the concentration of sugars, acids, and other dissolved solids in the fruit. A higher TSS content often indicates a sweeter taste and enhanced flavor profile. The elevated TSS in Pusa Amartara Pride can make it a desirable choice for those who enjoy a sweet and flavorful apple.

Shalimar Apple-1

Shalimar Apple-1 are apple varieties that were developed by the Sher-e-Kashmir University of Agricultural Sciences and Technology of Kashmir, India. These varieties are specifically bred to thrive in the unique climatic conditions of the Kashmir region.

Chaubattia Agrim

This variety is known for its suitability to the hilly regions of the state of Uttarakhand. It is selected for its adaptability to the local climate and soil conditions, ensuring better growth and productivity in that particular region.

Pusa Gold

Pusa Gold, a result of crossing Golden Delicious and Tydemann's Early Worcestor apples, is characterized by its early maturity. This means that the fruit of Pusa Gold reaches its full maturity earlier in the growing season compared to other apple varieties. Early maturity can be advantageous for commercial growers as it allows for an earlier harvest and potential market advantage.

Akbar

The Akbar apple variety is a result of crossing the Ambri and Cox's Orange Pippin apple varieties. It is specifically bred to enhance fruit quality. By combining the desirable characteristics of its parent varieties, Akbar aims to offer improved taste, texture, and overall fruit quality.

Firdous: Firdous is an apple variety that is developed by crossing three different varieties: Golden Delicious, Rome Beauty, and Malus floribunda. It is selected for its scab resistance. Scab is a fungal disease that can affect apple trees and reduce fruit quality. By incorporating the scab resistance trait from Malus floribunda and combining it with the desirable attributes of Golden Delicious and Rome Beauty, Firdous aims to provide growers with a variety that is both resistant to scab and offers appealing fruit qualities.

Shireen: Shireen is an apple variety resulting from crossing Lord Lam Bourne and Melba XR-1274078. It is specifically bred for its scab resistance. The scab resistance trait is derived from one of the parent varieties, Melba XR-1274078. By combining this resistance trait with the desirable attributes of Lord Lam Bourne, Shireen aims to offer growers an apple variety that is resistant to scab and maintains other desirable characteristics such as taste, texture, and appearance.

Redlane: The Redlane variety displays charming pink flowers and produces attractive and edible crab apples during the autumn season. This variety is self-pollinating, meaning it can fertilize itself without requiring additional pollinators.

Goldlane: Goldlane is a novel and distinct variety of the Malus domestica (Borkh.) apple tree. It showcases a columnar tree type with compact growth, primarily bearing fruit on spurs. Notably, it exhibits strong resistance against scab disease. Goldlane is also a self-pollinating variety, eliminating the need for external pollinators.

Sunlight: It is a white flower, a juicy, sweet red apple in the autumn. The growth of the plant is slow, and it does not possess self-pollinating capabilities. To facilitate pollination, it requires the presence of another apple tree nearby. Moonlight, Redlane, or Goldlane varieties are excellent options for cross-pollination. The circumference of Sunlight apple trees typically ranges between 80-100 cm.

Moonlight: Moonlight is a Czech apple tree variety introduced in 2010, known for its columnar growth habit. It produces medium-sized fruits that resemble Golden Delicious in shape and color. Moonlight exhibits resistance to both scab and powdery mildew diseases, making it a favorable choice for growers.

Propagation and Rootstocks

Apples are propagated using the asexual techniques of grafting and budding on rootstocks.

Rootstocks

Seedling Rootstocks

For growing seedling rootstocks, either crab apple (*Malus baccata*) seeds or seeds from self-pollinating apple cultivars like Golden Delicious and Granny Smith are utilized.

Clonal Rootstocks

Precocious, uniform, and resistant to some diseases and insect pests are clonal rootstocks. Developing in favor in recent years is size-controlling clonal rootstocks, particularly for the establishment of high density.

The rootstocks are either related species like *Malus sylvestris* (crap apple), *M. prunifolia*, or *M. sikkimensis* or their hybrid descendants, or seedling offspring of cultivated kinds. The primary objectives in producing rootstocks are either to dwarf the trees or to have resistance to soil temperatures above or below freezing, to woolly aphid, root knot nematodes, etc. The highest collar rot incidence was observed in M-

27 and M-26 and the least in indigenous root stocks (CITH-Apple rootstocks). Highest wooly aphid infection was found in M-9, B-9, M-27, EMLA-106 and ALNARP whereas least infestation noted in M-26 and indigenous apple rootstock (CITH-ARS-01, CITH-ARS-02, CITH-ARS-03, CITH-ARS-04, CITH-ARS-, CITH-ARS-09 and CITH-ARS-10) (Srivastava *et al.,* 2016). Malling Stocks were created in England at the East Malling Research Station. Rootstocks (MM) that are clonally propagated by either cutting or stooling were developed by the John Innes Horticultural Institute, Jointly Merton, and East Malling Research Station. These are a few of the significant rootstocks developed for certain purposes:

Rootstock	Category	Feature
Dwarfing	M9	Exclusively suitable for HDP in flat, irrigated locations
Semi-dwarf	M4, M7, MM106, M24	HDP-friendly and woolly apple aphid-resistant (*Eriosoma lanigera*)
Semi-vigorous	M-4, MM-111 and MM-104	Drought tolerance and woolly apple aphid resistant
Vigorous	M-2, M-13, Merton-793	Early fruiting and resistance to collar rot and woolly apple aphid
Ultra dwarf	M-27 (M13xM9)	Suitable for High-Density Planting

Propagation of Rootstock

Apple seeds need to be stratified (wet chilled) to emerge from dormancy. In alternate layers of wet sand at a temperature of 4-6°C from December to February, seeds stratify for 60–70 days. March is the time for sowing the stratified seeds in nursery beds, spacing them 8–10 cm apart from one another and 15-20 cm apart from one another. To prevent the desiccation of stratified seeds, the nursery beds are mulched with 10 cm of dry grass after sowing. When seeds begin to germinate, the mulch is removed. Regular cultural activities including hoeing, spraying fungicides and pesticides, irrigation, and weeding are all carried out. In one year, the seedling rootstock grows to a size that is graftable at 15 mm in diameter.

Clonal Rootstock

Commercial clonal rootstock propagation utilizes either mound layering or trench layering (Figures 1.3 and 1.4).

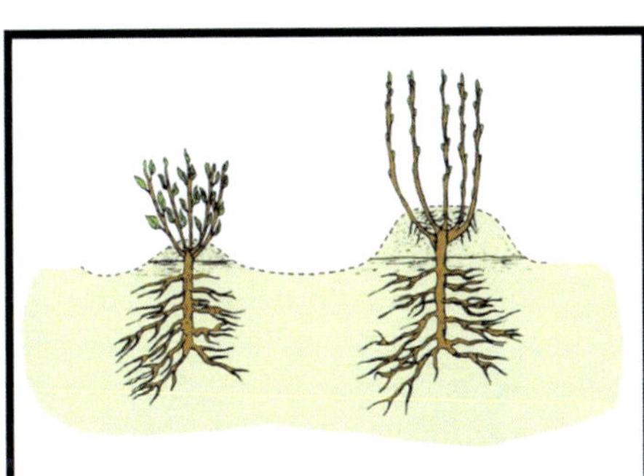

Fig. 1.3: Mound Layering

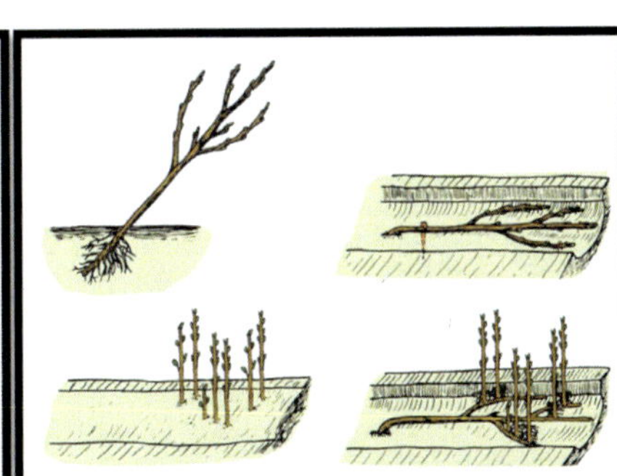

Fig. 1.4: Trench Layering

Propagation of Scion

Grafting and chip budding are the commercial methods of apple scion variety propagation. Scion wood from healthy, disease-free, true-to-type mother trees should be collected in January for grafting and budding. After properly marking the variety, the scion wood is harvested from one-year-old shoots and stored in moist sphagnum moss. These bundles of scion wood are kept cold until they are needed for grafting, or they are buried deep in the ground in a shaded area. With tongue and cleft techniques, February to March is the ideal time to graft apples. Chip budding is also possible in July and March (Figure 1.5).

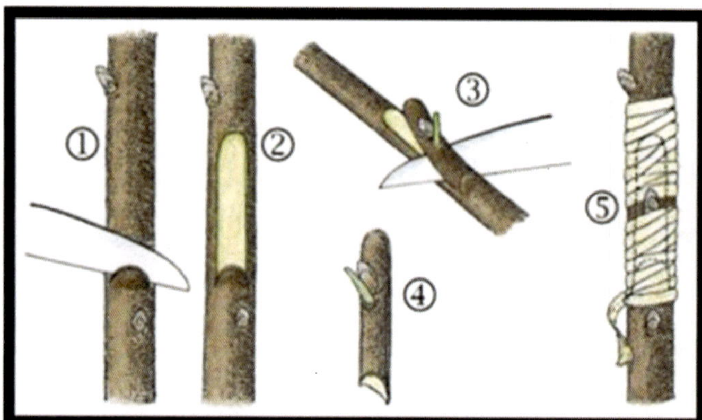

Fig. 1.5: Chip Budding

Planting and Orchard Establishment

Apples should be planted between January and February. The planting distance varies depending on the cultivar, the rootstock, and the soil fertility. The orchard layout needs to be prepared before planting. In level terrain, a square or hexagonal system of layout is used, whereas, in uneven ground, a terrace and contour system is used. Pits of 1x1x1 M size are dug after an orchard's planning and well before planting. At least a month before planting, the trenches are filled with soil with a mixture of 40–50 kg of well-rotten FYM and 1 kilogram of single super phosphate. After planting, tree basins are watered and mulched with 10 cm of dry grass, which helps maintain soil moisture and reduce weed growth. Since the majority of apple varieties are self-fertile, the orchard should be planted with the appropriate percentage of pollinizer kinds (25 or 33%) during planting time. Planting in an intense system is done at a size of 4M x 2.5M or smaller. Every two rows of the main variety require the usage of a full row of a pollinizer.

High-Density Planting (HDP) and Medium Density Planting (MDP)

Dwarfing rootstocks like M9, M4, MM106, and M26 are utilized commercially nowadays. To accommodate 2500 plants/ha, the grafts are planted at a close spacing of 2M x 2M. However, the HDP did not produce as much yield as the MDP, which allowed for a total of 1666 trees/ha (3M x 2M) of plants. MDP can produce an additional yield of 50% more than HDP. Additionally, the fruit quality and color in MDP are superior to HDP.

Ultra High-Density Plantation (UHDP)

Ultra High-Density Plantation (UHDP) is a method of apple cultivation that involves planting trees at a much higher density than traditional orchards. This can be achieved by using dwarf or semi-dwarf apple tree varieties, and by training the trees to grow in a specific shape, such as a central leader or a vase shape. The trees are planted much closer together than in traditional orchards, typically at a spacing of around 3-4 meters between trees, and 2-3 meters between rows. This results in a much higher tree density per hectare, with some UHDP orchards achieving densities of up to 4000 trees per hectare.

The main advantage of UHDP is that it allows for much higher yields per hectare, as the trees are planted closer together and are trained to grow in a specific shape that maximizes fruit production. This can lead to yields that are up to 3-4 times higher than traditional orchards. Additionally, UHDP orchards are much easier to manage, as the trees are smaller and more compact, and the orchard floor is much less shaded, making it easier to control weeds and pests. Another advantage of UHDP is that it allows for more efficient use of resources, such as water and fertilizer. As the trees are planted closer together, they compete for resources less, which means that they require less water and fertilizer to produce a given yield. This can result in significant cost savings for growers, as well as reduced environmental impacts.

UHDP apple orchards also have some disadvantages too. The main one is that they require more intensive management and pruning, as the trees are planted closer together, and the branches need to be trained and pruned regularly to maintain the desired shape and maximize fruit production. Additionally, the trees are more prone to diseases, pests and other issues, so they require more frequent monitoring and control measures. In conclusion, UHDP is a method of apple cultivation that involves planting trees at a much higher density than traditional orchards, resulting in higher yields and more efficient use of resources. However, it also requires more intensive management and pruning, as well as more frequent monitoring and control of pests and diseases.

Training and Pruning

The most crucial orchard management procedures for healthy canopy development and fruit production are training and pruning. When a plant reaches the bearing stage, training is done to shape or create a solid foundation for the trees to support the greatest yield. There are several training schemes for apples, including spindle bush, dwarf pyramid, and cylinder spindle, among others, but the modified central leader method is best for ordinary plantations.

Modified central leader system

Typically, a one-year-old whip plant that has no branches is planted during the dormant season. The plant's tip is automatically redirected 60 to 75 cm above the ground level immediately following transplanting. Most of the buds on the main branch will grow the following summer. The sprouting buds are trimmed off as soon as they appear to grow a clean stem up to 45 cm off the ground. The lowest of the three or four buds that are preserved is 45 cm above the ground and projects in opposite directions.

During dormant pruning, three to four major branches with a larger crotch angle are chosen if summer pruning is avoided. In a spiral pattern, the chosen branches should be placed 10-15 cm apart. Cut off any unattractive branches and branches that emerge lower than 30 cm from the ground. The chosen branches are pointing in the direction of a bud that is extending outward from the center of growth. The leader is moving backward until it is 30 cm above the final branch. 2-3 primary branches on the leader are chosen for the second dormant pruning and are spaced evenly apart. Two secondary branches that are developing outward should be chosen from the primary branches that were chosen the previous year. The chosen primary and secondary branches are pointed directly to the last third quarter of the growth. The secondary and tertiary branches chosen should be horizontally spreading, and branches that are growing upward or downward should be trimmed off. In the third year of training, undesirable branches are removed, and desirable side branches are headed back. The main branch should split off from a bud or weak shoot that will eventually become a secondary branch. The training should be finished by the fourth year.

Pruning

Maintaining a healthy balance between vegetative growth and spur development is the goal of pruning. The training is finished in the first 4-5 years following plant planting, however, pruning is continued after training for the duration of the tree's life. Thinning out and heading back are two fundamental aspects of pruning. All upright laterals and those growing inside the trees must be removed, and leaders and laterals must be headed back.

Apple fruit can also be seen on fruit buds on immature laterals in addition to spurs. To maintain a steady supply of fresh, healthy shoots, spurs, and branches, pruning should be done carefully. While pruning, some tree parts are removed while others are left alone. Unpruned areas will develop fruit buds whereas pruned areas will create shoot growth. Depending on the development, spur formation, and crop load, the laterals that were left unpruned in one year may be either shortened or left unpruned in the next year. The laterals must be heavily pruned until they are only 5 to 6 cm long after they have cropped and grown weak. The drastic shortening will encourage new growth from these stubs, which will provide a yield in two years. Every year, the entire cycle is therefore repeated to maintain normal growth, spur formation, and cropping. Dead, sick, and broken branches are chopped off during pruning, and Bordeaux or Chaubattia paste is applied to the cut surface to prevent fungus growth. In the dormant season, pruning is most effective (December to January).

Flowering

The study of flower growth is a significant field of study in plant sciences. In addition to being commercially significant, flowers, seeds, and fruit, the change from vegetative to reproductive growth is one of the crucial developmental stages in plant biology. The transformation of a bud from a vegetative to a floral stage marks the start of the process of developing fruit. Early in the summer is when this happens in apples, with the floral bud commonly differentiating around 100 days following bloom, however both earlier and later differentiation are possible.

Approximately 3-6 weeks following peak bloom, the commencement of flower primordia begins (June). Having five flowers, the inflorescence is fixed. In cymes, there are white, pink, or carmine flowers. Epigynous and hermaphrodite flowers are characteristic of most cultivars. The components of a flower include five petals, five sepals, 15–20 stamens, and a pistil with five carpals, each holding two ovules. inferior to the ovary (Figure 1.6).

Fig. 1.6: Stages of Flowering

Pollination, and Pollinizers

The majority of apple varieties are self-unfruitful, meaning they cannot develop fruit when fertilized by their pollen. As a result, cross-pollination and excellent fruit set require some compatible cultivars. Inadequate fruit set is frequently the result of a failure during the pollination Apple trees produce profuse bloom but do not set fruit, especially under unfavorable environmental circumstances. pollen generation, transmission, germination, pollen tube development, or ovule fertility throughout this period. Incompatibility and sterility are the two main reasons that apple trees fail to bear fruit. Poor pollination and fruit set are also caused by a wide range of climatic and edaphic conditions. When flowers are in bloom, low temperatures, rain, and overcast weather harm bee activity, pollen transport to the stigma, pollen germination, and ultimately fruit set.

Apple fruit set and cross-pollination can be increased by planting at least 25 to 33 percent of pollinizers, placing 5 to 6 honey bee colonies per hectare, and top-working 2-4 shoots of commercial cultivars with pollinizers (Hirst, 2017) the arrangement of bouquets.

Nutrition Management

The soil fertility, tree age, cultural practices, and crop load all affect however much manure and fertilizer are needed. The leaf and soil analysis should be carried out to determine the status of nutrients in the tree and soil to determine the actual nutritional requirement. Following extensive fertilizer analysis, the manure and fertilizer regimen for any fruit crop is recommended.

Manure and Fertilizer

All the minerals are necessary for apple trees. 500g of N, 250g of P, and 750g of K/tree can be applied, depending on the rootstock and the type of soil. Calcium deficiency causes the physiological condition known as "bitter pit" or "cork spot" in apple fruit. Bitter pits typically show as a little indentation in the skin toward the fruit's calyx. These regions form soft desiccated pits of collapsing tissues and turn brown. Fruits' colors change, and cracking appears. During the later stages of the fruit's development, 0.5% calcium chloride spraying should be used to manage it.

The deficiency symptom can be lessened by taking it four times at a 14-day interval. Additionally, a lack of boron results in inside fruit browning, corking patches on the flesh (internal cork), and occasionally cork tissue in the skin (external cork). During blooming and fruit set, spray sodium borate at a rate of 10g/lit. three times will help in resolving the issue.

Fertilizers containing calcium ammonium nitrate, single super phosphate, and muriate of potash are used to apply the N, P, and K, respectively.

Time and Procedure of Application

In December and January, FYM is applied along with the recommended dosages of P_2O_5 and K_2O. Two separate dosages of nitrogen are delivered. The first half of the N dose is applied one month before flowering (March), and the second half one month following the fruit set. When the soil is sufficiently moist, these fertilizers should be used. Broadcast the nutrients in tree basins 30 cm away from the trunk. However, band application of fertilizer is advised to prevent leaching and runoff losses of fertilizers in locations with very steep slopes and high rains. Soluble fertilizers can also be used more effectively by being applied by drip irrigation. Application of micronutrients, in particular boron, zinc, iron, and manganese, are crucial for normal plant growth, and apple deficiencies are common. Before flowering and in May–June, two foliar sprays of $ZnSO_4$ (0.5%) and boric acid (0.01%) might be used to correct it.

Water Management

For apples to produce fruits of a decent size and higher quality, the soil moisture level must be at its ideal, especially throughout fruit growth and development. Apples are primarily grown in hilly regions of India when the soil is undeveloped land and irrigation water supplies are insufficient. Furthermore, it doesn't rain much or at all in these places, especially during the crucial stages of growth.

When flowering, fruit set, fruit growth, and fruit development take place, April and July are when water requirements are at their highest. Water management is therefore crucial in apple orchards. Rainwater harvesting, preserving soil moisture in place, and drip irrigation are all examples of water management techniques. It is important to gather rainwater in water storage tanks for later use. The water should be used for irrigation, especially during times when there is a critical need for water. The drip irrigation method can be used in locations with limited water resources and can save more than 50% of the irrigation water used. From April to July, Apple should receive 6–8 irrigations spaced 7–10 days apart. Along with irrigation, hay or black polythene mulching is recommended for tree basins in March.

Intercropping

Intercropping in apple cultivation involves growing two or more crops together in the same field or orchard (Figure 1.7). This can be done by planting apple trees alongside other crops, such as vegetables, herbs, or grains, or by planting other crops in the spaces between apple trees. This method of apple cultivation is becoming increasingly popular as it can provide several benefits to growers.

Fig. 1.7: Intercropping

One of the main advantages of intercropping in apple cultivation is that it can increase yields. By growing multiple crops in the same field, growers can take advantage of the different growth patterns and nutritional needs of the different plants. This can help to improve soil health, reduce pest and disease pressure, and increase overall productivity. Additionally, intercropping can also provide a more consistent income stream for growers, as they can harvest multiple crops at different times throughout the growing season. Another advantage of intercropping in apple cultivation is that it can improve soil health. By growing a diverse range of plants, growers can help to improve soil structure, increase organic matter content, and reduce erosion. This can help to improve the overall health and productivity of the apple trees, as well as other crops. Intercropping can also help to improve pest and disease control. By growing a diverse range of plants, growers can create a more complex ecosystem that is less susceptible to pest and disease pressure. This can be achieved by planting apple trees alongside other crops that are less susceptible to the same pests and diseases, or by planting other crops that are known to repel pests and diseases.

Intercropping can also reduce the need for chemical inputs by creating a more balanced ecosystem that is less susceptible to pests and disease pressure. This can help to reduce the overall environmental impact of apple cultivation and can also help to improve the safety and quality of the apples.

Orchard Floor Management

Orchard floor management refers to the practices implemented to maintain and manage the ground cover in apple orchards. Proper orchard floor management is essential for weed control, moisture conservation, nutrient cycling, and overall orchard health. Here are some key aspects of orchard floor management in apple cultivation

Mulching: Mulching involves applying organic or synthetic materials on the orchard floor to suppress weed growth, conserve moisture, and regulate soil temperature. Organic mulches such as wood chips or straw can be used, or synthetic mulches like plastic films or landscape fabrics may be employed.

Cover cropping: Planting cover crops in the orchard provides multiple benefits. Cover crops help suppress weeds, improve soil structure, enhance nutrient cycling, and promote beneficial soil microorganisms. Common cover crops for apple orchards include clover, rye, and vetch.

Mowing and cultivation: Regular mowing or cultivation can be used to control weeds in the orchard. However, care should be taken to avoid damaging the tree trunks or surface roots. Cultivation tools like shallow tillers or weeders can be used to disturb the soil surface and control emerging weeds.

Weed Management

If weeds are allowed to spread throughout an orchard, soil moisture and minerals are depleted. As a result, it is recommended that apple orchards be managed using a clean basin management system. In this approach, the basin of trees is kept clear of weeds by using mulches, weedicides, or manual weeding. Implementing an integrated approach to weed management is recommended. This involves combining multiple strategies such as cultural practices, mechanical methods, and targeted herbicide applications to control weeds effectively while reducing reliance on weedicides alone. Pre-emergence and post-emergence weedicides can also be used to control herbicidal weeds. To effectively manage weeds in apple orchards, simazine at a rate of 4 kg/ha in March was followed by two spraying of glyphosate at a rate of 800 ml/ha each at monthly intervals in July and August. Simazine is a pre-emergent herbicide that provides effective control against grasses and broadleaf weeds. The quantity to be used should be based on the product label instructions and local regulations.

Monitoring and scouting: Regular monitoring of the orchard floor is important to identify weed species, detect potential weed problems, and determine the effectiveness of the management practices implemented. Scouting helps in early identification and timely intervention to prevent weed competition with the apple trees.

Important Practices

Fruit Thinning

Fruits that are small and of poor quality are produced when an apple tree bears heavily during the year. Thus, to control cropping and enhance fruit size and quality, careful thinning at the appropriate period of fruit development (the pea stage) is required. Both manual and growth regulator methods can be used to achieve the thinning. The best fruit thinning is achieved chemically, with foliar sprays of 20 ppm naphthalene acetic acid (NAA) applied at petal fall, as manual thinning of fruit lets is both time-consuming and economically unviable.

Fruit Drop: Apples blossom abundantly, but only a small portion of the flowers turn into fruit. Most of the flowers fade quickly after full bloom, with a small portion falling later. Fruit drop is a significant issue with apples. Fruit drop happens in three periods and ranges from 40% to 60% in the majority of commercial types.

- Early Drop is a natural occurrence caused by a lack of pollination within three weeks after petal fall. By ensuring efficient pollination, placing bee colonies, and planting additional pollinizer varieties in the orchard, this decline can be prevented.

- The June Drop is a significant drop caused by moisture stress. Utilizing irrigation water and mulching tree basins will help limit this decline.
- The Pre-Harvest Drop occurs before the harvest of fruits and the loss of farmers' income. It is triggered by a decrease in auxin levels and a rise in ethylene in the fruit. Checking this drop requires applying 10 ppm NAA 20–25 days before harvest.

Use of Growth Regulator

Breaking seed dormancy and enhancing seed germination, rooting in cuttings of clonal rootstocks, blooming fruit set, vegetative growth, fruit size and form, colour development, ripening and abscission, modifying apple harvest times and regulation of fruit drop in apples are all significantly influenced by growth regulators. These substances interact with the bio-chemical "machinery" of the plant after being ingested by plant cells, primarily through the leaves and fruit. They function by imitating naturally occurring plant hormones or by preventing the production of such hormones. These growth regulators play the following broad roles:

1. **Plant propagation:** Growth regulators promote the onset of rooting in clonal rootstock cuttings and the breaking of seed dormancy. At 100–200 ppm, GA_3 can activate seed germination and override physiological dormancy. Cuttings of clonal rootstocks can be rooted in by using IBA at 2000 ppm and NAA at 200 ppm (Tripathi *et al.,* 2006) and It is the best combination (1500 ppm IBA + 3000 ppm NAA) for rooting and growth of stool shoots of apple clonal rootstock Merton 793 through stool layering (Khatik and Sharma 2013).
2. **Effect on growth:** Growth helpers like auxin and GA_3 help in increasing vegetative growth, and growth retardants like PP333 retard the growth of trees and help in flower bud formation. A foliar spray of PP333 @ 500 to 1000 ppm reduces the vegetative growth of apple trees. Application of 250 ppm GA_3 stimulates vegetative growth but also enhances the alternate bearing phenomenon.
3. **Effect on fruit set and yield:** Fruit setting in warm and marginal places is a concern with apples. The fruit set and production of Delicious apples are improved by the use of triacontanol (20 ppm), miraculan (0.6 ml/L), paras (0.6 ml/L), or biozyme at 2 ml/L sprayed during bud swell and petal fall phases.
4. **Fruit drop:** In early maturing cultivars, pre-harvest fruit drop affects 40–60% of the fruit, which is more severe. The hormonal imbalance that causes the pre-harvest dropping can be managed by applying 10 ppm NAA (1 ml of Planofix in 4.5 L of water) one week before the anticipated fruit drop.
5. **Fruit thinning:** Numerous cultivars, like Red Gold, Starking Delicious, and Golden Delicious, are known to lead to heavy bear, resulting in smaller, low-quality fruits, thereby promoting the biennial bearing cycle. For fruit thinning and to sustain regular production of high-size and quality fruits, 10–20 ppm NAA spray 7–15 days following petal fall is most effective.
6. **Improvement of fruit shape:** The size, shape, and presence of fully formed calyx lobes affect apple market prices, particularly for apples from the Delicious Group. The shape of the apple fruits is improved by the application of promalin (30–60 ppm; GA3+7 + cytokinin) at the pea stage.

7. **Improvement of fruit colour and maturity:** Warmer climates in low-lying apple-producing places severely hinder the development of apple surface colour. Fruits with poor colour are offered for less cost. Ethrel application at 1200 ppm a.i. (2-chloroethylphosphonic acid) or 25 ppm NAA and 4.5 ml of ethrel per litre of water boost the apple's surface red colour.

Physiological Disorders and their management:

Several physiological disorders can occur in apple trees, which can negatively impact fruit quality and yield. Some common physiological disorders and their management include:

Bitter pit

This physiological condition affects the fruit's fresh market quality. The most vulnerable trees are those that are young and have just begun to develop fruit. Fruit picked at the appropriate stage of maturity is less likely to have a bitter pit than immature fruit. The fruit's flesh develops small, brown lesions that range in diameter from 2 to 10 mm (depending on the cultivar). The skin's subcutaneous tissue turns black and corky. The skin develops depressed patches on the surface at harvest or after time in storage (Figure 1.8). These spots show up as wet spots on the skin close to the calyx. After one or two months in storage, the spots typically become darker, more sunken than the surrounding skin, and completely developed.

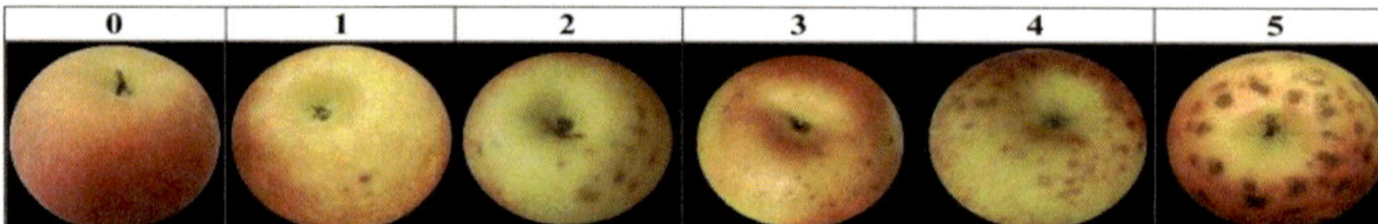

0	1	2	3	4	5
No signs of bitter pit.	Minimal signs of bitter pit. 1 to 5 small, light lesions.	Localized signs of bitter pit. 5 to 15 lesions, either small or medium.	Signs of bitter pit spreading. More than 15 lesions, either small or medium.	Signs of bitter pit surrounding calyx. More than 15 deep, dark medium lesions.	Deepening signs of bitter pit. More than 15, deep, dark large lesions.

Fig. 1.8: Different stage signs of bitter pit in apples fruit

Control

The prevalence of bitter pit is reduced by calcium sprays applied before harvest and calcium dip applied before storage. The plants should be sprayed 45 days before harvest, with a follow-up treatment 15 days afterward. Before storing, the post-harvest dip should be applied for 1-2 minutes.

Brown Heart

Large and overripe fruits are attributed to this physiological issue. It can also happen if the amount of CO_2 in storage rises above 1%. Brown staining of the flesh characterizes the symptoms, which typically start in or close to the core. Brown spots have clearly defined borders and might contain desiccated-induced dry cavities. The symptoms might range from a little area of dark flesh to the full body, with a healthy white flesh margin still present just below the skin. Early in storage, symptoms appear, and as storage time passes along, their severity may rise.

Control

It is best to refrain from picking overripe fruits. The fruits should be picked at their peak maturity if they are to be stored in a controlled environment (CA). To lower the incidence of brown heart formation, the CO_2 levels in CA should be less than 1%.

Cork Spot

The fruit's skin will first show signs of this physiological problem as a little flushed patch above the damaged brown spot. Normally, the damaged tissue is stronger than the healthy tissue. Cork spot development has occasionally been related to deficits in calcium and boron.

Control

Calcium and boron, in particular, are important for preventing this condition when nutrients are managed properly.

Scald

For apple growers, this physiological issue is a major issue. The type of apple, the environment, and cultural habits all influence susceptibility to this storage problem. Hot, dry weather before harvest, immature fruit during harvest, and high nitrogen and low calcium concentrations in the fruit all favor the incidence and severity of scald. This problem is also encouraged by inadequate ventilation in packing crates or storage spaces. The fruit warms up after being removed from the cold storage within 3 to 7 days, resulting in irregular brown patches of dead skin. The scald is not caused by the high temperatures; rather, they allow the symptoms of an earlier injury sustained during cold storage to manifest. When an injury is serious, symptoms may be seen in cold storage.

Control

Scald incidence is decreased by harvesting when the fruit is fully mature and by providing ventilation in cold storage. The administration of an antioxidant soon after harvest is the technique used to control scald the most frequently. DPA, or diphenylamine, is frequently employed. Although it can harm other apple varieties, ethoxyquin is also effective for select cultivars. For best control, antioxidants should be used within a week of harvest.

Water core

High leaf-to-fruit ratios, high amounts of boron and nitrogen in the fruits, low calcium content, excessive thinning, and fruit exposure to high temperatures all contribute to this physiological disorder. Large fruits are especially prone to this kind of problem. Disorder during the pre-harvest stage leads to the development of water-soaked areas in the flesh. These areas have a hard, glassy look and are only externally evident when the infection is quite bad. Fruits that have been severely impacted may smell and taste soured. On the apple's surface or close to the core, water-soaked regions can be found. In storage, symptoms that are mild to moderate may completely disappear.

Control

By avoiding delayed harvests, the incidence can be decreased most effectively. Fruit samples should have their water cores checked as they get closer to maturity.

Fruit should be picked before the water core has had a chance to fully develop. Fruit lots with moderate to severe water core symptoms should be marketed right away instead of being stored in a controlled atmosphere (CA).

Sun burn

This physiological condition is caused by the sun's extreme heat. Fruit on the tree's southwest side is typically harmed. Sunburn incidence can also rise with water stress. The fruits exposed to the sun develop white, tan, or yellow spots as the first signs. Areas of the fruit with extensive skin damage may become dark brown before harvest. These sections may sink and become spongy. Fruit that is exposed to the sun after being harvested can result in extremely bad sunburns.

Control

Avoiding abrupt exposure of fruit to extreme heat and sun radiation is the greatest technique of control. Pruning and training trees correctly are essential. Pruning throughout the summer must be done carefully to prevent getting too much sun. To lessen heat stress, pruned orchards should get regular irrigation. Once the damage has occurred, the sole remedy is careful sorting to remove the damaged fruit before packing.

Russeting

Fruit growers are often concerned about apple russeting in a humid atmosphere. Soon after the petal fall, rusting takes place. Russeting is more likely to occur in apple cultivars with thin cuticles. More frequently than on fruits that are still in the shade, it is seen on exposed fruits. Russeting can also result from frost during the flowering stage or the first stages of fruit development. The fruit peel breaks and develops cracks as a result of russeting.

Control

Russeting can be minimized by choosing clones that are less prone to it, as well as by providing sufficient irrigation, manuring, and pest control.

Fruit Drop

The majority of commercial apple types display three cycles of fruit loss: early loss, June loss, and pre-harvest loss. The absence of pollination and fruit competition are the causes of the early decrease, which is thought to be normal. The June decline is a result of environmental factors and moisture stress. Neither of these two declines results in significant economic losses nor are they successfully controlled by artificial means. Due to the decline in auxin levels, the full-grown marketable fruits abscise before the harvest, resulting in a significant economic loss.

Control: NAA (15 ppm) sprayed 20 days before the anticipated fruit drop or 20–25 days prior to harvest can be used to control the pre–harvest drop.

Blossom end rot: This disorder is caused by a lack of calcium in the fruit and can lead to a dark, leathery patch on the blossom end of the fruit. To prevent blossom end rot, growers should ensure that the trees have an adequate supply of calcium and that the soil pH is not too high. Additionally, growers should avoid over-fertilizing with nitrogen and should also be careful not to over-water the trees.

Frost injury: This occurs when the apple trees are exposed to freezing temperatures, causing damage to the buds, flowers and fruits. To prevent frost injury, growers should use frost protection methods such as wind machines, smudge pots, or irrigation to prevent frost from settling on the trees.

Major Insect-Pests of Apple and their management

Apple production and quality are poor compared to developed countries due to several factors including insects and diseases. Although a large number of pests attack apple trees, some of them are very serious and require special attention to control. Besides insect pests, mites are also involved in apple production and cause significant economic losses to commercial fruit growers. The main insect pests that attack apples are:

1. San Jose scale, *Quadraspidiotus perniciosus* (Comstock)

It was introduced to France in 1906 and has since been found on more than 32 host plants in India (Rahman and Ansari, 1941). Except for Antarctica, every continent has seen its spread. The insect can thrive on different fruits found in hilly areas but prefers to eat plants from the Rosaceae family, such as apples and pears.

Symptoms

The nursery's young plants became weak and ultimately died as a result of the small insects' constant sap-sucking. The ashy-grey scales that cover the leaves, twigs, fruits, and occasionally even the entire bark can be readily scraped off to reveal the orange-colored individuals underneath. The fruits have pink spots around the scales, and their market value has declined as a consequence (Figure 1.9).

Fig. 1.9: San Jose scale

Control

- Spray diesel oil emulsion plus Bordeaux combination (68 litres of diesel oil + 15 kg of copper sulphate + 3.75 kg of unslacked lime) that has been emulsified and diluted 5–6 times.
- When trees are entirely defoliated in the winter, spray 7.5 litres of any tree spray oil emulsion in 250 litres of water per ha.
- Spray 500 litres of water per ha with 1.25 litres of Malathion 50 EC or 6.25 litres of Chlorpyriphos 20 EC during the summer.
- To combat this insect problem, the parasite *Encarsia perniciosi* (Tower) may also be released to monitor the overwintering population.

2. Wooly Apple Aphid, *Eriosoma lanigerum* (Hausmann)

The damage to young apple trees was first reported in India in 1889 at Coonoor (Tamil Nadu) (Misra, 1920). The pest feeds on apple, crab apple, almond, hawthorn, pear, and quince throughout the year. The nymphs' bodies exude woolly filaments of wax, giving rise to the term "woolly aphids."

Symptoms

Fig. 1.10: Wooly Apple Aphid

The pest is active all year long. Although the winged form also attacks the trunk, branches, stems, twigs, leaf petioles, and fruit stalks, it predominantly targets the underground roots. Hotter and colder seasons, respectively, encourage upward and downhill migrations (Lal and Singh, 1947). Affected trees display a sickly appearance, lose vigour, and their growth, as well as their ability to bear fruit, are negatively impacted due to the drainage of sap by feeding. Young trees have roots that deteriorate to the point where even moderately strong winds can easily topple them. Fruits from trees with a lot of infestations are of poor quality because they are small, deformed, and have a bland flavor (Figure 1.10).

Control

- To stop aphid damage, use resistant rootstock like Golden Delicious, Northern Spy, and Morton Stock 778, 779, 789, or 793. (Atwal and Dhaliwal 1999).
- The foreign parasite *Aphelinus mali* can efficiently control the aphid population (Haldeman).
- During the summer, apply 800 ml of Malathion 50 EC to 500 litres of water per acre. Where the parasite *Aphelinus mali* is present, stay away from insecticides.
- Spray the plants with 1.25 litres of Diazinon 20 EC in 500 litres of water per hectare to protect them from aerial forms during leaf fall. Apply one litre of methyl oxydemeton 25 EC in 500 litres of water per hectare for regulating the root forms. The best time to make these sprays is in the winter.

3. Tent Caterpillar, *Malacosoma indicum* (Walker)

It is a major apple pest in north-western India, with the Kashmir Valley and the Shimla Hills seeing the most severity. The main pest is present on a variety of fruit, forest trees, and decorative plants, with apples being its favorite host (Figure 1.11).

Symptoms

Fig. 1.11: Tent Caterpillar

The caterpillar builds a tent-like home at a twig's fork and hides there during the day. The caterpillars concentrate on leaf lamina at night and feed voraciously, only leaving the midribs and little fragments of rigid veins behind. In cases of severe infestations, an orchard's apple trees may lose 40 to 50 percent of their leaves. The tents and leaf skeletons are obvious signs that this bug has attacked.

Control

- When pruning in December or January, destroy all egg bands to prevent the infestation of this insect.
- Cleaning the tents with a pole and some rags doused in kerosene and tied to its

end will kill the caterpillar. The larval-infested tree should be placed below an open container filled with kerosene so that any larvae that fall into it will also be killed.

- For every acre, apply 500 ml of Dimethoate 30 EC or 2.0 kg of Carbaryl 50 WP diluted in 500 litres of water.

4. European Red Mite, *Panonychus ulmi* (Koch.)

The European red mite damages the tree by eating its leaves, decreasing the chlorophyll, and boosting respiration. To do this, the mite mouth parts are inserted into the leaf cells to expel the contents. All motile stages feed on the leaves, preferably the bottom surface, yet when populations are great, both leaf surfaces are attached. The effect of extensive mite feeding is a characteristic of brown foliage that begins as a subtle cast to the green leaf and progresses to bronze in extreme cases. When Alternaria fungus develops after a mite infection, the damage is more severe. By laying overwintering eggs, a high mite population during the late season might indirectly degrade fruit even more.

Control

- In the autumn, when the leaves are falling, remove whatever branches holding mite eggs.
- When buds start to open in March, a winter spray of petroleum oil will help control winter eggs and serve as the first line of defense.
- Sprays for the summer are also useful. Malathion has limited persistence but could be helpful if a kill is needed soon after harvest.
- Other summertime sprays include Abamectin (1.8 EC) 0.01 percent, Propargite (57 EC) 0.05 percent, and Dicofol (18.5 EC) 0.02 percent. At the crop's post-bloom stage, fenazaquin (10 EC) 0.04 percent should be administered. Dimethoate (30 EC) 0.025 percent sprayed onto the apical area also controls mites.

5. Codling Moth, *Cydia pomonella* (Linnaeus)

It is a serious pest of apples. Caterpillars tunnel through the fruits by consuming the pulp, filling these tunnels with frass, and therefore destroying the fruit's suitability for human consumption.

Control

- To prevent the hibernating larvae from finding shelter, all the trash and weeds should be pulled out of the orchard and old trees.
- Before the larvae descend to the ground to hibernate, band the trees with grass ropes or sac (Jute) cloth folded three to four times. The larvae should then be collected and exterminated.
- To stop the spread of this hazardous pest, rigorously stick to quarantine regulations.
- The trees should be sprayed with Chlorpyrifos (20EC) 0.02 percent; the first spray, known as the calyx spray, should be applied at the time of the petals' fall and calyx's closure. Three to four cover sprays with Methyl oxydemeton (25EC) 0.02 percent should then be applied at intervals of 15 days and two months before the fruits are harvested.

6. Apple Root Borer, *Dorysthenes hugelii* (Redtenbacher)

The grubs girdle the roots, feed on the internal tissues, and consume the roots' roots. Because of this, the primary roots are badly harmed from the bottom up, and new trees die while older.

Control

- ones weaken and topple over in heavy winds. Avoid using dry, sandy soils for the planting of apple orchards to reduce the prevalence of this insect.
- Inter-culturing in the orchard aids in grub eradication.
- Use well-rotted FYM and thoroughly blend with the surrounding soil.
- It is crucial to treat the tree basins with Phorate granules at 100 grams a.i. per tree once an infestation has occurred.

7. Apple Stem Borer, *Apriona cinerea* (Cheverlot)

The grubs pierce the shoots and create circular galleries that descend to the tree's main stem and trunk. The grubs' eating may prevent the damaged trees from dying for many years, but it substantially reduces their vitality and output. The adult beetles consume bark, although they cause little damage.

Control

- Beetle and grub collection and extermination is a highly successful approach for reducing the pest population.
- No insecticide spray can be efficient since the pest feeds inside the woody areas. Finding the feeding openings, cleaning the channel, and injecting poisons are the only ways to get rid of this parasite.
- Cotton wicks soaked in Dichlorvos (0.1%), Dimethoate (0.03%), or Methyl -o-Demeton (0.025%) should be inserted into the living holes and sealed with mud.

Major Diseases and their management

1.Apple Scab (*Venturia inaequalis*)

Symptoms are visible on fruits and leaves. On the underside of the leaf, olivaceous dots first appear before darkening to black and becoming velvety. The dots have a feathery border and radiate outward on a new leaf. The lesions have a clearer shape on older leaves. The lesion could have a convex surface on one side and a concave portion on the other. Leaf blades that had a serious infection were bent, small, and deformed. Small, ragged, black circular blemishes might be seen on fruits. On fully developed fruits, a yellow halo can be observed around the lesions, and the centers of the spots turn corky (Figure 1.12).

Fig. 1.12: Apple Scab

Control: The following spraying plan is recommended for the efficient management of scab disease:

Stage	Fungicide/100 litres of water.
Silver tip-Green tip	Mancozeb (400 g)/Captan (300 g).
Pink bud	Contaf (30 ml)/ Baycor (50 g).
Petal fall	Bavistin (50 g)/Topsin M (50 g).
Pea stage	Mancozeb (300 g)/Captan (300 g).
Fruit development	Bavistin + Mancozeb (25+250 g).
15-20 days before harvest	Mancozeb (300 g).
Before leaf fall	Urea (5 kg).

2. Fire Blight (*Erwinia amylovora*)

Bacteria are the cause of this disease. On diseased plant components, the symptoms are noticeable as a unique fire-like look. The earliest symptom typically appears on leaves that become water-soaked, wilt, become brownish to black in color, and either fall to the ground or hang on to the tree. To twigs, the symptom had spread. In addition to spreading to the branches, terminal twigs wilt from tip to base. Fruits become sopped in water, change colour to brown, shrink, and then become black. The affected area might exhibit oozing. Infection is quite likely to spread to new shoots. Without becoming brown, the shoot tips fade and droop. On the stem, there is bacterial slime that is golden in color. On the outside of fruits, necrotic lesions and patches with seeping fluid are visible (Figure 1.13).

Fig. 1.13: Fire Blight

Control

- Handling fruit with care to avoid cutting yourself. The best control is achieved by dipping the fruits in aureofunginsol at 500 ppm for 20 minutes.
- As soon as fire blight is detected, the diseased trees and host plants should be collected and burned. Streptomycin sprays help treat the infection in apple blossoms in the spring.

3. Powdery Mildew (*Podosphaera leucotricha*)

A serious disease that impacts buds, new shoots, and leaves is powdery mildew. The disease manifests itself in dry weather conditions. This disease affects nursery plants more frequently. On both sides of the leaves and young shoots, there are patches of white, ash-like powder that are indicative of the disease. Leaves that are harmed become pale and curl up. Affected shoots are still immature and frail. Leaf drop and early fruit ripening may happen in cases of severe infection. Rusting is visible on the young infected fruit (Figure 1.14).

Fig. 1.14: Powdery Mildew

Control

- By pruning and removing the diseased plant sections, the incidence of the infection can be decreased. Bayleton (500 ppm) is sprayed on young seedlings in nurseries at intervals of 7 days to control the disease. Fungicidal sprays should be used in the field following the pruning of the afflicted plant portions.
- The disease can be efficiently controlled by spraying the crop with sulphur (0.3%), carbendazim (0.05%), or karathane (0.05%).

4. Bitter rot (*Glomerella cingulate*)

A faint, light brown discoloration appears beneath the skin. The discolouration grows into a cone-shaped area. The rounded, rough lesions degenerate. The lesions grew and now cover whole sections of fruit. Acervuli are created when tiny black specks form beneath the cuticle. Spores are observed in distinct rings that are arranged in pink masses.

Control

- In the field, apply Mancozeb 0.25%. Mancozeb 0.25% treatment to control the disease in storage

Maturity Indices and Harvesting

There are several different indices that can be used to determine the maturity of apples. These include:

DFFB: DFFB is a measure of the number of days that have passed since the apple tree has completed full bloom. This can be used as an indicator of the maturity of the apples, as the fruit will typically reach maturity around 90-180 DFFB.

TSS: Total Soluble Solids (TSS) is a measure of the sugar content in the apple. It is typically measured as the percentage of sugar in the juice of the apple. As the apple matures, the TSS will increase, indicating that the apple is becoming sweeter. A TSS of around 12-15% is considered optimal for most apple varieties.

Firmness: Firmness is a commonly used indicator of apple maturity. The firmness of an apple is typically measured in kilograms per square centimeter (kg/cm^2) using a device called a texture analyzer. Apples should be firm to the touch and not too soft.

Starch Conversion Index (SCI): This measures the conversion of starch to sugar in the apple and is often used to determine when an apple is at its peak sweetness.

Ethylene Production: Apples produce a hormone called ethylene as they ripen, and measuring the amount of ethylene can indicate the maturity of an apple.

Color: The color of an apple can also indicate its maturity. Red apples should have a consistent red color, while green apples should be a uniform green.

Size: Apples should have reached their optimal size before they are harvested.

Taste: Apples should be sweet and crisp, with a good balance of acidity and sugar.

Apples are climacteric fruits, indicating that their ripening does not occur at the same time as their maturity. At harvest, the fruits are still too green-red to be eaten. Fruits that are plucked before they are fully developed lack flavor and taste and shrivel when stored. Fruit that has been harvested when it is too mature has soft scald, internal decomposition, and a short shelf life. Several ripeness indexes are used to determine the ideal harvesting period.

Ultimately, the most accurate way to determine the maturity of an apple is to use a combination of these indices.

Table : Maturity indicators for significant apple cultivars

Cultivar	DFFB (Days)	Firmness (Kg/cm^2)	TSS (%)
Tydeman's Early Worcester	90 ± 4	7.8 ± 0.15	12.0-13.0
Red Cheif	110 ± 5	8.5 ± 0.4	14.0-15.0
Starking Delicious	120 ± 5	8.2 ± 0.4	13.0-15.0
Red Delicious	134 ± 5	8.5 ± 0.4	13.0-14.0
Golden Delicious	148 ± 5	8.4 ± 0.4	12.0-14.5
Granny Smith	180 ± 5	8.4 ± 0.4	13.0-14.0

Fruits should be harvested in two or three pickings because they do not fully ripen and mature on the tree at the same time. Fruit is harvested in such a way that bruises and stem punctures are prevented, and the pedicel must remain with the fruit. Apples are placed between the thumb, middle finger, and index finger, and with a fast twist of the wrist, the fruit and pedicel are easily plucked. Fruits are gently picked, put in a picking bag or basket, and taken to packaging facilities for grading and packing.

Yield

A four-year-old apple tree begins to produce fruit. A well-managed orchard of apples typically yields 10 to 20 kg per tree each year, depending on variety and season.

Pre and Post-harvest management

Fruits must be handled, graded, and stored properly after being harvested. The post-harvest losses of apple fruits have been estimated to be over 30%; however, these losses can be reduced by handling the fruits properly.

Precooling

Fruits should be stored in a cold, well-ventilated area to remove field heat as soon as they are picked. Methods like the use of air coolers, cold water sprinkling, fruit cleaning, and storing fruits overnight in a cool environment are used to remove field heat.

Grading

The fruits are graded based on their sizes and outward characteristics, such as color and shape. The apple fruits are mechanically or manually sorted according to size as follows:

Table: Grades Size of Apple

Grade	Fruit diameter(mm)
Super large	85 ± 2.5
Extra Large	80 ± 2.5
Large	75 ± 2.5
Medium	70 ± 2.5
Small	65 ± 2.5
Extra Small	60 ± 2.5
Pittoo	55 ± 2.5

Fruits are graded as AAA, AA, and A based on the fruit's color, shape, quality, and attractiveness. The packing of fruits affects their quality during storage, marketing, and transportation. Corrugated fiberboard (CFB) boxes with telescoping sides are typically used to package apples. Telescopic CFB cartons with trays typically measure 50.4 x 30.3 x 28.2 cm (outer jacket) and 50 x 30 x 28.2 cm (inner case). Fruit trays that change according to size grades are used to separate each layer of fruit in a carton.

Storage

Apples have a longer shelf life, which can be increased by creating ideal storage conditions. The cold storage delays fruit deterioration reduces microbial decay and guards against shriveling brought on by water loss. After harvest, apple fruits can be kept in cold storage for 4-6 months at a temperature of -1.1 to 0°C with 80-90 % relative humidity. Apple fruit can be kept in cold storage for 4-6 months at a temperature of -1.1 to 0°C and 80-90 percent relative humidity after harvest.

Processing

Before being packed for the fresh fruit market, over 30% of apple fruits are rejected as cull fruits. The removed fruits can be prepared as fruit pulp by processing and preservation, which can then be used to create a wide range of products with added value like jam, fruit leather, fruit toffees, and fermented goods like wine. Juice can be produced from fresh fruit.

Future prospects

Apple is regarded as a significant temperate fruit crop. It is a widely cultivated fruit that has significant economic and cultural importance. It is a major cash crop for many countries around the world, and it is also an important part of many traditional diets. In recent years, the apple industry has seen significant growth and innovation, with new varieties and growing methods being developed to meet the demands of consumers and growers alike. One of the key reasons for the continued growth of the apple industry is the fruit's versatility. Apples can be eaten fresh, cooked, or processed into juice, cider, and other products. They are also a good source of vitamins and antioxidants, making them a popular choice for health-conscious consumers. Additionally, apples are relatively easy to grow and store, which makes them a suitable crop for both large-scale commercial production and small-scale backyard gardening.

Another major factor that is driving the growth of the apple industry is the increasing demand for organic and locally-sourced produce. As consumers become more aware of the environmental and health benefits of eating fresh, natural foods, they are increasingly turning to organic and locally-sourced apples. This has led to a growing number of small-scale apple growers and orchards, many of which are focused on producing high-quality, organic apples for local markets. Overall, the apple industry is well-positioned for continued growth and success in the future. With increasing demand for healthy, natural foods, and a growing focus on organic and locally-sourced produce, the apple industry is well-positioned to meet the needs of consumers and growers alike. As technology and innovation continue to advance, the apple industry will continue to evolve and adapt to meet the changing needs of the market.

Crop improvement of apple fruit is a vital aspect of the future prospects of the apple industry. Advances in breeding, genetics, and biotechnology have led to the development of new apple varieties that are more resistant to disease, pests, and environmental stress, and have improved fruit quality, size and color. These improvements have led to increased yields, reduced costs and more efficient use of resources. One of the key areas of focus in apple crop improvement is disease resistance. Apple trees are susceptible to a variety of diseases such as fire blight, apple scab, and powdery mildew, which can cause significant damage to the fruit and reduce yields. Breeders are working to develop apple varieties that are resistant to these diseases, which will help to reduce the need for chemical pesticides and improve the sustainability of apple production. Another area of focus is improving fruit quality. This includes developing varieties with improved flavor, texture, and color, as well as extending the storage life of apples. This can be achieved through genetic modifications, traditional breeding, or by using precision agriculture techniques such as precision irrigation, nutrient management and machine learning algorithms.

In addition, biotechnology is also playing an important role in the crop improvement of apples. For example, genetic engineering can be used to produce apples that are resistant to disease and pests, or that have improved nutritional content. Biotechnology also offers the potential to develop new apple varieties with unique characteristics, such as increased tolerance to extreme weather conditions and climate change. Overall, crop improvement of apple fruit is a vital aspect of the prospects of the apple industry. With the continued use of new technologies and innovative breeding techniques, apple growers will be able to produce high-yielding, high-quality fruit that is resistant to disease and pests, and has improved nutritional content. These advancements will help to ensure the long-term viability and sustainability of the apple industry.

References

Agrawal, A., Gupta, S. and Kumar, R. (2019). Nutritional composition and health benefits of apples: An overview. *Journal of Food Science and Technology*, 56(2):469-478. doi: 10.1007/s13197-018-3535-9.

Anonymous, (2020). National Horticulture Board, Indian Horticulture Database. Ministry of Agriculture & Farmers Welfare, Government of India. www.agricoop.nic.in

Bal. J. S. (2014). Fruit Growing, Third Revised Edition, Kalyani Publishers, B-1/1292, Rajinder Nagar, Ludhiyana-141008, pp. 327-346.

Bose, T. K., Mitra, S. K. and Rathore, D. S. (2001). Temperate Fruits, Third Edition, Volume 2, Horticulture and Allied Publishers 27/3, Chakraberia Lane, pp. 1-122.

Boyer, J. and Liu, R. H. (2004). Apple phytochemicals and their health benefits. Nutrition Journal, 3(5):1-15. doi: 10.1186/1475-2891-3-5

Chadha, K.L. (2019). Hand Book of Horticulture. Second Edition, Volume 1, Indian Council of Agricultural Research, New Delhi. pp. 243-249.

Chattopadhyay, T.K. (2009). A Textbook on Pomology IV: Devoted to Temperate Fruits Crops, Second Edition, Volume 4, Kalyani Publishers B-1/1292, Rajinder Nagar, Ludhiyana-141008, pp. 1-45.

Hirst, P. M. (2017). Advances in understanding flowering and pollination in apple trees. *Achieving sustainable cultivation of apples*, 109-126.

http://nhb.gov.in/Horticulture%20Crops/Apple/Apple1.htm

http://www.agrimoon.com developed by Tamil Nadu Agricultural University (TNAU), Coimbatore, Tamil Nadu.

https://fdc.nal.usda.gov/fdc-app.html#/food-details/1102644/nutrients

https://www.hsph.harvard.edu/nutritionsource/food-features/apples/

https://www.newworldencyclopedia.org/entry/Apple#Origin_of_name

https://www. researchgate.net/ publication/282336735 _New_Apple_variety_released [accessed Jan 03 2023].

Indian Council of Agricultural Research. (2020). Database on Medicinal, Aromatic and Economic Plants. Retrieved from http://www.mpbd.info/plants/malus-domestica.php

Janick, J., Cummins, J. N. and Brown, S. K. (1996). Apples. In J. Janick (Ed.), Progress in New Crops (pp. 162-171). ASHS Press.

Khatik, P. C. and Sharma, D. D. (2013). Effect of IBA and NAA on stool layering in apple clonal rootstock, Merton 793. *Progressive Horticulture*, 45(2): 388-391

Kumar, V. and Kamra, S. (2018). Ready Reckoner for Pomology, First Edition, New Vishal Publications, West Patel Nagar, New Delhi. pp. 198-210.

Lal, R. and Singh, R. N. (1947). Studies on the life history and control of the woolly aphis, *Eriosoma lanigerum* (Hausmann), infesting apple roots. *Indian Journal of Entomology,* 9(1):77-98.

Misra, B. N. (1920). The woolly aphis of the apple in India. Memoirs of the Department of Agriculture in India, Entomological Series, 10(2):39-47.

Nagar, D., Garg S. and Rout, B. M. (2020). Glimpses of Fruit Science, First Edition, New Vishal Publications, West Patel Nagar, New Delhi. pp. 228-241.

Rahman, M. A. and Ansari, M. A. (1941). Host plants of the San Jose scale, *Quadraspidiotus perniciosus* (Comstock), in India. *Indian Journal of Entomology*, 3(1): 107-113.

Singh P. and Ramteke V. (2018). Comprehensive Fruit Science, First Edition, New Vishal Publications, West Patel Nagar, New Delhi. pp. 182-190.

Singh, V., Ali, Z. A. and Hassan, F. (2018). Health benefits of apple. In R. C. Ray & S. S. Singh (Eds.), Postharvest Physiology and Biochemistry of Fruits and Vegetables (pp. 433-455). Academic Press.

Srivastava, K. K., Singh, D. B., Kumar, D. and Mir, J. I. (2016). Screening of exotic and indigenous apple rootstocks for different horticultural traits. *SAARC Journal of Agriculture*, *14*(2):117-125.

Tripathi, P. C., Verma, M. K., Verma, R. K., Pandey, G. and Sofi, A. A. (2006). Response of NAA and root initiation and growth in apple root stocks. *Indian Journal of Horticulture*, 63(1):81-82.

2

Advances Production Technology of Pear

***Imran Ali*[1], *Vibhu Pandey*[2], *Jagraj Singh*[3], *Prashant Gautam*[2] and *Vijay Pratap Yadav*[4]**

[1]*M.Sc. Fruit Science, Department of Fruit Science, Sardar Vallabhbhai Patel University of Agriculture and Technology, Modipuram, Meerut-250110, India*
[2]*Ph.D. Research Scholar, Department of Fruit Science, Sardar Vallabhbhai Patel University of Agriculture and Technology, Modipuram, Meerut-250110 Uttar Pradesh, India.*
[3]*Ph.D. Research Scholar, Department of Vegetable Science, Sardar Vallabhbhai Patel University of Agriculture and Technology, Modipuram, Meerut-250110, Uttar Pradesh, India*
[4]*M.Sc.Ag. (Horticulture), National Post Graduate College Barahalgang Gorakhpur (Affiliated Deen Dayal Upadhyaya Gorakhpur University Gorakhpur), Uttar Pradesh, India*
Email: imranaligkp09@gmail.com

Introduction

Pear (*Pyrus communis* L.) is one of the most important temperate fruit crops after apple in the family Rosaceae. It is diploid and a member of the Pomoideae subfamily, having chromosome number 2n=2X=34. The pear is a tree and fruit in the Rosaceae family, which also includes apples and quince. The pear has a fibrous center and is a mildly sweet fruit. It is abundant in dietary fiber, plant components, and vital antioxidants. Pears come in a variety of shapes, including round, oval, and elongated, and they can be smooth or bumpy in texture. The fruit has a thin skin that can be smooth or rough, and it ranges in color from green to yellow to red. The flesh of the pear is white, yellow, or cream-colored and is juicy and sweet, with a slightly gritty texture. Later in the 19th century, better pear cultivars were introduced in India. Pears come in second place among temperate fruits in India in terms of both production and land area. Due to the pear's

Fig. 1.1: Pear Fruit

extensive climatic and soil tolerance, it is grown in both temperate and subtropical climate.

Importance and composition and uses

Pears are a good source of fiber and vitamin C, and they are relatively low in calories. Pear fruits are suggested for diabetic patients because of their low sugar content, which also lowers blood pressure and regulates heart rate. The 3.1 gm of dietary fibre it contains are very beneficial for intestinal health. About 27 to 41 mg of phenolics are present per 100 mg of fruit, making it a strong source of antioxidants. Moreover, pears have demonstrated vasodilator capabilities, and their components anti-arrhythmic effects often lower blood pressure and regulate heart rate. Fruits include the folic acid that makes up the vitamin B complex. It contains the glycoside arbutin, which has potential as a treatment for urinary tract infections. The body's uric acid is helped to leave the body with its help. The pears' leaves are effective in treating swelling and blisters. Fever, anaemia, and general malaise are also treated with their assistance. A destion can be made from the bark of pears. Drinks (such RTS, squashes), sweets, jam, and preserved fruits are among the common products that are manufactured from it (Reiland and Slavin 2015). The fruits are pyriform pome in shape and flesh of pear fruit contains the stone cell. It is developed from fleshy thalamus of monocarpellary or multicarpellary syncarpous inferior ovary called pseudo carp or false fruit. It's typical fruit with pleasant taste and flavour and is generally consumed as fresh cookies, dried or preserved. Pear fruits available in the world market belong either to *Pyrus communis, Pyrus pyrifolia* and their hybrid species, Pear stands 2nd after apples as the most frequently consumed fruit and the most economically important tree fruit in temperate zones of the world (Ahmed *et al.*, 2011). Pears has an important role in wood making industries, Pear wood is preferred for making wood instruments and furniture. Pear wood is also used to produce aromatic smoke for smoking meat or sobacco. It grows in cold climate and plants are generally deciduous. Deciduous fruits have a problem of low precocity and high vigour. There is a need to regulate the growth of deciduous fruit crops to maintain a balance between growth and fruiting (Sharma *et al.*, 2009). The three main types of carbohydrates contained in pears are sugars, starch, and cellulose. Brix-acid ratio in Patharnakh is 36: 1 in Le Conte is 47: I and in Baggugosha 60:1. Pears are typically eaten fresh, but they can also be canned, frozen, or dried. They are also used in cooking, particularly in the making of jams, jellies, and sauces. Pear fruits can be used to make a variety of delectable items, including pear preserves. The Patharnakh and Kieffer cultivars are recommended for canning.

Nutrient composition

According to the USDA, the nutritive value of an apple (100g of edible portion) is as follows:

Table: Nutritive value of apple fruit/100 g edible portion

Contents	Amount with Unit	Contents	Amount with Unit
Water	84 g	Vitamin C	4.3mg
Energy	57 kcal	Thiamin.	0.012 mg
Protein	0.36 g	Riboflavin	0.026 mg
Total lipid (fat)	0.14 g	Niacin.	0.161 mg
Carbohydrate	15.23 g	Vitamin B-6.	0.029 mg
Fiber, total dietary	3.1 g	Folate, DFE	7 µg
Sugars	9.75g	Choline, total.	5.1 mg
Potassium, K	116 mg	Calcium, Ca	9 mg
Sodium, Na	1 mg	Vitamin A, RAE	3 µg
Carotene, beta	27 µg	Iron, Fe.	0.18 mg
Cryptoxanthin, beta	11 µg	Magnesium, Mg	7 mg
Phosphorus, P.	12 mg	Alpha-tocopherol	0.18 mg
Lutein + zeaxanthin	29 µg	Vitamin K	4.4 µg
Sodium, Na	1 mg	Fatty acids, total saturated.	0.028 g
Zinc, Zn.	0.1 mg	Copper, Cu	0.027 mg

Source: USDA Nutrient Database: g = grams, kcal = kilocalorie, **µg** = micrograms; mg = milligrams

Origin and Distribution

The genus Pyrus has two domestication centres, one in China, Asia Minor to the Middle East, Caucasus Mountains, and one secondary site in Central Asia. The two domestication centres and primary origin of the genus Pyrus are located in these three regions. Pyrus species are dispersed over a wide range in nations spanning from the north to the north-west, west, and south Central regions (Sharifani *et al.*, 2008). The Asian pear (*Pyrus pyrifolia* (Burm) Nakai) was originated in China, where its cultivation dates back to 2500-3000 years. Asian pear has existed and being cultivated in the Himalayan region of India from very old times which might have been introduced from China by Chinese hostages of Lord Kanishka, while the European pear was introduced in Kullu district (H.P.) by European settlers during the 19th century. The first record of pear cultivation in Europe was in Ancient Greece where it was mentioned by Homer in the Odyssey, almost three thousand years ago deeming pears "a gift of God". In ancient Greek pears were differentiated by two separate words, "acras" and "apios" with the latter referring to the cultivated pear (Silva *et al.*, 2014). It ranks second only to the apple in terms of popularity, output, and vertical diversity among temperate fruits. A mildly sweet fruit with a fibrous centre is the pear. It contains a lot of dietary fibre, important plant compounds, and antioxidants. The improved pear cultivars were introduced to India, one of the major pear-producing nations in the world, in the latter half of the 19th century. Pears are the second most produced and most widely grown temperate fruit in India.

Due to the availability of low chilling, hard, semi-soft pear cultivars, its cultivation is restricted to the temperate and subtropical regions of Jammu & Kashmir, Himanchal Pradesh, Uttarakhand, Punjab, Haryana, Uttar Pradesh, Tamil Nadu, North

Eastern provinces, and Nilgiri Hills of Southern India. Semi soft cultivar of pear like Baggugosha and Le Conte which are popularly grown in lower hilly areas are shy and irregular bearer. New strains of semisoft pear like Punjab Beauty, Punjab Gold, Punjab Nectar and Punjab Soft collected from indigenous and exotic sources approved by P.A.U., Ludhiana are gaining popularity amongst the orchardists of Punjab in the low chilling pear growing areas. Generally, Pear cv. Patharnakh requires approximately 250 chilling hours. The Chinese sand pears (*Pyrus pyrifolia)* are widely grown in North Western Himalayan region including Jammu and Kashmir, Himanchal Pradesh and Uttarakhand. The maximum area under cultivation of Chinese sand pear existed in Kashmir valley. Therefore, it is locally known as "Kashmiri Nakh". It is also believed that Kashmiri Nakh is one of the naturalized indigenous cultivar grown since ancient time in India (Verma *et al.,* 2014). Under various environmental situations, it may differ dramatically in other Pear varieties. India's soil and agroclimatic conditions are well suited to the hard pear. Hard pear types with low chilling requirements in Punjab need 150–300 chilling hours, but high chilling kinds cultivated in areas with higher altitudes need 900–1000 chilling hours. European pears require a comparatively higher number of chilling hours Le. 1000-1500. Patharnakh pear has become an important and commercial cultivar in Punjab owing to its good keeping quality and standing of long transportation.

Area and Production

Pears are mostly grown in China (69%) on a global scale. Over 82% comes from China, India, Italy, Spain, Argentina, Algeria, the United States, and Turkey. Pear is the second-most abundant and produced temperate fruit in India. The major pear-producing states in India include Himachal Pradesh, Jammu and Kashmir, and Uttarakhand. However, pear cultivation is also practiced in other states such as Punjab, Uttar Pradesh, and Arunachal Pradesh. The area under pear in India is 42000 hectares with annual production of 276000 MT (NHB 2020)

Botany

Plant

A medium sized, upright growing tree, to 30 ft tall, generally 8-18 ft in cultivation. Size of pear trees is greatly influenced by rootstock and training method. The 2-4 inch long leaves are elliptic/ovate in shape with sharp points and slightly serrated or whole margins.

Flowers

White flowers with a diameter of about 1 inch and longer pedicels resemble apple flowers. A corymbose inflorescence with 5-7 blooms is present (also different from apple). The majority of cultivars need cross-pollination to produce commercial fruit. Several cultivars, including Bartlett, Orient, Baldwin, Kiffer, and Spalding, are partially self-fruitful. The primary pollinator is honey bees.

Fruit

A pyriform (European) or round (Asian) pome. The fleshy edible part, such in an apple, comes from hypanthium tissue. The five core seed chambers typically contain two seeds each, just like in an apple. The grit cells, also known as brachysclereids, which have thick walls and are lignified, are present in the flesh and are what give European pears their distinctive flesh texture. 1-2 pears are thinned and spaced 6" apart on each spur.

Taxonomy

Pear is classified into the following categories-

Rank	**Scientific Taxonomical Details**
Order	Rosales
Family	Rosaceae
Sub -family	Pomoideae
Genus	*Pyrus*
Species	*communis*
Basic chromosome	17
Somatic chromosome	34

There are 22 primary species, which are classified in to 4 groups on the basis of geographical distribution.

1. **European group:** *P. communis, P. nivalis, P. cordata, P. caucasica*
2. **Asian group:** *P. calleryana, P. betulifolia, P. dimorbhophylla, P. Koehnei*
3. **North Africa group:** *P. longipes, P. memorensis, P. gharbiana*
4. **East Asian group:** *P. pyrifolia, P. ussuriensis, P. hodoensis*

The important species which are either grown for commercial cultivation or for raising rootstocks are given below

1. ***Pyrus communis*** L.: European pear: Trees grows in upright manner. Bark light brown to grey in colour. Lenticells are present on the twigs Young shoots are light brown in colour Leaves above and light green under surface. It has persistent calyx, fleshy pedicels and pyriform shape in small, ovate, 5-10 cm long, margin serrated, base obtuse, apex acuminate Leaves are dark groom different form.
2. ***Pyrus pyrifolia*** (Burm F) Nakai, var Calta: Japanese or Chinese or Sand pear: Tree 9-10 metres in height, having spreading habit of growth bark light brown in colour with the presence of lenticells bark surface rough and splitted branches are in drooping habit pubescence present on young shoots leaves mostly ovale, 10-15 cm long, margin crenate, base round, apex narrowly acuminate leaves are dark green above and light green below, mibrib pale green. It has deciduous calyx, non-fleshy pedicel and fruit shape like that of apple.
3. ***Pyrus serotina*** Rehd: Shiara: Shiara is a wild relative of pear and is widely used as rootstock Tree 8-9 metres in height, spreading in nature. Leaves ovate, 8-10 cm long, pointed at the tip, modium in size. Spines long, stout and pointed. Leaf margin finely serrated. It has deciduces calyx and fruit shape like that of apple.
4. ***Pyrus pashia*** Hamilton: Kainth: This species is commonly used for raising of rootstock Trees having upright habit of growth, 9 metres in height. Branches are thicker than pear. Leaves elliptical, margin serrated, base and apex acute, 5-7 cm long. Spines present, long and pointed Fruits about 1 inch long, sub-globose and brown. It is found wild in the Himalayas to Norther China

5. ***Pyrus communis X Pyrus pyrifolia*** : Medium-vigor and spreading pear trees. The leaves are medium in size, 8 to 10 cm long, heart-shaped, larger at the base and narrower towards the apex, with a glabrous top surface and a light green colour. Fruits are small to medium-sized, pyriform, and have an appealing greenish yellow colour.

Other important species

The nashi pear (*P. serotina* Rehder): This sand pear is from China. Fruits are handsome in abundance; shelf life is better than European pears, poor as dessert but excellent for cooking. In India known by names like bada kainth, shiara, and zarenth. Important cultivars are Dong Guo Li, Su Li, Ya Li and Xue Hua Li.

Wild Himalayan Pear (*Pyrus pashia* Hamilton): This is Wild Himalayan Pear also known as Afghan pear native to southern Asia. Locally, known by batangi (Urdu), tangi (Kashmiri), mahal mol (Hindi) and passi (Nepal). Its fruit is edible and characterized as being pome. It looks like the russet apple and has an astringent but sweet taste when ripe.

Hybrid pears: These sand pears from Japan hybridize freely with the common pears, and several valuable hybrids are now widely and commonly grown e.g. Kieffer (*P. communis x P. pyrifolia*), Le Conte (*P. communis x P. pyrifolia*) and Garber. These hybrids are much stronger, more blight-resistant, more productive, and more rapid in growth, fruits are pyriform, better flavour than oriental.

Soil and climatic conditions

Pear grows best in deep, well-drained, fertile medium textured loamy soil. It is more tolerant to wet soils but less tolerant to drought than apple. Pear even do well on poorly aerated heavy soil with high water table which is heavy in texture for most of deciduous fruits. A pH range of 6.0 to 7.5 is desirable because Fe deficiency appears in alkaline soils.

At an elevation of 1,200 to 1,800 metres above sea level, where mean maximum temperatures of 15°C–25°C in winter and 20°–32°C in summer and mean minimum temperatures of 2°–10°C in winter and 10°–18°C in summer predominate, pears can be cultivated well. It can withstand temperatures as low as -26°C during the dormant season and as high as 45°C during the growing season. Pear production is negatively impacted by spring frost, and open blossoms are killed at temperatures of 3.3°C or lower. For a large number of pear cultivars to fulfill their chilling need and successfully blossom and bear fruit, 900–1200 hours below 7°C must be spent throughout the winter. Bartlett pear has a chilling requirement of about 1500 hours. Rainfall between 100 and 125 cm annually is regarded as acceptable. The best flavour and keeping quality in Bartlett pear develop in locations where summers are fairly hot and dry. Cool nights (10°C) during a month before harvest cause softening and premature ripening of this cultivar, thus impairing fruit size and quality.

Important Commercial Varieties

There are two main types of pear belongs to European or Common grows USA (95%), world (80%). Asian or Oriental are mainly grown in China/Japan (80%) and world (20%), fruits crispy, juicy, and sweet, globose to pyriform. Another group comprised of hybrid pears from cross between *Pyrus pyrifolia* and *Pyrus communis*. It is one of the most important fruits of temperate regions, being the species from which most orchard pear cultivars grown in Europe, North America and Australia have been developed.

European pear varieties: Beurré Bosc (Kaiser Alexander), Williams' Bon Chrêtien (Bartlett), Winter Nelis (Honey Pear), Abaté Fétel (Abbe Fetele), Beurré Hardy, Clapp's Favourit(Flemish Beauty x Williams), Concorde Patent/ Flemings (Conference x Doyenné du Comice), Conference, Doyenné Du Comice (Decana Del Comizio).

Asiatic pear varieties: Hanhong (Pyrus ussuriensis Maxim. x P. bretschneideri Rehd.), Ichiban Nashi, Shinsui, Kosui, Shinseiki, Hosui, Kikusui, Yoinashi, 20th Century ('Nijisseki'), Chojuro, Shinko, Niitaka, Ya Li, Tsu Li, Dasui Li, Shin Li, Okusankichi.

Interspecific hybrids: Leconte, and Keiffer.

Red colour sports: Red Anjou, Red d'Anjou, Corella, Butirra Precoce Morettini, and Red Bartlett.

Varieties based on maturity

Season	Varieties
Very early	Hood, Flordahome, Fla 39-49, Fla 57-75, Fla 58-45
Early	China pear, Beurre-de-Amanlis, Corella, Red Face, Forelle.
Mid-season	Beurre Hardy, Citron-do-Carme, Conference, Clapp's Favourite, China Sand Pear, Doyenne Bussoch, Genta Drauard, Harrow Delight, Fertility, William Bartlett.
Late	Beurro Bosc, Comice, Concord, Dr. Jules Guyot, Hardy, Passe Crassane, Vicar of Winkfield Williams, Winter Nelis.

High Hills: Bartlett, Red Bartlett, Max Red Bartlett, Flemish Beauty, Starkrimson, Conference, and Clapp's Favourite.

Mid, Low hills and Valley areas: Patharnakh, Kieffer, William, Lagoon, Fertility, Baggugosha, Punjab Beauty, Punjab Gold

Low chilling pear varieties : Patharnakh, Gola, Leconte, Keiffer, Smith, Baggugosha, China Pear, Pineapple, Baldwin.

Fig. 1.2: Pear Varieties

Varieties with their salient feature

The following varieties of pears are given along with their salient features

Bartlett. It is the popular commercial cultivar. The tree is prolific and regular bearer. The fruit is medium to large, ovate pyriform in shape and turn yellow in colour on ripening with thin skin. The pulp is soft, juicy, sweet and scented but keeping quality is poor. It is mid season variety and starts ripening from the first week of August. The fruit is suitable for dessert purpose, drying and canning

Conference. It is a regular and heavy bearer variety. The fruit is medium in size, pyriform in sape and pale yellow in colour at maturity. The pulp is salmon coloured, sweet and juicy. It is a mid seas, a variety and ripened from last week of July to first week of August.

Clapp's Favourite. It is a good and regular cropper. The trees are large, spreading, round topped and very productive. The fruits resemble Bartlett in size, shape, colour and flavour. Skin is pale lemon yellow with brown dots. The texture is fine, melting, buttery, juicy and flavoured

Flemish Beauty. The trees are modium in ze, vigorous and spreading with drooping branches Fruit is large, obovate or roundish, colour clear yellow, overspread on the exposed check with a dotted and marbled red blush. The flesh is tender, granular, juicy, sweet and aromatic with a slight pusky flavour. It ripens in September.

Patharnakh: It belongs to the *Pyrus pyrifolia* species. This type is the most widely used due to its robust bearing and excellent keeping qualities. The fruit is hard and sturdy, and it can be transported over long distances without going bad. The fruit has noticeable spots and is medium-sized and green in colour. The flesh is luscious, crunchy, and grittier. In the final week of July, it ripens. It is a heavy producer, producing 60 kg of fruit on average per tree.

Baggugosha: It belongs to the *Pyrus communis* species. The erect, healthy trees have an erratic bearing pattern. It contains tiny, yellowish-greenish fruits with a tapered stem end. The flesh is slightly rough and delicious. Punjab Beauty has an average yield of 60 kg per tree and ripens in August. The trees are regular in bearing, medium in vigour, and straight. The fruit has a medium-sized, reddish-yellow flush. More juicy and delicious than Le Conte and Baggugosha, the flesh is white. Fruits reach maturity in the third week of July. 80 kg is the typical yield per plant.

Punjab Nector. The trees are upright, medium in vigour and precocious. The fruit size is medium to large, ground colour is yellow green. The flesh is whitish, very juicy, TSS 13.6% and acidity 0.21%. When ripe, the mature fruits lose their roughness and become mellow. It mature in fourth week of July and yields 80 kg fruits per tree. Suitable for processing like ready to serve (RTS), nectar and squash

Punjab Gold. It is a sister line of Punjab Nectar. The trees are upright, medium in vigour and precocious. The fruit size is large, ground colour is golden yellow. The flesh is whitish, TSS 13.3% and acidity 0.22% It mature in fourth week of July and yields 72 kg fruits per tree. suitable for processing, such as nectar, squash, and RTS.

Le-Conte : It is a cross between *Pyrus communis* X *Pyrus pyrifolia* produced it. LeConte is a variety that requires little cooling. Medium-sized, upright, and spreading trees that have reached full maturity Trees produce a strong harvest but are vulnerable to blight. Its pyriform, tiny to medium-sized fruits have a greenish yellow colour and are small to medium in size. By the end of July, it repens, and each tree produces between 60 and 80 kg of fruit.

Kieffer: It is a cross between *Pyrus communis X Pyrus pyrifolia*. It is a variety requiring longer chilling hours than Le Conte and Patharnakh. Trees are extremely productive, flexible, robust, and disease-resistant. The fruit is medium to large in size, pointed at both ends, gritty, firm, and ideally suited for canning. Its colour is brownish red. Throughout August to September, it ripens. 100 kg is the typical yield per tree.

Propogation and rootstock

Propagation

Pear is typically cultivated on kainth (*Pyrus pashia*) or shiara seedling rootstock (*Pyrus serotina*). Between December and January, local residents gather kainth seeds, which are then directly sowed in the ground without stratification. Kainth seeds can be layered in damp sand for 35–45 days at 2–5°C for enhanced germination. For grafting or budding, seedlings older than a year are used. The most popular methods for growing pear plants are T-budding from April to September and tongue grafting from January to March.

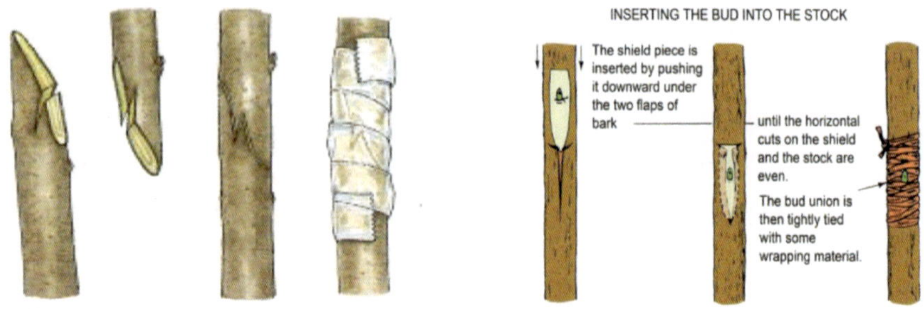

Above: Whip-and-tongue grafting sequence.

Fig. 1.3: Tongue grafting and T - budding

Several *Pyrus communis* F1 hybrids, including Old Home x Farmingdale, are propagated clonally and utilised as rootstock. Another closely related genus is quince (*Cydonia oblonga*), which has excellent compatibility with many commercial cultivars and is resistant to woolly aphids and worms but vulnerable to oak root rot, fire blight, frost, and too much lime in the soil. Quince clones can be easily multiplied using softwood and semi-hardwood cuttings. The quince clones QA, QB, and QC are a few that are utilised commercially. However, due to their incompatibility with quince, commercial cultivars of pears such as Bartlett, Bosc, and Eldorado require "Old Home" as interstock.

For greater fruit yield and quality, the chilled Doyenne du Comice scions grafted at 1.0 m height on 1.0 m long shoots of Pant pear -18 stock were discovered to be the best.

Rootstocks

Dwarfing	Quince-C, Quince-A, Pyrodwarf, Pyro-233, Ox F-97, OH X F 51, Oregon 211, Oregon 249, OH x F 34, OH x F 69, OH x F 87, OH x F 230, OH x F 233.
Cold hardy:	OH x F series
High temperature tolerant:	Oregon-211 and 249, Oregon-260, 261 and 264.
Drought tolerant:	Oregon 211 X 249, Oregon 260, 261 and 264.
High soil moisture tolerant:	Oregon 211 X 249, Oregon 260, 261 and 264.
Low soil pH tolerant	Oregon series
Resistant to powdery mildew	Oregon 211 and 249
Resistant to crown and root rot	Oregon Series X OH x F series
Resistant to crown gall	Oregon series and OH x F series
Resistant to canker/gummosis	OH x F series, Oregon 260, 261 & 264.
Nematode resistant	Oregon series

In plains, planting can be carried out at any time between December and mid-February. Yet on hills, planting often takes place in the late fall or the early spring.

Planting and Orchard Establishment

The field should be prepared a year before planting by cutting off the stems and roots of any existing trees and shrubs, levelling the ground, and creating a slight slope to allow any surplus water from heavy rains to drain. An initial spacing of 3 M x 2 M is recommended for a crop grown on its own rootstock (pears), which is modified to 6 M x 4 M after 4-5 years. Since quince has the effect of dwarfing the trees, 3.5 M x 1.1 M is sufficient for growing pear on quince. The pit should be 1M x 1M x 1M in size, and soil and compost are used to fill them. Late fall or early spring are both suitable times to start planting. The basin should be created and watered as soon as the plants are planted. On plains, planting can be carried out at any time between December and mid-February. Yet on hills, planting often takes place in the late fall or the early spring.

Planting and After Care

The optimal period to plant pear trees is from the middle of January to the first fortnight of February, before they begin to sprout new growth. Plants that are at least a year old and in good health should be utilised for planting. With good results, older plants that are 2-3 years old can also be planted. Before the plants are put in the field, the orchard site should be properly laid out. The trees may be planted at a distance of about 7.5 metres apart by which 180 plants are accommodated in one hectare by the square system of planting. The hexoganal technique, which allows 207 plants to be planted in a hectare, is another option for planting pears. Peer can also be placed 5 x 5 metres apart. Using quince rootstock, high density planting pear have been attempted. Planting distance of 3-3.5 x 1.2 m on Quince MA has been recommended for William's Basins should be pepared immediately after planting. The level of soil near the trunk should be kept slightly higher than the level of basins so that water may not come in direct contact of the trunk. Irrigation should be given immediately after planting so that soil may settle down properly around the roots. After two or three days, second irrigation should be given. Subsequent irrigations should be given as and when required. The irrigation interval upto March may be 15-20 days, but later on as the season warms up the interval may be reduced to 10-12 days upto June. Irrigation during rainy season is not required. The trunk of young plants should be white washed once in March, again in May and then in the rainy season to avoid injury from sun-burn and to prevent cracking of bark Farmyard manure should be mixed in the basins at the rate of 20 kg per tree in January-February, Irrigation should be applied immedately after it is thoroughly mixed in the basins. To check the attack of white ants, the trees should be given chlorpyriphos 2 ml per litre of water once in April and then after rains. The surplus shoots not required for building the framework of the trees should be removed. The shoots appearing on the rootstock should also be removed, otherwise they will have an adverse effect on the growth of plants. Trees should be protected from insect and pests by following the proper spraying schedule.

High Density Orcharding (HDP)

In pear, high density planting systems gave very high yields in varieties like Abbe Fetel, Conference and Doyenne du Comice. These all varieties found productive on dwarfing rootstock Quince and its various clones. Higher yields were mostly recorded at very high densities (5555 to 7936 trees/ha) which indicates that the planting in hedge-row pattern and trained with open centre systems like V-shape and Y-trellis facilitates maximum sunlight interception and results into maximum yields.

Orchard Cultural Practices

Training and Pruning

Proper training and pruning of pear trees is essential for the development of proper framework and to control the size of the tree for the convenience of horticultural operations, to maintain the vigour and growth of the tree, to spread the fruiting area uniformly on the tree, to secure fruit of Pear good size and quality and to encourage regular bearing. Pear trees can be trained according to modified leader system, free standing, pyramid, spindle bush, palmette and tatura trellis. For high density planting, tatura trellis system of training is followed. Typically, pears are trained using the modified leader method.

Training in the first year: In low-headed trees, the lowest branch is allowed to develop upto a height of 60 cm and accordingly top the plant is cut at about 90 cm. If high-headed trees are required, then the plant should be headed back at a height of 12 to 1.5 metres.

Training in second year: The largest shoot in the centre is kept as the leader, the beginning of the immature area is clipped back. Three to five laterals (major scaffold branches), spaced 15 to 22 cm apart, are maintained in various directions. By attaching these laterals to the ground with rope or thick thread, it is preferable to bend them downward. These laterals will be tied, which will promote the growth of secondaries.

Training in 3-4 year old trees: One or two of the new laterals that arise when the second season's growth is complete from the laterals left over should be chosen to build the secondary scaffolds framework. The laterals should be directed backward to promote the sprouting of buds that will give rise to tertiary branches.

Training in 5 year old trees: To an appropriately situated, outward-going lateral, the leader should now cut back.

Pruning of young bearing trees: Pear trees produce fruit on spurs, or stems with restricted growth. These spurs only grow a few millimetres per year yet produce fruit for 8–10 years. Consequently, for the first 10 years, only corrective pruning is done on young trees. Just cut off diseased, dry, and encroaching branches. Extreme pruning, which would result in the loss of fruiting spurs, should not be done during this time. When picking fruit, caution should be used to prevent injury to the spurs.

Pruning of older trees: The bearing wood of the elder trees needs to be rejuvenated and regenerated. The largest branches are completely removed to achieve this. To allow more light and greater colour, this helps to open up the trees. The renewal process takes 3 to 4 years to complete in stages. The trees once more sprout new fruiting spurs, which continue to produce fruit for an additional ten years. In early February or in January, pear pruning should be carried out.

Rejuvenation of old pear orchards. Around 20 years of age is when pear trees start to become less fruitful. Such trees can be revived to resume production. That can be accomplished by repositioning the three to four major scaffolds back by about 15 cm in the months of December and January. Remove all of the remaining scaffolds. Bordeaux paint should be applied to the cut edges. In March, the majority of the sprouts on these stubs will appear. Just one to two outgoing shoots from each stub should be kept in May, for a total of six to eight shoots per tree. Rejuvenated trees begin bearing fruit in their third year and produce a commercial crop in their fifth year.

Plant growth hormones

Crop regulation and pre-harvest management

- **Improving fruit set:** When applied to frost-damaged flowers, GA_3 treatments caused parthenocarpic fruits and enhanced fruit setting and fruit quality, whereas GA_3 applications made in a favourable environment had no positive impact. The most popular bioregulator for chemically triggering the development of parthenocarpic fruit is GA_3. Applying AVG to pears at full bloom or after bloom, respectively, enhanced fruit set. AVG is an inhibitor of the production of ethylene.
- **Fruit Thinning**: Chemical thinning in pear is intended to reduce the current season's crop load in pursuit of 1) inhibit fruit set to minimize green fruitlet hand thinning; 2) improve size and quality of surviving fruit; and 3) promote return bloom to encourage annual cropping. A well-managed chemical and manual thinning programme can boost the long-term profitability of many kinds, especially Bartlett, even though the majority of pear trees essentially self-regulate their yields without chemical intervention. When circumstances have been cool and/or wet during bloom, it may be prudent to thoroughly analyse pear set before using chemical thinners. Certain pear blocks can struggle to set commercially appropriate quantities of fruit. When pears are 5-15 mm in diameter, BA (MaxCel, RiteWay) may be used to decrease fruit set, enhance fruit size, and encourage return bloom. Reduce pear fruit set by applying NAD (Amid-Thin W) 5-7 days after petal fall. Pears can be thinned 15–28 days after full bloom with NAA (K-Salt Fruit Fix 200, K-Salt Fruit Fix 800, or Fruitone). When temps are between 70 and 75 F, use NAA products with a surfactant (wetting agent) for optimal results. For more specific information, please refer to the labels of each product. At this moment, there are no chemical bloom thinners approved for use in pears.
- **Fruit drop**: NAA is helpful for stopping pear pre-harvest drop. After application, NAA takes effect 2 to 3 days later and remains in effect for around 2 weeks. 5–7 days before harvest, apply.

Intercultural practice

Intercropping

In addition to providing a decent revenue from the land, wise intercropping also maintains the productivity of the land at a high level and controls weed growth. Wheat, peas, grammes, and seeji may be cultivated during the rabi season in new and revitalised pea orchards to generate some income during the non-bearing season. Intercrops such moong, mash, and toria can be grown during the Kharif season. Even in bearing orchards, crops like peas, grans, and senji can be cultivated during rabi and moong, mash, and sunflowers during kharif. During intercrops, additional fertiliser doses should be used.

It is also possible to interplant some fruit trees, such as peach, plum, papaya, and phalsa to act as fillers. However, the fillers must be taken out once the pear trees start to give fruit. Additional doses of the fertilizers should be added to filler trees according to their requirements, so that they do not compete for nutrients with the permanent trees. It should be kept in mind that once the peaches are planted as fillers, no intercropping with crops like berseem, patato, tomato, etc should be done during the rabi or kharif seasons. Otherwise the peaches as fillers or the main pear plants may suffer from the on slaught of root rot or any other devastating disease, because of the excessive watering to the intercrops, which is not required at any cost by the deciduous trees due to dormant condition in winter months.

Fig. 1.4: Intercropping

Weed Control

Weeds can he effectively controlled by spraying the herbicides namely hexpron or glycel or gramagne Spray hexuron 80 WP (diuror) at 4.0 kg/hectare as pre-emergence (first fortnight of March) after thorough cultivation so as to remove the established growing weeds/plant debris. Gylcd 41 SL (glyphosate) at 3 litres/hecture or gramoxone 24 WSC (paraquat at 3 tres hectare should be sprayed as post- emergence when the weeds are growing actively and to a height of 15-20 cm. Dissolve the herbicides in 500 litres of water per hectare so as to give a thorough coverage to the field/weeds. In case there is emergence of weeks during/after rainy season, another spray of glycel gramoxone at the same rate may be dane. Spray glycel gramosone during calm day so as to avoid spray drift to the foliage of the fruit tree.

Manures and manuring

To get the highest production, an ideal major nutrient dose for each tree is 600g N, 150g P, and 300g K. Typically, only in soils with limited P&K availability can pears respond to P and K. The 'P' won't be available at higher elevations where the soil pH will be less than 7. Similar to how it will be less "P" available when the soil pH is higher than 7 (an alkaline condition). Under these circumstances, adding more "P" will raise the yield. The Baggugosha variety of low-chill pears performed best when fertilised with nitrogen @ 60g/tree in two splits (2/3 in January and 1/3 in May), combined with a basal dressing of 40g each of phosphorus and potash.

Nutrient management

Optimum supply of nutrients is essential for a good yield and quality fruits. Soil pH plays an important role in the availability of various nutrients. In order to maintain the adequate availability of essential elements for best plant performance, it is necessary to know their deficiency, optimum and excess levels in the tissue as well as soil. Leaf sampling of 20 leaves from the middle of current year's shoot of average vigour and length from each of 10 selected trees is the best procedure. The suggested optimum levels (% dry matter) of nutrients for William pear are N (2.13-2.75 %), P (0.110-0.161%), K (0.80-2.16%), Ca (1.18-3.00%), Mg (0.317-0.419%), Mn (0.0068-0.022%), Fe (0.0028-0.024%), Cu (0.005-0.010%) and B (0.0010-0.0043%). Fertilizers and manures are applied relatively in large quantity as better irrigation facilities are available in plains. Throughout December and January, phosphatic fertilisers and farmyard manure are applied. Twenty to twenty-five days before to bud burst in the spring, nitrogen and potassium are applied. Nitrogen can be administered in the form of calcium ammonium nitrate (CAN) or urea, phosphorus as single superphosphate, and potassium as muriate of potash, as much as is practical. Two separate applications of nitrogen should be made, the first half in early February and the second half in April following fruit set.

Age	Quantity of nutrients given to each tree (gm)		
	Nitrogen	Phosphorus	Potash
Newly growing trees	50-100	50-100	100-120
Trees going to start bearing	150-250	150-200	150-200
Trees in full bearing	300-400	150-200	250-300

Water management

Pear trees can tolerate drought condition for quite long time. They can also withstand waterlogged condition more efficiently than any other temperate fruit crop. But for a good crop, timely irrigation and removal of stagnating water is necessary. In areas with convenient irrigation systems, trees should be watered every 10 to 15 days after they have produced some fruit. For optimum growth of fruit, shoots, and trunks in heavy soils, moisture delivery representing 50% or more of the maximum available moisture in the upper 1m is crucial. Lighter soils or those with a gravel substratum require irrigation that is applied more frequently and heavily. A loam soil should be wet 1.5-1.8 m deep. In hills, pear cultivation is mostly done under rain fed conditions but at many places irrigation facility may be available which is and additional advantage. Besides rain fall, 75-100cm irrigation may be applied annually in some regions. After harvesting in July-August, the trees should be irrigated at 20-days intervals through the end of October. After then, irrigation is not necessary until January, with the exception of when manures and fertilisers are added in December.

Flowering and Fruiting

The ovary is inferior, and the flowers are epigynous. Pear inflorescences are of the Corymb type. The Patharnakh pear has white, 3 cm or more in diameter, 4–12 flowers grouped together on slender pedicels that emerge with the initial foliage. The last week of February marks the beginning of blossoming, which lasts through the third week of March. In common pear, the flowers are 4-12 in umbel-like clusters on slender pedicels, white, appearing with the leaves, calyx persistent or rarely deciduous, stemens 15-20. The fruit is borne on spurs. It is pollinated through insects. The pollen is highly viable.

Pear fruit-setting begins towards the end of March. Fruit is carried on spurs. The fruit (Patharnakh) reaches harvest maturity after 155 days of fruit set. The fruit growth in terms of length and diameter shows three distinct phases and follows a pattern of the "Sigmoid Curve".

Fig. 1.5: Flowering and fruiting

Insect pest and their management

1. San Jose -Scale (*Quadraspidiotus perniciosus*); Hemiptera

San Jose Scale is an extremely important pest of Apple, Pear, Peach and plums. Nymph and female scales attack all above ground parts. Feeding site turns into a characters purplish red colour that initially growth of plants is checked but as scale increases in number plants may be die.

Control

- Prune the infested plant parts and destroy.
- Encourage the activity of parasitoids like *Prospaltella perniciosi and Aspidiotophagus* Sp.
- Winter spray with diesel oil emulsion at 8-12 1/ tree (diesel oil 4.5 1, soap 1 kg, water 54-721).
- Summer spray with Phosalone 50 EC 0.05% or Chlorpyriphos 20 EC 0.05%.

Fig. 1.6: Sanjose scale infected fruit of Pear

2. Pear psylla

The aphid-like creatures produce large, sticky droplets of honeydew that can cover the tree and fruit after sucking the sap from the leaves. Severe infestations can damage fruit buds, decrease the growth of shoots, and cause premature leaf drop.

Fig. 1.7: Pear psylla

Control

- Limiting nitrogen inputs will help prevent excessive tree vigour, and summer trimming will help lower the quantity of psyllas.
- Biological control: A number of natural enemies attack psylla, including parasitic wasps, adults and larvae of lady beetles, lacewings, and earwigs, as well as predacious plant bugs and minute pirate bugs.
- Spraying with Methyl Demeton @ 0.25% or Diazinon 20EC @ 0.2% is advised to control this pest three weeks after petal fall; repeating the spray in cases of severe infestation is crucial.

3. Colding moth (*Cydia pomonella*); Lepidoptera

Colding moth is major pest of pear and apple. The caterpillars bore into fruit and stop it from growing, which leads to premature ripening.

Control

- Control of caterpillars Bands of sacking (gunny bags) or corrugated cardboard can be tie round the tree trunks by late july till the end of October to provide alternative over wintering sites for the caterpillars.
- In Biological control *Trichogramma embryophagum* @ 20000 adults per 50 trees.
- The pesticides that can be use are Chlorpyriphos 0.02% @ 2ml/L or Methyl oxydemeton 25 EC @ 0.02%.

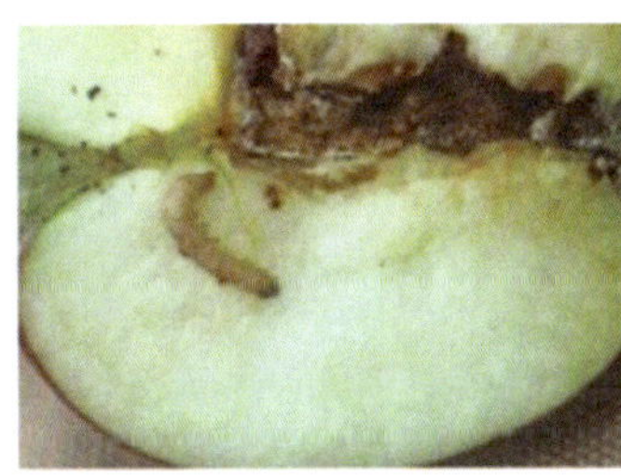

Fig. 1.8: Colding moth larva

4. Green peach aphid (*Myzus persicae*); Hemiptera

Several aphid species occasionally attack pears. Usually after pear bloom, when trees are growing rapidly, these aphids appear on foliage and shoots and establishing colonies. Green peach aphids can attain very high densities on young plant tissue, causing water stress, wilting, and reduced growth rate of the plant. Blemishes to the plant tissue, usually in the form of yellow spots, may result from aphid feeding.

Control

- Remove and destroy the damaged plant parts along with nymphs and adults.

Fig. 1.9: Green peach aphid (*Myzus persicae*)

- Use Natural Enemies like Parasitoid: *Aphelinus mali* and Predators: *Coccinella septumpunctata*, Syrphid Lygaeid bug, Coccinellid, Lacewing.
- Chemicals: Spray dimethoate 0.03% or methyl demeton 0.025% @ 2 ml/lit.

5. Aphids (*Schiazaphis sp.* and *Toxoptera sp.*)

Sometimes cause serious damage to the plants by sucking the cell-sap from the leaves and tender fruits. Aphid attacks the pear as soon as the sprouting starts (March-April).

Control

- They can be controlled by spraying Malathion 50EC 800ml with 500 liters water or 800 ml Rogor 30 EC (Dimethoate) in 500 litres of water.

Disease and their management

1. Pear scab (*Venturia pirina*) In early pear shoots, an infection is seen. The underside of the leaf is particularly affected by olive-dark mouldy sports. The patch turned grey and necrotic afterwards. The pedicel, calyx, and petals all acquire comparable olive markings. Blooms and petals that are harmed typically fall. On the fruits, dark-brown markings appear. The blemishes congregate, thicken, and crack.

Fig. 1.10: Pear scab

Control

- To check this disease, apply bulky organic manures of the trees.
- "Apply 0.2 percent" Captan at the dormant trees' pro-bloom stage and continue till petal fall at 10-day intervals. If the illness doesn't go away, it can take seven or eight sprays. To reduce ascospore discharge and subsequent disease infection, gather fallen leaves and carefully plough them under or burn them. To encourage healthy acration, prune the trees.
- Spray Mancozeb 300g / Captain 300g at fruit development stage.

2. Fire blight (*Erwinia amylovora*)

This is a bacterial disease that affects pears frequently and severely. Twig blight first affects young shoot tips before spreading to older twig parts. Twigs with blight initially look water-soaked before turning dark brown or black.

Control

- During the dormant season, blighted twigs and fire blight cankers should be pruned.
- Prevent irrigation, heavy pruning, and use of nitrogen fertilisers during bloom.

- Bordeaux mixture or other copper-based fungicides can be sprayed multiple times as petals open to help prevent new infections.
- The best control is achieved by dipping the fruits in aureofungisol at 500 ppm for 20 minutes.

Fig. 1.11: Fire blight (*Erwinia amylovora*)

3. Shoot-fruit blight and bark canker (*Phoma glomerata*)

Cankers appear on bud scars, wounds, twig stubs or in crotches. Cankers are elliptical, small circular, brown spots appear around a leaf-scar or superficial wound. As the canker enlarges, the centres become sunken with the edges raised above the surrounding healthy bark. The fungus perpetuates itself on the trees in bark cankers.

Control

- To control this disease, cankers on the trunk and in the crotches should be removed and the dead bark decorticated along with 2 cm of the healthy bark. All the dead wood and pruning should be destroyed.
- Cover the wounds with a disinfectant solution and after this apply Bordeaux paste to them. After a week cover the dried paste with Bordeaux paint.
- The whole operation should be immediately followed by spraying the pruned and canker treated tree with 2: 2: 250 Bordeaux mixture. Spray may be repeated in March, June and January.

4. Root-rot and Sap-wood-rot (*Polyporus polustris, Ganoderma lacidum, Schizophyllum commune*)

Gummosis of main trunk, limbs and secondary branches occur. The bark and wood of the roots rot and turn brown with white mycelial mat in the crevices. The affected trees begin to show symptoms of wilt, early leaf-fall and increase in fruit-set before collapsing. Rotting of roots leads to toppling over the trees, especially those with bulky tops.

Control

- Apply 10 g Bavistin 50% WP + 5 g Vitavax 75% WP mixed in 10 litres of water along the trunk and around the drip region twice; once before the monsoon (April-May) and once after the monsoon (September-October) to each fully developed tree (4-6 years). Provide a gentle irrigation to the trees right after the treatments. Fungicide dosage for young trees should be cut in half.

- Avoid severe hoeing to prevent damage to the roots, where the fungus enters the plant. Moreover, stay away from cultivating intercrops that demand a lot of water throughout the winter. Keep the soil from piling up around the tree's trunk.

Disorders

Internal Browning:- Development of Brown to dark brown water – soaked areas in the core or flesh occur during storage of Chinese pear. Cause of internal Browning is still unknown. Ripening- senescence predisposes the fruit to the disorder while it is in cold storage.

Control:- Harvest fruit while the most of the pears are still green on the tree, however a few at the top may start to acquire smooth surfaces and light yellow spots.

Fig. 1.12: Internal Browning

Core Breakdown:- A brown, soft breakdown of the core and surrounding tissues.

Control:- Fruit should be quickly cooled and held at the lowest non-freezing temperature possible to extend its shelf life after harvest (generally-2 to -1^{0}C, 28 to 30° F).

Black end in pear due to use of poor quality rootstock and Water imbalance.

Cork spot of pear:- Calcium deficiency

Calyx End rot:- Boron deficiency

Fig. 1.13: Core Breakdown

Maturity Indices

Maturity indices of pear are methods used to determine when a pear is ripe and ready to be harvested. The following are some of the common maturity indices used for pears given below

- **Fruit Color**: As pears mature, their skin changes from green to yellow, providing a visual cue of the fruit's ripeness.
- **Texture:** The flesh of the pear becomes softer as it ripens, indicating that it is ready to eat.
- **Sugar Content:** The sugar content of pears increases as they mature, making them sweeter. This can be measured using a refractometer or a Brix meter.
- **Ethylene Gas:** Pears release ethylene gas as they ripen, which can be used as a signal of maturity.
- **Stem Separation:** A mature pear's stem should separate easily from the fruit. If the stem is still attached, the fruit is not yet ripe.
- **Pressure Test:** To check for ripeness, gently press the stem end of the pear. If it gives slightly, the fruit is ready to eat. If it is too hard, it needs more time to ripen. If it is too soft, it may be overripe.
- **Other Indicators:** Overripe pears may have wrinkled skin, mold, or a strong, fermented odor.

Yield

Some of the parameters for harvest maturity of pear fruits include change in external fruit colour, decrease in firmness of the fruit flesh, and increase in total soluble solids content. The fruits should be picked and handled normally as soon as the first symptoms occur. The best annual fruit production from a tree is between 40 and 50 kilogrammes.

Harvesting and Post-harvest handling

Harvesting and post-harvest handling play a crucial role in maintaining the quality of pears and extending their shelf life. Harvesting of the pear fruit at proper stage of maturity, is very important. The standards of maturity fruit growth in pears includes a change in the colour of the fruit's exterior, a softening of the fruit's flesh, and an increase in the concentration of total soluble solids (TSS). The fruit of Baggugosha and Punjab Beauty reaches harvest maturity after 145 days and of Pathamakh after 155 days of fruit-set. The size and weight of the fruit continued to increase till harvest in August. During the last stage of harvest, the fruit's colour turns bright yellow. The best months for harvesting pear varieties are July and August. Preharvest spray of ethephon at 100-300 ppm hasten maturity of fruits by two weeks. At 15.6–21°C with RH 80–85%, the majority of pear cultivars mature and produce their best dessert qualities. Picking the fruit should be done carefully to avoid damaging it. Unripe fruits that are harvested at their peak of ripeness can be kept for up to 5 months at a temperature of –1°C.

Picking should be done carefully to avoid damaging the fruit-bearing spurs. Once the spurs are broken damaged, that will take long time to become fruit-bearing spurs again and will ultimately affect the fruit yield in the coming bearing seasons.

After harvesting, pears must be carefully handled to prevent damage, as they are highly susceptible to bruising and other types of physical injury. Pears are typically stored at a temperature of 0-2°C with a relative humidity of 90-95%, which helps to reduce the rate of respiration and delay the onset of senescence (Gao *et al.*, 2019). Various post-harvest treatments such as controlled atmosphere storage, modified atmosphere packaging, and the use of ripening agents have been investigated to extend the shelf life of pears. Controlled atmosphere storage involves storing pears in an atmosphere of reduced oxygen and increased carbon dioxide levels, which helps to delay ripening and preserve fruit quality. Modified atmosphere packaging involves storing pears in a package with modified gas composition, which can also delay ripening and improve fruit quality. The use of ripening agents such as ethylene gas can also be used to accelerate the ripening of pears, allowing them to be sold and consumed when they are at the desired level of ripeness (Watkins 2019). In conclusion, proper harvesting and post-harvest handling are essential for maintaining the quality and extending the shelf life of pears. Further research is needed to investigate new and innovative post-harvest treatments to improve pear storage and preservation.

Future Prospect

Pear is considered to be a valuable crop with a lot of potential for the future. The increasing demand for pears, easy cultivation, low maintenance and the ability to be grown in a sustainable manner make it a desirable crop for growers. With the use of new technologies, such as precision farming, pear cultivation can become more efficient, profitable, and environmentally friendly.

Crop improvement of pear as a future prospect is a key area of research and development that aims to improve the productivity, profitability, and sustainability of pear cultivation. Pear breeding programs focus on developing new cultivars that have improved resistance to pests and diseases, better fruit quality, and higher yields. This can be achieved through traditional breeding methods, such as crossbreeding, or through more advanced techniques such as genetic engineering and gene editing. One important area of research is the development of new cultivars that are resistant to fire blight, which is a bacterial disease that can cause significant damage to pear orchards. This can be achieved through traditional breeding methods, such as selecting for resistance to fire blight in wild pear varieties or through genetic engineering and gene editing, which allows for the precise introduction of resistance genes into commercial cultivars. Additionally, there are also efforts to develop new cultivars that are better suited for different climates and growing conditions, as well as those that have improved fruit quality and storage life. This can be achieved through the development of cultivars that are better adapted to specific regions, have improved resistance to pests and diseases, and have improved fruit quality characteristics such as size, shape, color and texture.

Another important aspect of crop improvement is the use of precision farming techniques to increase efficiency and reduce costs. This includes using precision irrigation and fertilization systems, as well as using sensors and other technologies to monitor tree health and optimize growing conditions. This can help to reduce water and fertilizer use, improve yields and reduce the environmental impact of pear cultivation. In addition, new technologies such as gene editing, can be used to improve the nutritional value of pears, making them a healthier and more desirable fruit for consumers. This can be achieved by introducing genes that increase the content of beneficial compounds such as antioxidants, or by reducing the content of compounds that are associated with negative health effects, such as sugar. In conclusion, Crop improvement of pear as a future prospect is an important area of research and development, aimed at developing new cultivars that are resistant to pests and diseases, have improved fruit quality, higher yields and can be grown in different climates and growing conditions. Additionally, precision farming techniques and new technologies such as gene editing can be used to improve the efficiency, profitability, and nutritional value of pears. These efforts will contribute to the profitability and sustainability of pear cultivation in the future, making pears more accessible and more desirable for the consumers.

References

Ahmed, M., Anjum, M. A., Shinwari, K. Z, Awan, M. S. and Rabbani, M. A. (2011). Assessment of fruit quality parameters of Pyrus germplasm collected from Azad Jammu and Kashmir (Pakistan). *Pakistan Journal of Botany*, 43(2): 971-981.

Anonymous, (2020). National Horticulture Board, Indian Horticulture Database. Ministry of Agriculture & Farmers Welfare, Government of India. www.agricoop.nic.in.

Bal. J. S. (2019). Fruit Growing, Third Revised Edition, Kalyani Publishers, B-1/1292, Rajinder Nagar, Ludhiyana-141008, pp. 324-336.

Bose, T. K., Mitra, S. K., Rathore, D. S. (2001). Temperate Fruits, Third Edition, Volume 2, Horticulture and Allied Publishers 27/3, Chakraberia Lane, pp. 1-122.

Chadha, K. L. (2019). Hand Book of Horticulture. Second Edition, Volume 1, Indian Council of Agricultural Research, New Delhi. pp. 243-249.

Chattopadhyay, T.K. (2009). A Textbook on Pomology IV: Devoted to Temperate Fruits Crops, Second Edition, Volume 4, Kalyani Publishers B-1/1292, Rajinder Nagar, Ludhiyana-141008, pp. 46-68.

Gao, Z., Liu, H., Duan, X., & Jiang, Y. (2019). Postharvest biology and technology of pears. *Postharvest Biology and Technology*, 156, 110935.

http://eagri.org/eagri50/HORT181/lec31.pdf.

http://www.kiran.nic.in/pdf/Agri-Kaleidoscope/Horticulture-resources/PEAR.pdf.

https://naldc.nal.usda.gov/download/CAT88913041/PDF.

https://www.horticulture.com.au/globalassets/hort-innovation/resource assets/ap12002 planting-and-management-systems-for-new-blush-pears.pdf.

Muthukumar. P. and Selvakumar. R. (2017). Glaustas Horticulture (Pear) . New Vishal Publication New Delhi. (Edition-2) 175 -177.

Reiland, H. and Slavin, J. (2015). Systematic review of pears and health. *Nutr. Today* 50: 301-305.

Sharifani, M., Hemmati, K. and Fallahi, SHE. (2008). Evaluation of useful botanical traits for Iranian Pyrus species. *Acta Horticulturae* 769: 185-188.

Sharma, R. R. and Krishna, H. (2017). Text Book of temperate fruits, CBS Publisher &Distributers Pvt Ltd.

Sharma, S., Rehalia, A. S. and Sharma, S. D. (2009). Vegetative growth restriction in pome and stone fruits- a review. Agric Rev 30(1): 13-23.

Silva, G. J., Medeiros Souza, T., Barbieri, R. L. and Costa de Oliveira, A. (2014). Origin, Domestication and Dispersing of Pear (Pyrus spp.). Advances in Agriculture, Article ID 541097.

Singh, P. and Ramtake, V. (2018). Comprehensive of fruit science (Pear). New Vishal Publication New Delhi. (Edition-1) 191.

Verma, M. K. (2014 A). Pear Production Technology. Training manual on teaching of post-graduate courses in horticulture (fruit science). 2014c.

Watkins, C. B. (2019). Pears. In Postharvest physiology and biochemistry of fruits and vegetables (pp. 325-350). Elsevier.

Verma, M. K., Lal, S., Mir, J. I, Bhat, H. A. and Sheikh, M. A. (2014 B). Genetic variability among "Kashmiri Nakh" pear (Pyrus pyrifolia): a local variety grown in North Western Himalayan region of India. *African Journal of Biotechnology* 33:3352-3359.

3

Advances Production Technology in Peach

Pavan Prajapati*, Piyusha Sharma and Neha Kumari

PG Research Scholar, Department of Horticulture, Rajasthan College of Agriculture, MPUAT, Udaipur, Rajasthan
**Email- pavansantha6@gmail.com*

Introduction

Peach (*Prunus persica* (L.) is an important deciduous fruit crop in temperate regions. The basic chromosome number 8 and somatic chromosome number 16. It belongs to family Rosaceae family, subfamily Prunoideae. It has a deciduous nature and can reach heights of 5 to 10 m. It is classified with the almond within the genus Prunus in the subgenus Amygdalus that is distinguished from the other subgenera by the corrugated seed shell.

Fig. 1.1: Peach Fruit

The majority opinion among botanists today is that they are Chinese in origin and were transported along the silk road in ancient times, most likely by around 2000 BC, to Persia and the Mediterranean region. Introduction of the cultivated peaches probably took place in the latter half of the 19th century.

China, Japan, Iran, Turkey, and other Mediterranean nations have historically produced significant amounts of peaches. Peaches have been cultivated in these nations for thousands of years. The United States, Canada, and Australia have all grown in importance more lately. Due to insufficient summer heat, oceanic climate regions like the Pacific Northwest and the British Isles are generally unsuitable for peach growth, though they can occasionally be grown trained against south-facing walls to capture additional heat from the sun. In the British Isles, trees can thrive in a protected, south-facing location and produce both flowers and a plentiful supply of fruit.

It is grown in India in the Himalayan mid-hill region, which stretches from Jammu and Kashmir to the Khasi Hills at 1,000–2,000 metres above mean sea level. Low cooling peaches are grown in the sub-mountainous areas of Punjab, Haryana, Delhi, and western Uttar Pradesh. It is also being grown to a small extent in the highlands of south India and in the north-eastern area of the country.

Peach production worldwide totals 25,003,345 MT (FAO, 2019) [Note: statistics on global production includes both nectarines and peaches] On around 3.5 million acres, peaches and nectarines are commercially grown in 71 different nations. In India, peach is cultivated at an area of 18000 ha with annual production of 111000 MT (Anon, 2020).

In addition to being a well-liked fruit, peaches are well-known in China, Spain, and Vietnam for the numerous folktales and customs connected to them. According to Chinese mythology, peaches were eaten by the immortals because of their mystical ability to grant everyone who ate them a long life because they are a symbol of longevity.

Besides fresh and canned fruit, peaches don't have many specific marketing specializations. Clingstones are almost exclusively canned, whereas freestone peaches are offered fresh. Nectarines are marketed fresh in about 90% of cases. The breakdown of utilization is as follows: Fresh- 45-55%, Frozen- 5%, Canned- 35–40%. 9.5 pounds are consumed per person annually when peach and nectarine are considered. It contains a lot of vitamins, minerals, sugar, and proteins.

Composition and uses

Peaches and nectarines are best for processing and fresh market purpose. Peaches are good source of sugars (sucrose), proteins, vitamins (carotene, thiamine, riboflavin and niacin) and minerals (potassium, sodium, calcium, magnesium, iron & zinc). The pectin substances about 0.86 percent while tannin is only in minute quantity. Nectarines are riches in organic acids than peaches, the dominant acid being malic followed by citric acid, respectively. Total soluble solids in peaches range from 10 to 13 percent and acidity 0.4 to 0.6 percent.

Botany and Taxonomy

It is a little tree with a spreading canopy that grows quickly; in cultivation, it is typically 6–10 feet tall. In average, trees only live for 15 to 20 years, and on places where peach farming has already occurred, their lifespans are even shorter. The lanceolate leaves measure 7–15 cm long and 2-3 cm wide. Light pink to carmine or purplish flowers with a diameter of 1 to 1.5 inches. A hypanthium surrounds the perigynous ovary. The blooms, which have five petals and a diameter of 2.5 to 3 cm, are produced in the early spring before the leaves. It self-pollinates but is typically grown with pollinators. The peach is a drupe, which means it has a single large seed enclosed in a hard wood structure called the "stone" or "pit," yellow or pale flesh, a delicate scent, and velvety skin that is prone to bruising. The seed is 1.5–2 cm thick, oval-shaped, and crimson. Stone fruits include peaches, cherries, plums, and apricots (drupes). It grows totally from a superior ovary and consists of the exocarp, which is the outside layer, the mesocarp, which is the intermediate layer and becomes fleshy, and the endocarp, which hardens to produce the seed stone. Fruits have a velvety skin with yellow or red colour, pulp with white or yellow colour and a compressed, pitted or furrowed stone. The fleshy mesocarp may adhere to the stone (clingstone) or may seperate from it (freestone). The kernal may taste bitter or sweet depending on the variety.

The scientific name for peach is *Prunus persica* (L.) Batsch. All commercial cultivars belong to *P. persica* (L.) Batsch. The wild species of peach are *P. davidiana*, *P. mira*, *P. fergamensis* and *P. kansuensis*. The species, *P. behmi* is a natural hybrid of almond and peach. This species occurs extensively in the dry temperate zone of Kinnaur, India and is commonly used as rootstock for peach, almond and plum. The nectarine is a botanical variety of peach (*P. persica*) and is classified as *P. persica* var. nucipersica.

Species

1) ***Prunus persica:*** It is a diploid spp. (2n=18). Most commercial species of peach grown in India.
2) ***Prunus persica*** **(L.) Batsch var. *nucipersica* (Nectarine):** A natural mutant of peach, nectarine originated in China around 2000 years ago. It is a hairless or fuzzless peach which has smooth skin like plum but, the trees do not differ from peach.
3) ***Prunus persica*** **(L.) Batsch var. *Platycarpa* (Saucer Peach or Flat Peach):** the fruit of flat peach is flattened and thicker than the stone length having a calyx-like eye or broken cavity at apex. The saucer shape of flat peaches makes it more convenient to eat.
4) ***Prunus davidiana (*****Father David's Peach):** It is originated in Northeast China. It is drought tolerant hence, use as a rootstock; however, it is sensitive to nematodes.
5) ***Prunus kansuensis*** **(Rehd).:** A wild species found in Northeastern China and generally used as a seedling rootstock. Fruits are astringent and very poor in quality.
6) ***Prunus mira*** **(Poit & Turpin):** It is a natural hybrid of almond and peach. The species is found extensively in the dry temperate zones of India. Locally known as Behmi and used as rootstock for peach, almond and plum.

Climate and Soil

The best climate for peach cultivation is one that is humid, has a dry summer, and a chilly winter. Additionally, it is winter hardy and sensitive to low temperature damage. Buds that are swollen are harmed at -6.5°C. To disrupt bud dormancy throughout the winter, peaches need to chill for 500–800 hours. Areas between 900 and 1600 meters above mean sea level can readily meet their chilling requirements. Some types that don't need much chilling need 250 to 300 chilling hours. In submontane areas of Punjab, Haryana, Uttar Pradesh, and Jammu, where summertime temperature can reach 45°C, these types are successfully produced. Spring frosts, hailstorms, excessive humidity and desiccating winds in the summer, a low minimum winter temperature, and chilling hours are the limiting variables in peach cultivation.

It grows best in deep sandy loam soil that is rich in organic materials. Peach is particularly prone to water logging and demands ideal drainage. The pH of the soil should range from 5.8 to 6.8.

Commercial Cultivars

Peach cultivars can be grouped into table cultivars and canning cultivars based on their use.

Table cultivars: It should be freestone, yellow fleshed and with a regular bearing habit. Some of the world known table cultivars are Alexander, Elberta, J.H. Hale, Babcock, Cardinal, Dixigem, Redtop, Redhaven, Rio Oso Gem, Candor, Red Globe and Baifeng.

Canning cultivars: It should be yellow-fleshed, clingstone with a small non-spliting pit, uniform size, mature uniformly, devoid of red colour at the pit. However, in case of dehydration, the fruits may be freestone, white fleshed and sweet. Some of the suitable canning cultivars are Certex, Carolyn Dixon-1, Halford, Fortuna, Crawford's Early, Golden Bush, Steuart, Vivian and Veteran.

In addition to the above two groups, the peach cultivars which do not require more cold for breaking dormancy can also be grouped as low chilling cultivars. Some of the common low chilling cultivars are Early Amber, Early Grande Flordabelle, Flordagold, Flordaking, Flordasum, Flordaqueen, Rio-Grande, Sharbati and May Gold.

Nectarine cultivars: Nectarine cultivars are mostly preferred for table purposes. Some of the best known cultivars are Annqueen, Charokee, Late Le Grand, Nectared, Sun Grand, Sunlite, Sunred, Sunrise and Sunripe.

Some of the important peach cultivars are described below:

Alexander: The fruits are large, round, freestone, skin greenish white covered with brilliant red, flesh greenish white, juicy, sweet and aromatic.

Babcock: It was introduced in 1933 at California. The trees are vigorous and productive. Fruits medium, freestone, skin light pink with red blush, flesh white and very sweet.

Baifeng: Origin is in Japan but has become an important mid-season cultivar in China. Medium fruits, clingstone, flesh white, self-melting, excellent quality.

Candor: It originated from cross Redhaven × Early Red Free. Medium fruits, freestone, skin bright red or yellow, flesh yellow, fine textured, excellent flavor. It is very promising cultivar for the hills because it ripens before the onset of monsoon.

Dixigem: It resulted from a cross between Admiral Dewey and St John in 1944. Tree needs about 950hours chilling. Fruits medium, freestone, skin bright yellow with red blush, flesh light yellow with good quality.

Early Amber: Fruits are medium, clingstone, yellow with red blush skin, flesh yellow, very good quality.

Elberta: It is mainly use for canning purposes. Fruit large, freestone, skin smooth, pale yellow with red splash; flesh yellow, firm, juicy sweet and of excellent quality.

Flordasum: Medium sized fruits, freestone, round shaped with red blush skin, flesh yellow, acidic sweet.

J.H. Hale: It is an old peach cultivar. Self-unfruitful (male sterile) cultivar, fruit large, round, freestone, skin and flesh yellow, firm textured and of excellent quality.

May Gold: Fruits medium, clingstone, skin yellow with red blush, flesh firm, melting with good flavour.

Nectared: Mid-season, medium fruits with good colour and quality.

Redhaven: It is cross of Halehaven and Kalhaven introduced by Stanley Johnston at Michigan in 1940. Regular cropper, chilling requirement of 950 hours.

Rio Oso Gem: It has originated in California, resembles J.H. Hale and ripen a week before Elberta, Large fruits, freestone, attractive skin colour, flesh yellow, fine textured and good quality.

Sharbati: It is chance seedling selected at Saharanpur. Late cultivar fruits ripen in the second week of June. Fruits are medium, clingstone, skin greenish yellow with red tinge on the base of the fruit, flesh soft, juicy, aromatic and white. TSS -13.5%.

Sunred: Fruits are small, flesh slightly adhering to the stone, bright red skin, flesh yellow, excellent quality for dessert.

Sunrise: Fruits are medium, flesh slightly adhering to the stone, skin thick mahogany red, flesh yellow, firm, crisp, mildly acidic.

Veteran: Medium to large fruits, skin yellowish, fruit good quality, flesh soft, suitable for both fresh and canning purposes.

Propagation

Commercial peach propagation methods include budding or grafting onto seedling or clonal rootstocks. For growing seedlings, wild peach rootstock seeds are used. In the damp sand, seed is first stratified for 10 to 12 weeks at 4-5°C or below. Pre-treating seeds with specific agrochemicals and plant bio-regulators, such as thiourea (0.5%), GA (200ppm), or BA (100ppm), not only promotes seedling growth but also shortens the time needed for stratification and increases seed germination. The stratified seeds are sown in well-prepared beds at a depth of 5 cm, 20 cm apart in rows. To prevent seed desiccation, seed beds are irrigated after sowing and mulched with dried grass that is 6 to 10 cm thick. In March, seed germination begins. The clonal rootstocks are multiplied through mound and trench layering. Commercial techniques for peach propagation include T budding and tongue or cleft grafting. T budding in the hills between May and June and tongue grafting in February are advised. In plains, budding occurs between June and September, and grafting takes place between November and January.

Rootstocks

Peach seedlings are typically utilized as the rootstock along with plum, apricot, and almond seedlings. Since wild species generate more robust and resilient seedlings than produced kinds, wild peach seedlings are favored in the hills. To grow the rootstock for peach growth in the plains, seeds from the sharbati, sufaida, and wild apricot species are employed.

Commercially available clonal rootstocks include GF-677 and sloh hybrid. GF-677 is useful on alkaline because it is resistant to chlorosis. The other significant clonal rootstocks include Siberian C (cold-hardy), St. Julien hybrid No. 1 and 2, Damas GF-677, and Nemagaurd and Shalil (nematode-resistant).

Some rootstocks are following with their characteristics

Bailey: Best in the northern states, Cold hardy rootstock with overall good performance.

Guardian: Slightly more expensive, used extensively in the eastern USA, and in areas where peaches were planted previously, rootstock with resistance to peach tree short life syndrome.

GF677: Highly vigorous, not well adapted for eastern USA, peach-almond hybrid for adaptation to high pH soils, highly vigorous.

Halford: Good rootstock for northern states.

Lovell: Fair resistance to peach tree short life and moderately invigorating rootstock.

Nemagaurd: Susceptible to other nematodes and peach tree short life. Invigorating rootstock with resistance to root knot nematode.

Siberian C: Short-life and poor in the southern states, cold hardy rootstock used only in the northern tiers of United states and Canada.

Planting

In peach, one year old materials are preferred for planting. Plants are spaced at 4 X 4 M in square system. In recent years high density planting is practiced in commercial orchards in view of the short life span of the trees. Early spring planting is better than planting in autumn. The tree base should be whitewashed to protect the bark from sun burn after the planting is completed.

Training

The suggested training programme is open centre (vase-shaped). Nursery trees can be pruned back 18 to 24 inches above ground level and often have no lateral branches. Below this cut, scaffold branches will sprout. These scaffold branches must be grown either high enough to accommodate management procedures like weed control, fertilisation, and irrigation line maintenance, or low enough to make activities like pruning, thinning, and fruit harvesting easily accessible from the ground.

Nursery trees with greater than a half-inch diameter may have already developed lateral branches. Using pruning shears, remove most of these branches, leaving three to five branches that are evenly spaced in a north, south, east and west direction. During the first spring and summer, trees should be managed to produce as much vegetative growth as possible. Leave major pruning for the winter months when the trees are dormant. Open vase system is considered good method of training in peach.

Pruning

Pruning's primary goal is to keep the balance between vegetative development and fruiting. Because it bears fruit laterally on the growth from the previous season, peach production necessitates frequent and rigorous trimming. Once it has borne fruit, it is understood that it will never again conceive. Therefore, pruning is done to remove the unproductive sections that would later create new fruiting branches in the next season.

Peach pruning involves two fundamental steps: thinning and heading back of shoots. It is recommended that shoots thin out by 40–50% and headed back 75% under mid-hills conditions. To yield the most fruit of the highest quality, pruning should be done to allow for yearly growth of 25–30 cm under mid-hill conditions and 30–70 cm under subtropical climates. When pruning, it's important to remove any branches that are dead, sick, or damaged.

Dormant Pruning in First Year

The aim is to choose branches during the first-year dormant pruning that will grow into the major structural branches for a tree with an open centre. Choose three to four limbs, if not previously chosen, evenly distributed around the trunk at a 45-degree angle and about six inches apart vertically. Trim these branches 2 to 3 feet away from where the scaffold branches first appeared. Leave little sprouts along these branches to protect the scaffold limbs from sunburn. Eliminate overly robust upright branches that can obstruct the development of the scaffold branches.

Dormant Pruning in Second Year and Later

Choose two to three healthy lateral branches that are developing away from each of the main scaffold branches during the second dormant season for peaches. Choose lateral branches that are not above one another. Cut back the secondary lateral branches from the major scaffold branches by 20 to 36 inches. Remove any more branches. To provide the scaffold branches some shade, leave a few small branches and twigs, especially on the tree's north and east sides.

In the third and fourth dormant seasons, the same fundamental trimming techniques are applied. By keeping the center open, you'll encourage the growth of new fruiting wood, enhance the quality of the fruit, and make it possible for employees to collect fruit without using ladders. Open access to the interior canopy for fruit thinning later in the season must be considered when doing dormant pruning. Peach trees should not grow vertically over a height of 7 feet. Pickers can easily harvest as a result of not needing ladders and operating from the ground.

High Density Planting

Due to its great yield efficiency in high density planting systems, this plant training technology is particularly well-liked. Trees are planted with a 5 or 6 x 1 m spacing. At the time of planting, a plant that has been growing for a year is headed back to 20 cm above the ground. In order to create a V-shaped canopy in the following growing season, two limbs or branches are chosen in opposing directions and trained over the interrow area at an angle of 60° from the horizontal.

Role of Bio-Regulators

The application of potassium gibberellins between 80 and 100 ppm in late August and September increases the bud hardiness and gibberellic acid (200 ppm) increased the hardiness in peach near bloom. Application of ethephon increased the peach flower bud resistance to low temperature stress by delaying the bloom for one week. Removal of seed testa and GA treatment reduced stratification period enhancing the germination.

The rooting in semi-hardwood cuttings of peaches was found to be improved with IBA at 1500-3500 ppm and by basal wounding or with the help of misting noted that soft wood cuttings of peach treated with IBA at 20°C to 30 C improved rooting and their subsequent development but either rise (40°C) or fall (15°C) in the temperature reduced the rooting capacity.

Weed Managemnet

Manual weeding is used to control peaches, but it is time- and money-consuming. Due of their uneven shape, peach roots are frequently harmed by continuous ploughing. Therefore, using herbicides is a great alternative. In order to effectively manage broad-leave weeds as well as common grasses in new orchards, pre-emergence applications of Diuron at 800 gm-1 kg/acre or even post-emergence applications of Glyphosate at 6 ml/acre in 200 litres of water during February and March can be used.

Irrigation

Fruit growth in peaches and nectarines has a double sigmoid growth pattern. Between the initial quick fruit growth following fruit set (stage I) and the ultimate rapid fruit growth before maturation (stage III), there is a moderate fruit growth phase (stage II) that is accompanied by pit hardening. Peach and nectarines have higher water requirement during rapid fruit growth during stage I and III. At stage one which coincides with spring the soil moisture levels are near optimum but, require one or two irrigations in plains. The rapid fruit growth during the stage III coincides with hot and dry summer. Fruit drops and low fruit size and quality are the results of water stress during the dry and hot summer. In the hot and dry summer, frequent irrigations are necessary for fruit development and maturation (April-June). As a result, it's crucial to irrigate peach orchards during the summer at intervals of 3–4 days in light soils and once a week in heavy soils. Minimum two irrigations are required during the period of fruit development in hilly region. Irrigation methods like drip and sprinkler are more efficient methods of irrigation, which save water and enhance productivity and quality of produce. Drip irrigation at 100% crop evapotranspiration along with black polyethylene mulch (100 micron) is recommended for peach crop for mid-hill zone of Himachal Pradesh as it helps in conservation of more moisture compared to surface irrigation and yield increases as compared to surface irrigation.

Nutrition Management

The peach has been found to have a relatively high requirement for nitrogen and potassium. The entire farmyard should be fertilized in December and January, and if irrigation facilities are available, half of the N should be applied in the spring before flowering and the other half a month later.

When the tree is 1-2 years old, apply FYM@10-15kg/tree, UREA@150-200gm/tree, SSP@200-300gm/tree and MOP@150-300gm/tree. when the tree is 3-4 years old, apply FYM@15-20kg/tree, UREA @500-700gm/tree, SSP@500-700gm/tree and MOP@400-600gm/tree. When the tree is 5years and above, apply FYM@25-30kg/tree, UREA@1000gm/tree, SSP@1000gm/tree and MOP@800gm/tree.

Flower and Fruit Drop

Early fruit drop, June fruit drop, and pre-harvest fruit drop are three phases that the peach varieties go through. Due to fruit competition and a lack of pollination, the early drop is thought to be normal. The June decline is brought on by environmental factors and moisture stress, and it can be stopped by retaining soil moisture by mulching or irrigation. The mature marketable fruits abscise prior to harvest due to a decrease in auxin levels or a rise in ethylene levels in fruit, causing the pre-harvest drop to result in significant economic losses. Before the anticipated fruit drop or 20–25 days before harvesting, NAA (10 ppm) is applied to check for fruit drop loss.

Insect pest and their management

Peach leaf curl aphid (*Brachycadus helichnysi*): Peach's most serious pest causing damage to floral and vegetative buds resulting in smaller, distorted, pale, curled leaves. Spraying systemic insecticides like dimethoate at 0.02 to 0.03 per cent during leaf emergence controls the pest.

Fruit flies (*Bactrocera zonata*): It is a destructive pest of tropical peach. The fly punctures and lays eggs inside the fruit during maturity to ripening. Due to rotting, infested fruits develop brownish patches and eventually drop. Spraying with 0.03 per cent dimethoate at fruit maturity effectively controls the pest.

Stem borers (*Sphenoptera lafertei* or *S. dadkhani*): It is a serious pest of peach. The infested plants show bark spliting and gum oozing. The leaves become pale yellow, plants become weak and may die in severe infestation. Adult beetles feed on foliage and lay eggs. The young grubs bore into the stem making small chambers in the wood under the bark. Borers can be controlled by cleaning the holes and inserting cotton wick soaked in gasoline or methyl parathion (0.2%) and then plugging them with mud. It can also be controlled by spraying of acephate 75 SP 2g/litre or malathion @ 0.1%.

Twig borer (*Anarsia lineatella*): The worm bores the young twigs and fruits. The infested twigs dry and show wilting and fruits start rotting. To control the pest, the infested twigs and fruits should be removed and destroyed followed by spraying with malathion (0.05%) .

Nematodes: The dominant nematode groups are root lesion (Pratylenchus), RKN (Meloidogyne), dagger nematode (Xiphinema) and ring nematode (Criconemoides). Some nematode species act as a vector of viruses. Use of resistant rootstocks is the most convenient and reliable method for controlling nematode problem in peach. Some of the root-knot nematode resistant rootstocks are Nemaguard, Okinawa, S-37, Shalil and Nemared. Other control measures include the application of nematicides, soil fumigation, growing of nematode resistant cover crops and high irrigation regime.

Disease and their management

Leaf curl (*Taphrina deformans*): Leaf curl is a common fungal disease of young peach plants. High temperature and high humidity favor the incidence of the disease. The affected leaves are deformed, thickened, puckered, rolled and reddened and become brittle and waxy. As surface causing the leaves to turn brown and shed prematurely.

The leaf curl incidence can be the disease advances, chlorophyll disappears, and a powdery coating develops on the upper controlled by spraying carbendazim.

Powdery mildew (*Sphaerotheca pannosa*): The fungus infects leaves, young shoots and twigs. The young leaves show coating of thick layer of mycelium while white patches develop on older leaves. In severe infection, the fruits first develop pink colour and later change to dark brown. It is a serious fungus of peach and can be effectively controlled by spraying carbendazim or with sulphur.

Bacterial spots (*Xanthomonas campestris* pv. *pruni*): The disease shows symptoms on leaves, twigs and fruits. On leaves or fruits, small water-soaked areas develop and eventually drop. The disease forms cankers on twigs which are the sources of primary infection in the subsequent growth phase. To check the spread of the disease, the cankerous growth should be pruned and burnt, and then be sprayed with antibiotic like terramycin. In general, healthy vigorous plants are less succeptible to the disease.

Silver leaf (*Chondrostereum purpureum*): It is a serious fungal disease of deciduous temperate trees. The affected leaves show hay-gray colouration resembling the color of lead. Slanting cut on wood exhibits brown discoloration. It can be controlled by pruning of affected branches.

Viruses

Numerous virus diseases are known to occur in peach orchards. Peach yellows and little peach are two important viruses causing economic loss. Some other virus diseases on peach are red suture, asteroid spots, golden net, line pattern, mottle, willow twig, peach yellow bud mosaic, peach wart, ring spot and stem pitting. Effective virus disease management practices include sanitation measures like removal of diseased parts, control of vectors such as leaf hoppers, aphids and nematodes and follow certified bud-wood programme.

Physiological Disorders

Split pit and gumming: During pit hardening, sometimes fruits show splitting at the joint of dorsal and ventral side of the fruit. This splitting is accompanied by exudation of gum which may fill the entire pit cavity surrounding the kernel. Seeds of the affected fruits are usually abortive. Splitting and gumming make the fruits unfit for consumption. Prolonged dry period followed by sudden rain resulting in fluctuation of atmospheric temperature and humidity causes fruit splitting.

Sunscald: Constant exposure to intense sunlight may cause sun scalding on tree trunk, young twigs and branches. To protect the plants from sun scalding, providing shade by wrapping straw or hay around the trunk and branches and painting the exposed surface with lime paste may be practiced.

Yield

Mature productive peach trees generally produce 7 to 10 tonnes of fruits per hectare. However, in high density planting a yield of 78 MT/ha may be obtained. Fruit yield in peach, as in other stone fruits, is governed by various morphological parameters. Although a very scanty work on yield forecasting was negative correlation between blooming intensity and yield in peach, and a significantly positive correlation between percentage of fruit set and number of fruits which ultimately contributed to yield.

Harvesting and Post-Harvest Mangement

It is crucial to choose peaches at the right stage of ripeness because this affects the fruit's quality and shelf life after harvest. Days after full bloom, calendar dates, fruit size, hardness, ground colour, sugar, and TSS: acid ratio are some of the markers used to determine fruit maturity. Days needed from flowering to maturity vary depending on the cultivar and the location (Alexander, 86 days; July Elberta, 101 days; Babcock, 122 days and Elberta, 127 days). Three to four pickings are required during the harvest of peach fruits since they do not mature consistently. The typical method of fruit harvesting is hand picking. Peach fruits should be harvested depending upon the place of market. For distant markets, harvest the fruit at firm mature stage while for local market nearly ripe fruits can be picked.

During harvesting, collect the fruits in a basket/carton to prevent injury and bruising to the fruit. Immediately after harvesting transfer the fruits to a shady place. To remove the field heat, dip the fruits in cold water for 10-15 minutes and dry them under shade. Fruit ripening is slowed down by hydro-cooling, which also prolongs the shelf life of peach fruits. The end of April to May (Flordaprince, Shan-e-Punjab, Earligrande, and Punjab Nectarine) and June to July are the peak harvesting months for several peach varieties (Red Haven, Glo Haven, July Elberta and Shimizu Hakuto). After being planted in the field for two years, peaches begin to bear fruit. For roughly 20 years in highlands and 12–14 years in subtropics, the trees give fruit. 100–125 kg of yield is possible per tree.

Being highly perishable fruit peach cannot be stored for a larger period under room conditions. It is also sensitive to low-temperature injury. Proper packaging before storage is must to extend the availability period. Before packaging, sorting of fruits is necessary to remove the injured, damaged and undersized fruits from the lot. Corrugated fibre board (CFB) boxes are typically used for packing. Effective peach fruit storage lasts 25 days at a temperature of 0-3.5°C and a relative humidity of 85-90%. It has been discovered that a number of antisense substances, including $CaCl_2$, $Ca(NO_3)_2$, GA, and boric acid, can increase the shelf life of peach fruits. Physically mature peach fruits of the Shan-e-Punjab variety should be dipped in $CaCl_2$, 2 per cent, then packaged in 4 kg CFB boxes to preserve quality for 30 days in cold storage (0-1°C, 90-95% RH), with the least amount of deterioration and weight loss.

Modern Technology

The major countries with highly mechanized farming operations are Japan, Taiwan and Korea. After that India also adopted mechanization farming to strengthen temperate fruit production. In recent year it developed a Self-Propelled Trunk-Shaker and Catch Harvester, Canopy shaker, Intelligent Automated Fruit Harvesting Robots, Self-Propelled harvesters etc. With the picking manipulator and automated mechanization coupled with artificial intelligence, this machine's progress is highly successful at capturing fruit and can reach every potential drop position in the workspace, improving the selectivity, precision, and robustness of farming. It lowers labour expenses, improves harvest timing, allows for selective harvesting, boosts operation efficiency, and maximizes profitability. By boosting harvest production quickly, mechanization of Indian horticulture for harvesting will assist the growers maintain their competitiveness in the long run. As a result, the temperate fruit business urgently needs to embrace integrated mechanization strategies.

Future Prospects

Peach fruit is a delicious and nutritious addition to any diet. It is a great source of vitamins A and C, as well as potassium and dietary fiber. Peaches are also low in calories and fat, making them a healthy option for those looking to maintain or lose weight. Peaches are also high in antioxidants, which can help to protect against cancer and other diseases. The fruit is also considered to have anti-inflammatory properties. In addition to being a healthy food, peaches are also versatile and can be eaten fresh, canned, or frozen. They can be used in a variety of recipes, such as pies, jams, and smoothies. Overall, peaches are a tasty and healthy choice for anyone looking to improve their diet.

Peaches are perishable goods and more prone to quick senescence, deterioration, and rotten development. The rate of respiration and transpiration both increase as the temperature rises, and if produce isn't properly stored, it quickly dries out, wilts, and spoils. Amounting to Rs. 5000 crores, it has been estimated that 20 to 30% of horticulture produce is wasted before consumption due to ineffective harvesting, handling, storage, transportation, and marketing techniques. There are two methods that could be used to solve this issue. One is the construction or expansion of cold storage facilities in the regions that produce fruits and vegetables. An alternative strategy is to turn the fruits into a variety of goods such as juices, concentrates, fruit bars, baby meals, pulps, etc. that may be kept for a long time and add to the value of the product. Marketing of horticulture produce is presently facing great exploitation from intermediaries. The best way to market the horticulture produce will be to minimize the intermediaries in the marketing channel. The awareness among growers, new concepts of online marketing, involvement of educated youth with the horticulture sector and such like factors are likely to influence the marketing of horticulture produces and in future years it is expected that besides better quality retention of the produces, the producer will get higher share of consumer's rupee.

Crop improvement of peach fruit involves a variety of techniques to enhance the yield, quality, and disease resistance of the fruit. One common method is breeding and selection, where new varieties are developed by crossing different types of peaches with desirable traits. Another approach is the use of genetic engineering, where specific genes are inserted into the peach tree to improve certain characteristics, such as disease resistance or fruit size. Another important aspect of crop improvement is proper orchard management. This includes proper pruning, fertilization, irrigation, and pest control to maintain the health of the trees and optimize fruit production. Another technique that is used in crop improvement is the use of tissue culture which involves the growth of plants from small pieces of plant tissue in a laboratory environment. This method allows for the mass production of genetically identical plants that can be used to establish orchards. Overall, crop improvement of peach fruit requires a combination of breeding, genetic engineering, orchard management, and tissue culture to enhance the yield, quality, and disease resistance of the fruit.

References

Anonymous, (2020). National Horticulture Board, Indian Horticulture Database. Ministry of Agriculture & Farmers Welfare, Government of India. www.agricoop.nic.in

Bose, T.K., Mitra, S.K. and Sanyal, D. (2001): Fruits: Temperate fruits, ed. 3 vol. 1, Naya Udyog, Calcutta, pp. 331-437.

Chadha, K.L. (2019): Handbook of Horticulture, ed. 2 vol. 1, ICAR, New Delhi, pp. 391-395.

Chattopadhyay, T.K. (2019): A Textbook on Pomology- Temperate Fruits, ed. 2 vol. 4, Kalyani Publishers, Noida, pp. 206-221.

Dhillon, W.S. (2013): Fruit Production in India, ed. 1, Narendra Publishing House, Delhi, pp. 457-475.

Kumar, V. and Kamra, S. (2018): Ready Reckoner for Pomology, ed. 1, New Vishal Publications, New Delhi, pp. 220-226.

Selvakumar, R. and Muthukumar, P. (2018): Glaustas Horticulture, ed. 2, New Vishal Publication, New Delhi, pp. 103-108.

Sharma, R.R. and Srivastava (2004): Plant Propagation and Nursery Management, ed. 1, International Book Distributing Co., New Delhi, pp. 43-242.

Sunda, N.R. (2020): Basic Agriculture, ed. 3, Vishva Bhawan Publication, Jaipur, pp. 377-450.

4

Advances Production Technology of Plum

[1]Charu Shekhawat and [2]Deeksha Gautam

[1]Research Scholar, Department of Horticulture, Agriculture University, Kota, Rajasthan
[2]Research Scholar, Department of Fruit Science and Horticulture Technology, Odisha University of Agriculture and Technology, Bhubaneswar-751003, Odisha
Corresponding Email: Charushekhawat95@gmail.com

Introduction

Plum, scientifically known as *Prunus domestica*, is a popular fruit-bearing tree belonging to the Rosaceae family. Renowned for its delicious and versatile fruits, plum cultivation has a long history that spans different cultures and regions. Plums are enjoyed worldwide both in their fresh form and processed into various culinary delights. Because of its temperate character, plum is a significant crop in hills, but it is just a modest fruit crop in Punjab and the north Indian plains. Because it requires little freezing and has a shorter juvenile phase, it is ideal for planting as a filler fruit plant in mango, litchi, and pear orchards. There are a few solid-block plum plantations in the Amritsar, Fazilka, Ambala, and Kurukshetra regions as well. Due to the trees' abundant flowering, high fruit set, and early ripening habits of varieties suitable for plains, they are extremely productive. A fully developed tree can typically produce 50–60 kg of fruit. Plum trees are deciduous and typically bear fruits during the summer season. The fruits come in a wide range of colors, including shades of purple, red, yellow, and green, with variations in size and flavor. This diversity in plum varieties allows for a delightful selection for consumers, catering to their individual preferences.

Fig. 1.1: Plum Fruit

History

Plums are widely cultivated over the world, and there are numerous varieties that may grow in a variety of soil types and climates. The common European plum, *P. domestica*, is at least 2,000 years old and likely originated in the area near the Caucasus and the Caspian Sea. The Damson plum (*P. insititia*) is another Old world plum species that most likely originated in Europe or Asia. According to historical texts, the area near Damascus was where these plums were first cultivated. The Japanese plum was initially cultivated in China thousands of years ago, but it underwent substantial development in Japan before being exported to the rest of the world. Japanese plums are the most popular fresh plum marketed commercially because they have a longer shelf life than the majority of European species.

Origin & distribution

Despite being a native of China, the plum was first grown commercially in Japan and America. Due to its cultivation in Japan, where cultivars later spread to other countries, it is known as the Japanese plum. Plums are currently grown in every country with a temperate climate.

Area & Production:

The area and production of plums in India have witnessed growth and expansion in recent years. Plum cultivation is predominantly concentrated in states such as Himachal Pradesh, Jammu and Kashmir, and Uttarakhand, which offer favorable climatic conditions for its cultivation. Regarding the area under plum cultivation, it has steadily increased over time. As of the available data up to 2020, the total area dedicated to plum cultivation in India was around 24,000 hectares(Anon 2020). Himachal Pradesh accounted for the largest share of the plum cultivation area, followed by Jammu and Kashmir and Uttarakhand. This expansion of the cultivation area indicates the growing interest and recognition of the economic potential of plum production in the country. In terms of production, India has experienced a positive trend in plum yields. The country produced approximately 84,000 metric tons of plums in 2020 (Anon 2020). Himachal Pradesh emerged as the leading plum-producing state, contributing a significant portion to the total production. Other regions, such as Jammu and Kashmir, Uttarakhand, and certain parts of North-East India, also make substantial contributions to the overall plum production in the country.

Importance, nutritional composition and uses

Plums hold significant importance as a fruit due to their nutritional composition and versatile uses. These delicious fruits are packed with essential vitamins, minerals, and antioxidants, making them a valuable addition to a healthy diet. Plums are a rich source of vitamin C, which boosts the immune system, promotes collagen production for healthy skin, and acts as a powerful antioxidant. They also contain vitamin A, which supports vision, immune function, and cell growth. Plums provide dietary fiber, which aids in digestion, regulates blood sugar levels, and promotes a feeling of fullness. In terms of minerals, plums are a good source of potassium, which plays a vital role in maintaining heart health, regulating blood pressure, and supporting proper muscle and nerve function. Additionally, plums offer small amounts of minerals like copper, manganese, and magnesium, which are important for various bodily functions.

The uses of plums extend beyond their nutritional value. They can be enjoyed fresh as a healthy snack, or incorporated into a wide range of culinary creations. Plums are commonly used in making jams, jellies, and preserves due to their natural sweetness and vibrant flavor. They can be added to baked goods such as pies, tarts, and cakes, imparting a delicious fruity taste. Plums are also a great addition to salads, lending a burst of sweetness and tanginess. Furthermore, they are used in the production of beverages like juices, smoothies, and even alcoholic drinks such as plum wines and brandies. Overall, plums are not only a tasty and versatile fruit but also an important part of a balanced and nutritious diet. Their nutritional composition, including vitamins, minerals, fiber, and antioxidants, contributes to their health benefits. The various uses of plums in both sweet and savory dishes make them a valuable ingredient in the culinary world, adding flavor, color, and nutritional value to a wide range of recipes.

Botany

Prunus salicina is a member of the Rosaceae botanical family, which also includes peach and pears. While hills have smaller trees than the plains, both have medium-sized trees. The bark on twigs and trunks is typically crumbling from the old scaffolding and is a dark, blackish-brown color.

The leaves are oblong and have sharp points. Lamina green on top and light green on the underside describe the serrated leaf margins. It blooms during the first two weeks of February. Bronze flowers are mounted on tiny spurs with closely spaced rings. Thus, flowers are close together, and the inflorescence is known as a cymose. Fruit is a drupe (stone) with a fleshy mesocarp and a thin edible exocarp. The seed is enclosed in the stony hard endocarp.

Taxonomy

The taxonomy of plum is as follows:

Rank	Scientific Taxonomical Details
Kingdom:	Plantae
Phylum:	Magnoliophyta
Class:	Magnoliopsida
Order:	Rosales
Family:	Rosaceae
Genus:	Prunus
Species:	Prunus domestica

Source: Rehder (1940)

The subspecies cross easily, so that numerous intermediate forms can be found: their sweetness and tartness may vary, their colors varying from bluish-purple to red, orange, yellow or light green.

P. domestica ssp. domestica – prune plums, zwetschge (including ssp. *oeconomica*)

P. domestica ssp. insititia – damsons and bullaces, krieche, kroosjes, perdrigon and other European varieties

P. domestica ssp. intermedia – egg plums (including Victoria plum)

P. domestica ssp. italica – gages (greengages, round plums etc. including spp. *claudiana* and rotunda)

P. domestica ssp. pomariorum – spilling

P. domestica ssp. prisca – zibarte

P. domestica ssp. syriaca – mirabelle plums

Plum types

Moyer plum - The Moyer plum is one sort of plum that has a red, sour-flavored skin surrounding its delicious, luscious yellow flesh. The common plum tastes sweet. It has purple skin and luscious, amber-colored to yellow-orange flesh.

Damsons plum - Plum damsons Popular plum with a tart flavor that has a dark purple skin and a yellowish green interior.

Plum Elephant Heart - The luscious, juicy red flesh of this variety of Japanese plum is covered in dark red to purple flecks.

Plum greengage Green - plum variety with a mouthwatering honey-sweet flavour.

Myrobalan Plum - Small, round plums with a red or yellow cherry-like appearance.

Santa Rosa berries

Red plum

In high elevation plum growing regions like Santa Rosa and Mariposa, there are numerous plum cultivars.

Soil and climatic requirements

Plum trees thrive in well-drained soils that are rich in organic matter. Sandy loam and loamy soils are considered ideal for plum cultivation as they provide good drainage while retaining sufficient moisture. Soil pH between 6.0 and 7.0 is suitable for plum trees (Singh *et al.*, 2014). Regular soil testing should be conducted to assess nutrient levels and pH, and appropriate soil amendments should be applied accordingly.

Plums are adapted to a wide range of temperatures, but they require a certain amount of chilling hours to break dormancy and promote flowering. Different plum cultivars have varying chilling requirements, typically ranging from 300 to 1200 hours below 7°C (Mukherjee *et al.*, 2016). The suitable temperature range during the growing season is between 15°C and 30°C for optimal growth and fruit development. Temperatures below 7.2°C only need to be chilled for a shorter period of time. It can withstand both frost and hot summers. Some cultivars, like Santa Rosa, require 700–1000 chilling hours to grow, although cultivars with lower chilling requirements 250–300 chilling hours are favoured for plains. Plum trees require a moderate amount of water, and a well-distributed rainfall pattern is beneficial for their growth. In India, plum cultivation is successful in regions with an annual rainfall ranging from 600 to 1200 mm. Excessive rainfall during the flowering and fruiting stages can cause fruit cracking and disease incidence. Proper drainage systems should be in place to prevent waterlogging. Plums can be cultivated in a wide range of altitudes in India.

In general, they are grown at altitudes between 500 and 2500 meters above sea level. Higher altitudes provide cooler temperatures, which can be advantageous for chilling requirements and flavor development in certain cultivars.

The microclimate within the orchard plays a crucial role in plum production. Sheltered sites that are protected from strong winds are preferred to prevent wind damage to the trees and reduce water stress. North or east-facing slopes are advantageous as they receive early morning sunlight, which helps in drying dew and reducing disease incidence.

Commercial Varieties

European plum (*Prunus domestica*) is a widely cultivated fruit tree in India, and several commercial varieties are grown for their unique characteristics and culinary uses.

European plum varieties

Italian Prune: Italian Prune is known for its medium-sized, oval-shaped fruit with a dark blue skin and sweet flesh. It is often used for drying and culinary purposes due to its high sugar content and rich flavor. Italian Prune is self-sterile and requires cross-pollination for fruit set.

Giant Prune: Giant Prune produces large, round fruit with a purple skin and juicy flesh. It is suitable for fresh consumption and can also be used for drying and processing into prunes. Similar to Italian Prune, Giant Prune requires cross-pollination for fruit set.

Reine Claude (Green Gage): Reine Claude, also known as Green Gage, is a small to medium-sized plum with a greenish-yellow skin and sweet, aromatic flesh. It is highly prized for its unique flavor and is popular for culinary uses and desserts. Reine Claude is self-sterile and requires cross-pollination.

Yellow Egg: Yellow Egg is characterized by its large, egg-shaped fruit with a golden-yellow skin. It has a sweet flavor and firm flesh, making it suitable for fresh consumption and processing into jams and preserves. Yellow Egg plum trees are self-sterile and require cross-pollination.

Golden Transparent: Golden Transparent is a dwarf European plum variety that is self-sterile. It produces small, yellow plums with a rich apricot flavor. Due to its dwarf nature, it is suitable for small gardens and containers.

Imperatrice: Imperatrice is a commercial European plum variety known for its medium-sized fruit with a dark purple skin and juicy, sweet flesh. It is highly valued for its excellent flavor and is suitable for fresh consumption as well as processing into various culinary products.

Lombard: Lombard is another popular European plum variety cultivated for its medium to large-sized fruit. It has a deep purple skin and firm, flavorful flesh. Lombard plums are known for their sweet taste and are often used for fresh eating, canning, and cooking purposes.

President: President plums are characterized by their large size and attractive dark purple to black skin. They have juicy, sweet flesh with a rich flavor. President plums are commonly consumed fresh and are also suitable for processing into jams and jellies.

Victoria: Victoria plums are highly regarded for their large, oval-shaped fruit. They have a reddish-purple skin with yellow flesh that is sweet and juicy. Victoria plums are versatile and can be enjoyed fresh, used for cooking, and processed into various products such as pies and preserves.

Starking Delicious: Starking Delicious is a well-known European plum variety that produces large fruit with a deep purple skin. The flesh is firm, juicy, and has a pleasant sweet flavor. Starking Delicious plums are favored for fresh consumption and are also suitable for canning and baking.

Demistons Superb: Demistons Superb plums are medium-sized with a dark purple skin. They have juicy flesh with a balanced sweet-tart flavor. These plums are valued for their eating qualities and are commonly enjoyed fresh or used in cooking and baking.

Golden Drop (Yellow Egg Plum): Golden Drop, also known as Yellow Egg Plum, is a variety with large, oval-shaped fruit. It has a yellow skin and flesh with a rich apricot flavor. Golden Drop plums are sought after for their distinct taste and are often used for fresh eating, desserts, and preserves.

Japanese plum varieties:

Beauty: Beauty is a popular Japanese plum variety known for its large, round fruit with a dark purple skin. It has sweet and juicy yellow flesh with a pleasant flavor. Beauty plums are highly prized for their attractive appearance and are commonly consumed fresh or used in various culinary preparations.

Santa Rosa: Santa Rosa is a widely grown Japanese plum variety recognized for its medium to large-sized fruit. It has a reddish-purple skin and firm, sweet, and flavorful flesh. Santa Rosa plums are versatile and can be enjoyed fresh, dried, or used in jams, jellies, and desserts.

Mariposa: Mariposa is a Japanese plum variety that produces medium-sized fruit with a red skin and yellow flesh. It has a sweet and tangy flavor with a hint of acidity. Mariposa plums are often consumed fresh, used in baking, or processed into jams and preserves.

Alubukhara: Alubukhara is a self-unfruitful Japanese plum variety, which means it requires cross-pollination from another compatible plum variety for fruit set. It produces small to medium-sized fruit with a reddish-purple skin and firm, juicy flesh. Alubukhara plums are commonly used for making preserves, pickles, and chutneys.

Kelsey: Kelsey is a Japanese plum variety that produces large fruit with a reddish-purple skin. It has sweet, juicy, and flavorful yellow flesh. Kelsey plums are popular for fresh consumption and can also be used in cooking and baking.

Satluj Purple: This cultivar cannot bear fruit on its own and needs Kala Amritsari to pollinate it. To achieve good fruit set, Kala Amritsari plum should be planted as a backup plant in backup rows. The tree has an upright growth style and is of average vigour. The fruit weighs 25–30 g and is of average size. When the epicarp ripens, it turns red in colour. The flesh of the mesocarp is firm and golden in colour. Juice has an acidity of 0.5 to 0.7 percent and a TSS of 12 to 14 percent. Fruits are often consumed fresh. It is a poor producing variety that ripens early. Early May is when it ripens, and each tree produces 35 to 40 kg of fruit.

Kala Amritsari: It is a high yielding, locally adapted cultivar. The trees are strong and have numerous branches. This variety is the most popular on the plains. Fruits are medium in size, spherical and oblate with two depressions. As it ripens, the epicarp (skin) turns a dark purple colour. The flesh, or mesocarp, is yellow and has pulp that is luscious. Fruits have a mildly acidic flavour. Juice has an acidity level of 1.D-1.3% and a TSS of 13 to 15%. Fruit matures in the latter half of May. About 55 kg is the average output per tree. Jam and squash are commonly made from its fruits.

Titron: Although it is a self-fruitful cultivar, using Alucha Early Round as a pollinizer will boost the output. Compared to kala Amritsari, titron trees are smaller. Additionally, the fruit is smaller than Satluj Purple and Kala Amritsari. A thinner epicarp than kala Amritsari exists. Yellow and less luscious is the meat. Due to the very low acidity, the flavour is subacid. Fruit production per tree is 30–35 kg. It works great for creating jam as well as serving food. TSS of 13.5% and acidity of 0.4%.

Kataruchak

Kataru Chak is a self-fruitful Japanese plum variety, meaning it can set fruit without the need for cross-pollination. The cultivar was created in the Punjabi village of Kataruchak in the Gurdaspur area. The cultivar can produce some of its own fruit. Where both kataruchak and Kala Amritsari had been planted, the yield per tree rose. Because the fruit's epicarp (skin) has a white bloom (purbescence), it sells for more money than Kala Amritsari. The trees at Kala Amritsari are just as healthy. The fruit is heart-shaped, larger in size, and purple in colour. It ripens right after Amritsari kala. Fruit production per tree is 45–50 kg. Fruit, jam, and squash are all suitable for use in food.

Plum hybrids

Plums are linked to other drupe fruits including nectarines, apricots, and peaches. Natural hybrids of several plum and apricot varieties include plumcots, apriums, and pluots. Here is a basic explanation of these hybrid plums.

Plumcots: Plumcots are an apricot and plum hybrid fruit that is half plum and half apricot in composition. These fruits resemble apricots in shape but have plums' sweet flavor and skin color.

Apriums: These drupes resemble apricots in texture and have more apricot than plum content. They have orange flesh that is delicious, just like apricots.

Pluots: The stone fruit known as pluots is more plum-like than apricot. They resemble red apricots and taste strongly of plums.

Recent verities of Plum: Black Ruby, Friar, Black Beauty, Black Splendor, El Dorao, Black Amber, French Prune

Cultivars

Numerous cultivars have been selected for garden use. The following have gained the Royal Horticultural Society's Award of Garden Merit:

1. 'Blue Rock
2. 'Blue Tit
3. 'Czar
4. 'Imperial Gage
5. 'Jefferson
6. 'Laxton's Delight
7. 'Mallard
8. 'Marjory's Seedling
9. 'Opal
10. 'Oullins Gage
11. 'Pershore
12. 'Victoria

Honey Sweet: Transgenic Variety of Plum

Honey Sweet is a transgenic variety of plum that has been genetically modified to enhance its sweetness and flavor. This variety is the result of scientific advancements in biotechnology and genetic engineering techniques. The transgenic modification in Honey Sweet plum involves the introduction of specific genes that regulate the production of natural sugars and flavor compounds. These genes are typically sourced from other plant species or organisms with desirable traits. The purpose of developing Honey Sweet plum is to create a variety that offers an exceptional level of sweetness and a unique flavor profile. This transgenic variety aims to provide consumers with a plum that is even more enjoyable and satisfying to eat.

Propagation

The vegetative methods of propagating plums include budding and grafting on rootstocks. These can also be multiplied through occasional misting of both leafy soft wood cuttings and hard wood cuttings. Hardwood and semi-hardwood cuttings that have been treated with IBA (2000–5000 ppm) for improved rooting Wild apricot seeds are stratified for 45 to 50 days at a temperature of 3-5°C to break the rest before being used to raise seedlings. The stratified seeds are planted in nursery beds 6–10 cm deep, 25–30 cm apart, and 10-15 cm apart from one another.

In a year, the seedling grows to a graftable size. Own-rooted Kala Amritsari plants are frequently utilized for planting in Punjab. Commercial plum clonal rootstock multiplication involves stacking. Most plum clonal rootstocks have been found to be challenging to root. Rooting is aided by the application of 2500ppm IBA to stool shoots. Additionally, hardwood cuttings are used to propagate clonal rootstocks. During the dormant season, the hardwood cutting should be taken and submerged in a 2500 ppm IBA solution for 30 seconds. Once rooted, these cuttings are placed in a mist propagation chamber.

Propagation of scion

In February, clonal and seedling rootstocks with a thickness of 0.8 to 1.2 cm are grafted using the tongue and cleft technique. Seedlings that are not graft-ready in February should be budged in June or July using the T- and chip procedures. In January, scion wood from healthy, disease-free, true-to-variety trees should be obtained for grafting. The one-year-old shoots are used to harvest the scion wood. They are correctly labelled for variety and packaged in damp sphagnum moss. Until the grafting is complete, these bundles of scion wood are kept in cold storage or buried deep in the ground in a shaded area. The ideal period for grafting these fruits is February in lower elevation and in March at higher elevations. Chip budding can also be done in March, July and September.

Rootstock

Seedling rootstocks: Although seedlings of peach, apricot, and almond can be utilized, plums are primarily cultivated on plum rootstocks. In India, wild apricot seedlings are primarily used to propagate plum (Chuli)

Myrobalan B, Myrobalan 29C, Mariana GF 8/1, Mariana 2624, and Mariana 8-6 are clonal rootstocks for plum. Pixy (Dwarf), St. Julien A, GF-667, Mariana 9-25, Mirabi, Brompton, Damas GF 1869

Rootstock influence

By affecting canopy design, nutrient uptake, blooming, yield, and fruit quality, rootstocks play a crucial role in the production of various fruit crops. Additionally, it can deal with biotic and abiotic challenges such nutritional stress, salinity, temperature stress, and soil pathogens. Most of the commercial fruit crops grown in India have rootstocks with significant commercial potential that have not yet been fully utilized. Finding a rootstock with the greatest number of desirable traits is therefore a crucial step. Other nations use peach *(P. persica*), apricot (*P. armeniaca*), almond (*P. amygdalus*), and some hybrids, like Marianna plum *(P. cerasifera P. munsoniana*), as rootstocks for peach, apricot, and almond trees. Plum (*P. domestica*) selections like "Brompton" and "Common plum" from; St. Julien, "Common Mussel" and meet the set of climatic requirements for the tree's intended growth.

Planting

When the plant is dormant, between December and January, plums are planted. The location of an orchard should be completely cleared of bushes and weeds before planting. The plantation land should be ploughed. The orchard area should be adequately mapped out around two months before planting. Terraces in hilly areas should be kept inward to enable soil conservation. An orchard should be laid out using a terrace or contour system on sloppy terrain, whereas a square method is used in flat terrain. The planting distance varies depending on soil richness, rootstocks, and variety. Plums are typically planted six metres apart and six metres apart.

The semi-dwarfing clonal rootstock-raised plum plants in high density plantations should be planted at a distance of 4 m by 4 m.

Pits measuring 1 m x 1 m x 1 m are excavated between October and November. Fertile top soils mixed with 40 kg of well-rotten FYM and 1 kilogram of single super phosphate should be used to fill pits. The pits are saturated with 10 liters of chlorpyriphos (4 ml/L) solution to prevent any insect damage. To prevent any scion roots, the graft union should be kept 10-15 cm above the ground at the time of planting. Young plants should receive regular irrigation, and the basin area is mulched with dried grass that is 15 cm thick.

High density planting

High density planting is when you put as many plants as you can into a given space in order to generate the most profit per tree volume without compromising the fertility of the soil. The first high density orchards were established in Europe around the end of the 19th century, and ever then, low density traditional orchards have been declining. The fundamental idea behind an HDP is to maximize return on investment for each unit of input while utilizing vertical and horizontal space as efficiently as feasible. Planting very densely, 500–10,000 plants per hectare, depending on the fruit crop Optimal high density is between 1,500 and 10,000 plants per hectare. Medium high density is between 500 and 1500. (3) Extremely high density: 10,000–100,000 plants per hectare.

Meadow orcharding: Meadow grassland is another name for planting with a very high density 10,000–100,000 plants/ha severe top trimming is used to maintain tree form and is done similarly to mowing grassland.

Training and pruning

Pruning and training depending on the growth patterns of the individual varieties, plums are typically taught using an open centre system or a modified central leader method. The cultivars with spreading growth patterns, like the Japanese plum, ought to be trained to have an open centre system. No matter the variety, HP follows the open centre system. But on the plains, where there is abundant sunlight, trees should be taught using a modified central leader method with four to five scaffold branches.

Open center system

The plant is pruned back to 40–60 cm above the ground after planting. Along with the Central leader, the tree also produces 3 to 6 laterals over the growing season. Three to five wide-angle, evenly spaced scaffold branches should be chosen for the first winter pruning, and the rest branches should be cut off. Also entirely erased is the Central leader. The chosen branches will return to the centre of the growth. 2-3 secondary branches are chosen on the parent branches during the second dormant pruning. After trimming, the primary and secondary leaders' tips should be roughly 30–40 cm apart from one another, therefore placement should be the main factor in choosing secondary branches.

Pruning

Thinning and heading back of shoots are two fundamental aspects of pruning in plums. The majority of plum types have spurs on wood that is two years old. These spurs have a lifespan of 5 to 6 years. Pruning is required each season to allow for some spur regrowth. The amount of trimming is done in a way that encourages a 25–50 cm yearly shoot growth. For proper fruiting, it is advised to thin branches by 25–30% and head back sprouts by 50–75% on bearing plum trees. Dead, diseased, and broken branches ought to be removed during pruning.

From full bloom to four weeks after petals fall, thinning is typically done by hand, mechanically, or chemically. Chemical thinning has lower thinning costs, higher fruit size, and greater fruit quality advantages over manual or mechanical thinning. Fruit thinning in plums between full bloom and pit hardening has been advised using Naphthalene acetic acid (10-50 ppm), 3-Cyclopiazonic (100-300 ppm), 2,4,5-T (2-25 ppm), Sevin (1000-2500 ppm), Dinitro-ortho-cresol (1000-2500 ppm), and Paclobutrazol (1000-2500 ppm).

Intercropping

For the first five years, inter crops may be cultivated in a solid block of plums. There would be minimal room for crop growth if the plum were planted as filler. Consequently, inner-crops can only be raised for 1-2 years. Vegetables or legumes like peas, grains, or moong should be grown with prudence. Intercrop water requirements should promote plum growth and fruiting.

Nutrition management

Nutrition management is a critical aspect of plum production in India as it directly influences tree health, fruit quality, and overall yield. Proper nutrient management ensures that plum trees receive adequate amounts of essential elements for their growth and development. It involves the careful application of fertilizers and organic amendments to meet the specific nutritional requirements of the trees.

Organic Amendments: Farmyard manure (FYM) is an important organic amendment that enriches the soil with organic matter, improves its structure, and enhances nutrient availability. The recommended quantity of FYM varies depending on soil fertility and can range from 10-20 kg per tree per year (Singh *et al.*, 2016).

Macronutrients: Nitrogen (N), phosphorus (P), and potassium (K) are the primary macronutrients required by plum trees. Nitrogen promotes vegetative growth, and the recommended quantity ranges from 100-150 grams per tree per year (Smith *et al.*, 2018). Phosphorus enhances root development and flower formation, with an application rate of 50-75 grams per tree per year (Jones & Johnson, 2015). Potassium contributes to overall tree health and fruit quality, and the recommended quantity depends on soil test results.

Micronutrients: Micronutrients, such as iron, zinc, manganese, copper, and boron, are essential for plum trees, albeit in smaller quantities. These nutrients are typically required in the form of foliar sprays or soil applications, based on soil and leaf analysis results. The precise quantities and application methods should be determined through soil testing and consultation with agricultural experts. Spray 500 litres of water with 1 1/2 kg each of zinc sulphate, ferrous sulphate, and un-slaked lime in March, just as fruits start to ripen and become pea-sized.

Weed Control

Several weeds appear in plum orchards. Particularly prevalent weeds include motha, Khabal grass, baru grass, parthenium, and dab. Plant basins can have weeds eradicated by hoeing. Spraying Roundup (Glyphosate) at a rate of 10 ml per liter as needed, especially during the wet season, might be used to manage weeds in the vacant space.

Water management

Water management is a crucial aspect of plum production in India as water availability and efficient irrigation practices directly impact tree growth, fruit development, and overall yield. Proper water management ensures that plum trees receive sufficient water during critical growth stages while minimizing water stress and optimizing water use efficiency.

Various irrigation methods can be employed in plum orchards to efficiently deliver water to the trees and maintain soil moisture levels. Drip irrigation is a commonly used method in plum production due to its precision and ability to deliver water directly to the root zone. Research by Choudhury *et al.*(2019) suggests that drip irrigation can significantly improve water use efficiency and enhance plum tree performance.

Developing an effective irrigation schedule is essential to meet the water requirements of plum trees at different growth stages. Factors such as soil type, tree age, weather conditions, and evapotranspiration rates should be considered. Research by Kumar *et al.*(2020) emphasizes the importance of adjusting irrigation frequency and duration based on tree water needs, soil moisture monitoring, and climatic conditions. Because they have shallow roots and grow quickly, plums require enough moisture during the growth season. Numerous variables, including the kind of soil, the climate, and the species of key fruit trees, may affect the watering schedule. During the months of April, May, and June, frequent irrigation at weekly intervals may be used. To prevent flower and fruit drop, watering should not be applied during the full bloom and ripening stages. There is no need for irrigation during the wet season. In the months of September, October, and November, the gap could be extended to 20 days. Irrigation should not be administered in December or January.

Mulching

Applying organic mulch around the base of plum trees can help conserve soil moisture, reduce weed competition, and regulate soil temperature. Mulching with materials such as straw, wood chips, or compost can improve water infiltration and retention in the root zone, thus reducing irrigation requirements. Research by Gupta *et al.*(2017) demonstrates the positive effects of mulching on soil moisture conservation and plum tree growth.

Fruit Set and Pollination

The majority of native American plums are not self-fruitful and require the interplanting of pollinators in the orchard. The pollinators also help the partially self-fruitful cultivars like Santa Rosa and Beauty set fruit. Regarding fruit sets, European plums can be divided into self-fruitful variations, where 30% of the flowers produce fruit, and self-unfruitful variants, where only 1.5 to 2% of the flowers produce pollinator-required fruits. Strong beehives per acre should be planted at least every third tree in every third row to ensure proper pollination and a healthy harvest. If the cultivars don't generate much pollen, grafting a pollinizer branch on will help.

Fruit thinning

Fruit thinning is an essential practice in plum production to achieve optimal fruit size, improve fruit quality, and maintain tree health. Plum trees often set more fruits than they can adequately support, leading to smaller-sized fruits, increased competition for nutrients, and potential branch breakage. Fruit thinning helps to regulate fruit load, ensure proper nutrient distribution, and promote uniform fruit development. Thinning can be done manually or through chemical methods. Manual thinning involves selectively removing excess fruits by hand, leaving an appropriate distance between the remaining fruits to allow for proper growth. This method requires careful observation and judgment to determine the ideal number of fruits to retain on each branch. Chemical thinning involves the application of plant growth regulators or chemical sprays to reduce fruit set. These chemicals disrupt natural hormonal balance, resulting in the shedding of excess fruits. Chemical thinning should be carried out with caution, following recommended rates and timings specified for plum trees, as excessive thinning can lead to poor fruit set or tree stress. Thinning should ideally be done during the early stages of fruit development when fruits are still small. It is important to consider factors such as tree vigor, crop load, and prevailing weather conditions when determining the thinning intensity. Regular monitoring of fruit size and tree health throughout the growing season is crucial to adjust thinning practices if necessary (Mitcham 2006; Watkins & Schupp 2016).

Physiological disorders and their management

Physiological disorders are non-infectious disorders that affect the growth and development of plum trees and their fruits. These disorders are often caused by environmental factors, nutrient imbalances, or other internal plant processes. Understanding and managing these disorders is essential for successful plum production. Here are some common physiological disorders and their management strategies (Kumar *et al.*2021; Watkins & Schupp 2016).

Fruit Cracking: Fruit cracking occurs when the fruit skin splits or cracks, often due to fluctuations in moisture levels. To manage fruit cracking, maintaining consistent soil moisture is crucial. Adequate irrigation and mulching help regulate soil moisture, preventing sudden changes that can lead to fruit cracking.

Blossom-end Rot: Blossom-end rot is characterized by a dark, sunken area at the blossom end of the fruit. It is caused by calcium deficiency or irregular calcium uptake by the fruit. Proper calcium nutrition, achieved through soil amendments or foliar sprays, helps prevent blossom-end rot.

Internal Browning: Internal browning, also known as brown heart, is the browning of the flesh within the fruit. It is caused by a calcium imbalance and high temperatures during fruit development. Proper irrigation, shading, and maintaining adequate calcium levels through foliar applications can help manage internal browning.

Fruit Splitting: Fruit splitting occurs when the skin of the fruit cracks and separates into distinct sections. It is often triggered by rapid fruit growth and fluctuations in moisture levels. Consistent irrigation, avoiding excessive nitrogen fertilization, and providing adequate tree support help prevent fruit splitting.

Sunburn: Sunburn is the scorching or browning of fruit due to excessive exposure to sunlight. Providing proper tree canopy management, such as pruning to maintain adequate shade, and using reflective mulches or shading materials can help protect fruits from sunburn. Management of physiological disorders involves a combination of cultural practices, such as proper irrigation, nutrient management, and canopy management, along with environmental control measures. Regular monitoring of the orchard, early detection of symptoms, and timely implementation of appropriate management strategies are essential for mitigating physiological disorders in plum production.

Major diseases and their management

Bacteria canker *(pseudomonas syringae)*

It is most serious disease od plums caused by bacteria, *Pseudomonas syringae* and *Xanthomonas pruni*. Gums frequently ooze out of fruits, bark, or outer sapwood. Early growth, a cold and damp early spring, and frosty temperatures during bloom all contribute to the prevalence of the bacterial canker. Spraying copper fungicides like copper oxychloride (0.3%) in the fall and winter can help suppress the disease.

Powdery Mildew (*Podosphaera tridactyla*): Powdery mildew is a fungal disease that can affect plum trees. The symptoms of powdery mildew in plums include the appearance of a white, powdery coating on the leaves, stems, and fruit. Infected leaves may become distorted, curled, or stunted, and may exhibit yellowing or browning. The disease can also impact the quality of the fruit, causing discoloration, deformities, or cracking. Control powdery mildew by applying fungicides like sulfur (2% concentration) or potassium bicarbonate (2% concentration) during the growing season (Michailides & Sibbett 2013).

Peach Leaf Curl (*Taphrina deformans*): Manage peach leaf curl with copper-based fungicides (0.5% concentration) applied during the dormant season, prior to bud break. Proper sanitation practices, including the removal of infected leaves and debris, are also essential.

Bacterial Gummosis: There are no effective chemical controls for bacterial gummosis. Focus on orchard sanitation, removing and destroying infected plant material, and maintaining tree health through proper cultural practices like nutrition, irrigation, and pruning.

Crown Gall (*Agrobacterium tumefaciens*)

They spread infection through wounds. Galls are caused by the hyperplasia and hypertrophy of cells, which are triggered by the presence of microorganisms. The infection can be prevented by using nursery stock that is free of gall and gall-resistant rootstocks. Prevention is key, and it involves using disease-free planting material and avoiding wounds during pruning or other activities that can facilitate bacterial entry.

Brown rot (*Monilina fructicola*): Affects both the flower and the fruit. Browning and bloom dying are symptoms. On fruits, tiny brown patches grow into a sizable, brown, mushy, and watery region. On immature branches, the illness causes wilting and cankering up to the shoot level. To manage brown rot, apply fungicides such as captan (0.2% concentration) or thiophanate-methyl (0.1% concentration) during the bloom and fruit development stages. Proper sanitation, including the removal and destruction of infected fruit, is essential.

Plum pox or sharka: The agents that transmit Sharka or PPV as well as through budding and grafting material are called aphids. Symptoms include shallow fruit depressions and chlorotic foliage. The spread of disease can be controlled and halted by maintaining adequate orchard hygiene, removing, and burning the diseased trees. The application of an organophosphorus insecticide can suppress aphids. There are no effective chemical treatments for plum pox. Management strategies involve planting virus-free planting material and controlling aphid vectors that spread the disease.

Major insect pest and their management

Insect pests can cause significant damage to plum trees and fruits. Here are some common insect pests of plums, and their management strategies (Jayanthi & Manimekalai 2018; Santharam & Indira 2021).

San Jose Scale (Quadraspidiotus perniciosus): San Jose Scale is a sap-feeding insect that can cause fruit damage and reduce tree vigor. Insecticides containing active ingredients such as chlorpyrifos (0.05% concentration) or petroleum oil (2% concentration) can be used to control San Jose Scale infestations.

Blossom Thrips (*Frankliniella spp.*): Blossom thrips can damage plum flowers and reduce fruit set. Insecticides such as spinosad (0.02% concentration) or acetamiprid (0.05% concentration) can be applied during the bloom stage to manage blossom thrips.

Plum Weevil (*Conotrachelus nenuphar*): Plum weevils are destructive pests that cause damage to developing fruit. Insecticides like carbaryl (0.1% concentration) or malathion (0.2% concentration) can be used to control plum weevil populations.

Plum Scale (Lecanium spp.): Plum scale insects are small, immobile pests that suck sap from plum trees, leading to weakened growth and fruit damage. Systemic insecticides containing imidacloprid (0.05% concentration) or acetamiprid (0.05% concentration) can be applied as soil drench or foliar sprays to manage plum scale infestations.

Plum Fruit Moth (*Grapholita spp.*): Plum fruit moth larvae tunnel into plum fruits, causing internal damage and rendering them unmarketable. Insecticides such as phosmet (0.15% concentration) or spinetoram (0.02% concentration) can be used to control plum fruit moth larvae.

Borers: Borers, such as peach tree borer and flatheaded apple tree borer, attack the trunks and branches of plum trees, causing damage and potential tree death. Borer infestations can be managed by applying insecticides containing chlorpyrifos (0.05% concentration) or bifenthrin (0.05% concentration) to affected areas of the tree trunk.

Termites: Termites can infest and damage the wood of plum trees, affecting their overall health and stability. Termite management typically involves the use of termiticides like chlorpyrifos (0.1% concentration) or imidacloprid (0.05% concentration) applied as soil treatments around the tree base.

Maturity indices

Maturity indices are used to determine the optimal harvest time for plums, ensuring that they are picked at their peak ripeness and flavor. Here are some commonly used maturity indices for plums (Watkins & Schupp 2016).

Color: Plum fruits undergo color changes as they ripen. Depending on the variety, ripe plums can exhibit different colors, such as red, purple, yellow, or green. Color development is often a good indicator of fruit maturity, with fully developed and vibrant hues indicating ripeness.

Firmness: Plum firmness is an important indicator of maturity. As plums ripen, they generally become softer. A gentle squeeze can be used to assess the firmness of the fruit. Ripe plums should yield slightly to pressure but should not be overly mushy or too firm.

Sugar content: The sugar content of plums increases as they ripen, resulting in sweeter and more flavorful fruits. Measuring the sugar content, often expressed as the percentage of soluble solids or Brix, can provide a quantitative assessment of plum maturity.

Tasting: One of the simplest and most reliable methods to assess plum maturity is through tasting. Sampling a few fruits from the tree and evaluating their flavor can help determine if the plums have reached the desired sweetness and taste.

Harvesting & Yield

Depending on the variety, plum fruits are considered mature when they have reached the correct size and developed the proper color. In Punjab, plums are picked starting in the second week of May. Fruits should be picked when ripe and firm for local marketing. Fruits are chosen while firm but the skin has only developed 50% of its colour for marketplaces that are far away. Plums should be picked off their pedicels in order to protect the fruit from damage. The age of the tree and the management techniques used will both have a significant impact on yield. A plum tree that is fully matured often produces 40–80 kg of fruit.

Post-harvest Handling

Post-harvest handling of plums involves practices aimed at maintaining fruit quality, extending shelf life, and reducing post-harvest losses. Some key considerations for post-harvest handling of plums include:

Sorting and Grading: Plums should be sorted and graded based on their size, color, and quality. This helps in uniform packing and ensures that only high-quality fruits reach the market. Different grades of fruits are packed in different boxes and labelled.

Three standard sizes as per "Ag-Mark" System of grading plum are as follows:

Grade	Fruit size (mm)	Box Size (cm)	No. of layers	No. of Fruit per layer
Special	42 and above	37 × 16.5 × 16.5	3	28–32
Grade	1 36–42	37 × 16.5 × 16.5	4	38–43
Grade	11 Below	36 37 × 16.5 × 16.5	4	50–56

Cleaning and Sanitizing: Plums should be cleaned to remove any dirt, debris, or surface contaminants. Sanitizing solutions can be used to reduce the risk of microbial contamination.

Packaging: Plums are often packed in protective containers or trays to prevent bruising and damage during transportation. Packaging materials should be sturdy, well-ventilated, and designed to maintain fruit quality.

Storage: Due to their perishability, plums have an extremely short shelf life. According to research conducted in India by HP, plums may be kept at 0°C and 80–90% humidity for one to two weeks. The fruits can be preserved for a period of two to three months using the CA storage method, which has been used abroad to maintain 2-3% oxygen and 2-8% CO_2.

Handling and Transportation: Plums should be handled with care to minimize mechanical damage. Proper loading, unloading, and transportation practices should be followed to prevent jostling and impact-related injuries to the fruits.

Quality Monitoring: Regular monitoring of fruit quality during storage and distribution is crucial. Any signs of decay, mold, or deterioration should be promptly addressed to minimize losses. By implementing proper post-harvest handling practices, the shelf life of plums can be extended, ensuring that consumers receive high-quality fruits and reducing economic losses for growers.

Future Prospect

Crop improvement holds immense future prospects for plum production in India. Through the application of modern breeding techniques, the development of improved plum varieties can be achieved. The primary goals of crop improvement include enhancing traits such as yield, fruit quality, disease resistance, and adaptability to different agro-climatic regions. Crop improvement efforts can focus on developing plum varieties with improved taste, texture, and aroma to cater to consumer preferences. Disease-resistant varieties can be developed to reduce the reliance on chemical pesticides and enhance sustainability in plum production. Additionally, breeding for improved shelf life and post-harvest handling characteristics can extend the marketability of plum fruits.

The use of advanced techniques like genetic engineering and genomic selection can expedite the breeding process by targeting specific traits and accelerating the development of new varieties. Integration of molecular markers and genomics can aid in the identification of desirable traits and facilitate more efficient breeding programs. Furthermore, research and development in plum crop improvement can contribute to the conservation and utilization of genetic diversity within the plum germplasm. This can help in the preservation of unique and valuable genetic resources, allowing for the development of region-specific plum varieties and addressing specific challenges faced in different parts of India.

References

Anonymous, (2020). National Horticulture Board, Indian Horticulture Database. Ministry of Agriculture & Farmers Welfare, Government of India. www.agricoop.nic.in

Bal. J. S. (2014). Fruit Growing, Third Revised Edition, Kalyani Publishers, B-1/1292, Rajinder Nagar, Ludhiyana-141008, pp. 374-385.

Bose, T. K., Mitra, S. K., Rathore, D. S. (2001). Temperate Fruits, Third Edition, Volume 2, Horticulture and Allied Publishers 27/3, Chakraberia Lane, pp. 241-275.

Chadha, K.L. (2019). Hand Book of Horticulture. Second Edition, Volume 1, Indian Council of Agricultural Research, New Delhi. pp. 400-403.

Chattopadhyay, T.K. (2009). A Textbook on Pomology IV: Devoted to Temperate Fruits Crops, Second Edition, Volume 4, Kalyani Publishers B-1/1292, Rajinder Nagar, Ludhiyana-141008, pp. 65-87.

Choudhury, R.K., Singh, R.K., & Naskar, S.K. (2019). Advances in irrigation scheduling techniques for fruit crops. In: Singh, B., & Kumar, P. (Eds.), *Advances in Precision Farming Techniques in Horticultural Crops* (pp. 251-267). Springer.

Gupta, D., Bhardwaj, V., & Gupta, R. (2017). Mulching in fruit crops: A review. *International Journal of Current Microbiology and Applied Sciences*, 6(3), 288-298.

http://ecoursesonline.iasri.res.in/mod/page/view.php?id=96188

http://ecoursesonline.iasri.res.in/mod/page/view.php?id=96188#:~:text=Rootstock%20%3A,5000ppm)%20for%20better%20rooting

http://www.kiran.nic.in/pdf/Agri-Kaleidoscope/Horticulture-resources/PLUM.pdf.

http://www.kiran.nic.in/pdf/Agri-Kaleidoscope/Horticulture-resources/PLUM.pdf

https://leafyplace.com/types-of-plums.

https://www.researchgate.net/publication/313611588_Physiological_Disorders_and_Their_Management_in_Stone_Fruits

https://www.researchgate.net/publication/343275361_Production_Technology_of_Peach_Plum_and_Apricot_in_India

https://www.yourarticlelibrary.com/fruits/plum-cultivation-in-india-production-area-climate-harvesting-and-fruit-handling/24693

Jayanthi, P. D., & Manimekalai, R. (2018). Pest management in fruits. In: Crop Protection: Principles and Practices (Vol. 2, pp. 162-163). Scientific Publishers.

Jones Jr., J.B., & Johnson, R.E. (2015). Nutrient deficiency symptoms and plant tissue analysis. In: Plant analysis handbook IV. Micro-Macro Publishing.

Kumar, P., Singh, P., Kumar, V., & Singh, R. (2020). Smart irrigation scheduling for fruit crops. In: Singh, P., Singh, R., & Kumar, P. (Eds.), Climate Change and Sustainable Horticulture (pp. 221-240). Springer.

Kumar, P., Singh, R., & Bhardwaj, V. (2021). Physiological disorders in plum. In: Handbook of Horticulture: Fruit Crops. ICAR-Indian Agricultural Research Institute.

Michailides, T. J., & Sibbett, G. S. (2013). Diseases of Temperate Zone Tree Fruit and Nut Crops. UCANR Publications.

Mitcham, E.J. (2006). Pruning and thinning stone fruit trees. University of California Agriculture and Natural Resources. Retrieved from https://anrcatalog.ucanr.edu/pdf/8044.pdf

Mukherjee, S. K., Mondal, S., Ghosh, S., & Ghosh, P. (2016). Plum cultivation in India. In: S. Mukherjee, S. Kundu, & S. Dasgupta (Eds.), Plum: Production and Postharvest Management (pp. 15-26). Apple Academic Press.

Rehder, A. (1940). Manual of Cultivated Trees and Shrubs, 2nd edition. The Macmillan Company, New York.

Santharam, G., & Indira, V. (2021). Fruit insect pests and their management. In: Horticultural Insect Pests and Their Management (pp. 264-266). Daya Publishing House.

Singh, B.; Sirohi, N.P.S. (2013) Protected cultivation of vegetables in India: problems and future prospects. *Acta horticulture*. P339-342.

Singh, J.P., Kumar, A., & Singh, G. (2016). Soil fertility management for sustainable crop production. In: Soil Science: Step to Development. Daya Publishing House.

Singh, R. P., Kundu, S., & Kundu, A. (2014). Soil management in horticultural crops. New Delhi: Daya Publishing House.

Singh. M., Niwas. R. (2015). Thermal Time Utilization of Plum in Semi-Arid Region of Gangetic Plain. Science research, 3(1): 19-24.

Smith, R., Smith, S., & Green, T. (2018). Fertilizing fruit crops. University of Nevada Cooperative Extension Fact Sheet. Retrieved from https://extension.unr.edu/publication.aspx?PubID=1684

Watkins, R., & Schupp, J.R. (2016). Fruit disorders - physiological. Penn State Extension. Retrieved from https://extension.psu.edu/fruit-disorders-physiological

Watkins, R., & Schupp, J.R. (2016). Fruit thinning. Penn State Extension. Retrieved from https://extension.psu.edu/fruit-thinning

Watkins, R., & Schupp, J.R. (2016). Harvesting and storing plums. Penn State Extension. Retrieved from https://extension.psu.edu/harvesting-and-storing-plums

5

Advances Production Technology of Apricot

Sanjay Chetry*[1], Shruthi S[1], Manjula Rakshi[1] and J. Auxcilia[2]

[1]Ph.D. Scholar, Department of Fruit Science, Horticultural College and Research Institute, Tamil Nadu Agricultural University, Coimbatore, Tamil Nadu
[2]Professor (Hort.), Directorate of Extension Education, Tamil Nadu Agricultural University, Coimbatore, Tamil Nadu
**E-mail: sanjaychetry47@gmail.com*

Introduction

Prunus armeniaca L. is a highly nutritious fruit belonging to the Rosaceae family and subfamily Prunoideae. It is known for its attractive appearance, pleasant flavor, and aroma. The name "apricot" is believed to have originated from the Latin word "praecocia" meaning "early matured" or the Arabic word "albarquq" meaning "short ripening period." Apricot is considered to be an important temperate fruit crop due to its high nutritional and health benefits. Apricots are rich in a multitude of vitamins and minerals, including Retinol (Vit. A), Thiamine (Vit. B1), Riboflavin (Vit. B2), Ascorbic acid (Vit. C) and Tocopherol (Vit. E) as well as calcium, iron, sodium, phosphorus, potassium, and magnesium. It also has low levels of fats and carbohydrates, making it an important food in human nutrition (Batu *et al.*, 2007; Karatas and Kamisli, 2007). Research has indicated that incorporating apricots into one's diet may counteract the risk posed by several ailments, including cardiovascular disease, cancers, cataracts, and other degenerative diseases (Oomah and Mazza, 2000).

In India, commercial cultivation of apricot takes place in the states of Himachal Pradesh, Jammu and Kashmir, Uttarakhand, and North eastern hill state. It has broad altitudinal adaptability, thriving between 900 and 2000 meters above mean sea level. Apricots can be cultivated in a broad spectrum of climatic conditions, from the frigid winter conditions of Siberia to the subtropical climates of North Africa, as well as from the arid lands of Central Asia to the moist areas of Japan and eastern China. As per data from FAO, United Nations in 2018, apricot was cultivated in 68 countries worldwide, covering an area of 5.49 hectares, with a total production estimated to be approximately 3.84 million MT. The Mediterranean Sea, encompassing nations such as Turkey, Spain, Italy, France, and Greece, currently dominates the world's apricot production. Turkey leads as the top producer of apricots, yielding 0.75 million tons (2018), with Uzbekistan, Iran, Algeria, Italy, Spain, Pakistan, France, Japan, and

Ukraine following behind. The combined output of these top 10 apricot-producing nations constitutes 70.37 percent of the world's total apricot production.

Apricot is a perishable fruit that is known for its shelf life of roughly 4-5 days, making it frequently consumed in the form of dried fruit. Countries like Turkey, Iran, China, the US, Australia, and South Africa are among the primary producers of dried apricots. Notably, Turkey alone accounts for 80% of global production and 70% of exports of dried apricots (Ercisli, 2009).

Apricot is particular in their climatic requirements and certain cultivars are known for their tendency towards inconsistent yields, attributed to the crop's limited adaptability range. Unlike other fruit crops where a small number of cultivars are globally grown, each apricot cultivar is typically limited to specific geographical regions with specific ecological conditions (Layne *et al.*, 1996). Historically, apricot has remained unaffected by overproduction and is frequently sold at a premium price, both in their fresh and processed forms (Hormaza *et al.*, 2007). It is a significant commercial crop in temperate regions, ranking third after plum and peach, due to its nutritious and therapeutic properties which earned it the moniker "golden fruit" (Alajil *et al.*, 2021).

History

The early explorers considered apricots to be valuable treasures, comparable in value to exotic spices and precious gems. While some assert that apricots have been present in India since 3,000 BC, it is more likely that the fruit arrived in India from China during or after the Han Dynasty (206 BC-220 AD). Bertrand Russell, in his book "*Useless Knowledge*", stated that during the reign of Afghan King Kanishka (78AD-144AD), Chinese captives brought apricots to India.

Apricots are believed to have originated in China. There exists Chinese written evidence, cited by De Candolle in 1886, of apricot cultivation dating back to the end of III Millennium BC. In contrast, apricot cultivation in Central Asia was introduced much later, around the I-II Millennia BC (Sinskaya 1969). Modern excavations in the southern region of Turkmenistan and Uzbekistan offer no indications of the utilization of fruit and nuts in western Central Asia before 1500 BC (Miller 1999).

Local Names of Apricot known in India

Zardalu, Khubani and Khumani (Hindi); Saara Paruppa and Sarkkarai Badami (Tamil); Jaldharupandu (Telegu); Mutta Pazham and Sheema Pazham (Malayalam) and Chuli (Ladakh).

Origin and distribution

Vavilov (1951) has enlisted three centers of origin: The Chinese Centre (Mountainous regions of northern and western China), The Central Asian Centre (mountainous areas extending from the Hindu Kush to Kashmir), and The Near Eastern Centre (mountainous regions extending from northeastern Iran to the Caucasus and central Turkey). Vavilov determined that the Near East region serves as a secondary center of origin for apricot varieties.

Apricot is an important fruit crop and is mainly grown in the dry temperate and mid-hill regions of India. Original apricots are from China whereas the wild apricot (Zardalu) is native to India. The best conditions for apricot fruit plants are moderate winters and little water. It is a Mediterranean crop that requires hot, dry summers with little humidity.

A recent investigation of single nucleotide polymorphisms (SNPs) in apricot has disclosed that the domestication of the cultivated apricot was the outcome of two distinct events. The European cultivated apricots were found to have diverged from the wild populations of Northern Central Asia and led to the differentiation of four groups: the Mediterranean countries, Continental Europe, North America, and North Africa (Groppi *et al.*, 2021).

The apricot's dissemination from China extended westward throughout central Asia, Iran, and Transcaucasia region, before reaching Italy in the Ist century, England in the 13th century and by 1720 to finally North America. The apricot is also thought to be native to Armenia as is evident from its scientific name *Prunus armeniaca* where it was cultivated since ancient times. Presently, apricot cultivars have pervaded regions of the world with climates suitable for their cultivation.

Several alternative paths have been suggested for the spread of apricots from the Middle East to different regions:

i. The dissemination of apricots extended from the Middle East to Egypt and Northern Africa, eventually reaching Spain (Khadari *et al.*, 2006)
ii. Another path for the distribution of apricots ran north from the Black Sea region, either originating from Turkey or directly from Iran.
iii. The dissemination of apricots through the Danube River valley and Germany was significantly influenced by Roman soldiers and Turkish landowners.
iv. The frequent introgression of plant specimens from the primary centers of origin is suggested as the source of extensive diversity in Italian germplasm according to a molecular study by Geuna *et al.*in 2003.

Area and production

The leading countries that produce apricots globally are Turkey, Iran, Italy, Spain, Pakistan, France, USA, Afghanistan, Czechoslovakia. Apricot growing areas in India range about 2303 hectares. In India, commercial apricot cultivation takes place in the hill regions of Himachal Pradesh, Jammu and Kashmir, Uttarakhand and in some limited areas of the North Eastern hills of Arunachal Pradesh, Nagaland, Manipur, Meghalaya and Mizoram. Shimla and Kinnaur districts of Himachal Pradesh seem to be the home for wild apricots with tremendous biodiversity for different quantitative and qualitative characteristics. With the introduction of low-chilling varieties, it is now being cultivated in the plains of northern India (Amritsar, Hoshiarpur, Pathankot, Patiala districts of Punjab). In India, Ladakh is the top producer with a total production of fresh apricots of 15,789 tonnes and dried apricots of 1,999 tonnes in 2019 (Stobdan *et al.*, 2021). Himachal Pradesh and J&K are India's top two apricot-producing states after Ladakh. Apricot is also cultivated in Uttarakhand and North-Eastern hilly region of India.

Fresh apricot production

The global annual production of fresh apricots was 4,260,466 tonnes from 2017 to 2019. Turkey is a major player in apricot production, with other notable countries including Uzbekistan (12.2%), Iran (7.6%), France (7.1%), Italy (6.0%), Algeria (5.5%), Spain (3.8%), Afghanistan (2.9%), and Pakistan (2.8%), comprising 68% of the world's apricot production. In terms of export, Spain leads the way with 21.1% in 2018 and 2019, followed by Suriname (17.0%), Turkey (14.4%), Uzbekistan (10.5%), Italy (7.8%), France (5.2%), and Greece (4.9%). Germany is the top importer of fresh apricots, accounting for 15.3% of the global trade in 2018 and 2019, with other significant importing countries including the Russian Federation (12.9%), Iraq (9.2%), Kazakhstan (8.4%), Pakistan (5.4%), France (5.0%), and Italy (4.5%).

Dried apricot production

Global annual export of dried apricots was estimated at 162,635 tonnes during 2018 and 2019. The export of dried apricots has seen an increase of 21.5% from 2010 to 2019. Turkey is the largest exporter of dried apricots, accounting for 59.7% of global exports. Other significant exporters include Afghanistan (14.9%), Tajikistan (6.7%), Uzbekistan (5.1%), Kyrgyzstan (3.9%), Belarus (2.4%), Spain (2.3%), Netherlands (1.0%), France (0.9%), and Germany (0.8%). The United States is the leading importer of dried apricots, accounting for 9.5% of global trade in 2018 and 2019.

Nutritional composition

Uses

Apricots can be consumed in various forms, including fresh, dried, and frozen can be transformed into various processed items such as jam, jelly, syrups, and juices. They are nutritionally dense, low in calories and contain a wide array of essential vitamins and minerals. Additionally, apricot kernels are abundant in high-quality oil and can be used for culinary purposes, in cosmetic products and as a source of benzaldehydes and activated carbon.

Apricots are versatile fruit that can be consumed in a variety of forms, including fresh, dried and frozen. They can also be processed into various products such as jam, jelly, syrup, or juice. Apricots are nutrient-dense, offering high levels of essential vitamins and minerals while being low in calories

Nutritional composition of Apricot per 100 g (Moustafa and Cross, 2019)

	Raw Apricot	Dried Apricot
Nutrients		
Water	86.35 g	30.89 g
Sugars	9.24 g	53.44 g
Fibers	2.0 g	7.30 g
Proteins	1.40 g	3.39 g
Lipids	0.39 g	0.51 g
Minerals		
Calcium	13.0 mg	55.0 mg
Iron	0.39 mg	2.66 mg
Magnesium	10.0 mg	32. 0 mg
Phosphorus	23.0 mg	71.0 mg
Potassium	259 mg	1162.0 mg
Sodium	1.0 mg	10.0 mg
Zinc	0.2 mg	0.39 mg
Vitamins		
Retinol (Vit. A)	1926 IU	3604 IU
Thiamine (Vit. B1)	0.03 mg	0.015 mg
Riboflavin (Vit. B2)	0.04 mg	0.074 mg
Niacin (Vit. B3)	0.6 mg	2.589 mg
Pyridoxine (Vit. B6)	0.054 mg	0.143 mg
Folate (Vit. B9)	9.0 μg	10.0 μg
Ascorbic acid (Vit. C)	10.0 mg	1.0 mg
Tocopherol (Vit. E)	0.89 mg	4.33 mg
Vitamin K	3.3 μg	3.1 μg

- **A good source of pro-vitamin A** (abundant in apricot) controls the immune system, thereby safeguarding the skin.
- Great source of dietary fiber that benefits the digestive system. The fruit has high fiber content (equal amounts of soluble and insoluble fiber are present overall), which aids in lowering the level of harmful cholesterol and simultaneously raises good cholesterol in the body, protecting the heart.
- Potassium in the fruit balances the electrolyte levels in our body, maintaining the health of our cardiac muscles.
- Apricots are known for their high content of phytochemicals including polyphenols and carotenoids, contributing to their unique taste, hue, and nutritional significance (Dragovic-Uzelac *et al.*, 2007).
- Apricots are a fruit that is rich in carotenoids, with a β-carotene content ranging from 2.00 to 20.77 mg/100g (Ali *et al.*, 2011). β-carotene acts as an antioxidant by neutralizing oxidative damage to cells caused by ROS. Additionally, carotenoids are thought to play a role in preventing degenerative disorders and promoting overall health (Bramley, 2003).

- Historically, apricot has been employed for centuries as traditional medicine in China and by the people living in the mountainous regions of Himalayas. In Chinese medicine, apricot is believed to have properties that aid in regenerating bodily fluids, purifying the body and alleviating thirst. The kernels are also utilized to enhance the respiratory system and reduce coughing (Kan and Bostan, 2010).

Botany

Apricot trees are small to medium growing up to a height of 8-12 m and are deciduous with either upright or spreading growth habits. The trunk of the apricot tree can reach a diameter of 40 cm with its bark having reddish to gray-brown color and smooth twigs. The leaves are simple, alternate, ovate measuring 5-9 cm in length and 4-8 cm in width with a rounded base, a pointed tip with a margin finely serrated, glabrous above and pubescent beneath. Flowers are solitary, pinkish to white with deep red sepals (5 in number) appearing before the leaves with a size of 2 - 4.5 cm. They feature five petals that are white to pinkish, and numerous stamens that stand erect with an ovary being perigynous. The apricot fruit resembles a tiny peach and its fruit type is a drupe, measuring 1.5 - 2.5cm in diameter (some modern cultivars may be larger). It is yellow to orange in color, often with a red tinge on the side that is exposed to the sun. The surface of the fruit can be smooth and hairless (glabrous) or have very short hair (pubescent). The seed is solitary and encased in a hard stony shell referred to as a stone, with a grainy, smooth texture and three ridges running along one side. Fruits are velvety, more or less round with orange to yellow or white colored, flesh is sweet, orange to white enclosing a smooth to rigid and flattened stone. Trees are precocious with the ability to produce fruit as early as the second or third year after planting.

Taxonomy

The apricot is a widely cultivated fruit that belongs to the Rosaceae family, which comprises approximately 2830-3100 species and 100 genera (Hummer and Janick 2009). The Rosaceae family encompasses a complicated taxonomy, including various levels of classification, and is constantly subjected to ongoing debate and revision due to recent genetic findings leading to new taxonomic rearrangements (Potter *et al.*, 2007, Shi *et al.*2013). The Rosaceae family is now being divided into three subfamilies: Rosoideae, Amygdaloideae, and Dryadoideae (Potter *et al.*, 2007). The subfamily Amygdaloideae, also known as Prunoideae, includes the genera *Pyrus*, *Malus* and *Prunus*, which contain several common stone fruits. The genus *Prunus* has recently been reorganized into 3 subgenera (Shi *et al.* 2013), with almonds and peaches now falling within the subgenus *Prunus*, which is further subdivided into sections. The section Armeniaca includes the common apricot (*Prunus armeniaca*) and the Japanese apricot (*Prunus mume*).

The genus *Prunus* has been recently reclassified into three subgenera (Shi *et al.*2013), with almonds and peaches now placed under the subgenus *Prunus*, which is further divided into different sections. The common apricot (*Prunus armeniaca*) and the Japanese apricot (*Prunus mume*) belong to the section Armeniaca.

The number of apricot species varies according to the classification system used, with six different species namely *Prunus brigantina* Vill., *Prunus holosericeae* Batal

(Tibetan Apricot), *Prunus armeniaca* L. (Common Apricot), *Prunus mandshurica* (Maxim), *Prunus sibirica* L., and *Prunus mume*. Charmagz, Halman, Rachkaikarpo, Nari and Shakarpara are popular apricot cultivars cultivated in J&K (Singh, 2016).

Soil and climate

Apricots are grown well in mid-hilly regions of India which requires coolers climatic conditions. Yellow apricots thrive well in warm weather and are cultivated at elevations ranging from 1000 to 1500 meters above sea level. Southwestern and Northeastern parts of India are highly suited for the growing of Apricots. Apricots can be cultivated in a wide range of soils. Though mostly it preferred a well-drained soil with high organic matter content for its better growth and higher yielding capacity. Micro-climate is a significant factor in the production of apricots that possess characteristics and features for fresh consumption. Generally, the highest quality fresh apricots in Trans-Himalayan Ladakh, India are produced in villages located at elevations above 9,500 feet. A total of 93 villages are located at this elevation. Fruit sweetness increases with increasing altitude. Every 100-meter increase in elevation results in a 1.2 °Brix increase in TSS of the fruit. Raktsey Karpo, a cultivar of Apricot was found to have significantly higher TSS (28.1 ± 3.8 °Brix) compared to cultivars with brown seed coats (Angmo *et al.*, 2017)

Commercial varieties with salient features

In India, apricot is cultivated in varying climatic condition ranging from mid to high hills. Varieties appropriate for the mid-hills are not suitable for the high hills or the arid temperate zone. In India, there are around 100 varieties of apricot. The majority of them are of exotic origin.

Apricot varieties recommended for cultivation in different states	
State/Union Territory	**Varieties**
Himachal Pradesh	Mid Hills: New Castle, Early Shipley and Shakarpara High Hills: Kaisha, Nugget, Royal, Suffaida, Charmagaz and Nari Dry Temperate: Charmagaz, Suffaida, Shakarpara and Kaisha
Uttarakhand	Charmgaz, Kaisha, Moorpark, Turkey, Ambroise, Early Shipley, Chaubattia Alankar, Chaubattia Madhu, ChaubattiaKesri and Bebeco
Kashmir	Turkey, Australian, Charmagaz, Rogan and Shakarapara
Ladakh	Halman, Rakchakarpa, Tokpopa, Margulam, Narmu and Khante
New Promising Varieties	Early maturing: Baiti,Beladi Late-maturing: Farmingdale, Alfred

Apricot genotypes were classified based on kernel taste into two groups *viz.*, sweet kernel and bitter kernel.

Sweet Kernel type	:	CITH-AP-1, 2, 3, 4, 5, 8, 9, 10, 11, 12, 13, 14, 15, 16, 17, 18, 19, 21, 23, 24, 26, 28, 30, 32, 33, 34, 35, 36, 37, Harcot, Balcota, Erani, Chinese, Australian, Tokpopa, Heartly, Afghani, Sateni, Yerevani, Communis Holy, Nari, PAS-1, PAS-2, PAS-3, PAS-4
Bitter kernel type	:	CITH-AP-6, 7, 20, 22, 25, 27, 31, Rival, Tilton, Communis, Turkey, Fairmedcester, Viva Gold, New Castle

Source: CITH, Annual Report, 2015-16

Salient features of apricot varieties

Varieties	Salient features
Gold Sweet Apricot  minnetonkaorchards.com	Tree's springtime canopy of pink-hued and white blooms attracts pollinators. Fruit skin is golden with delicate blush and the leaves are heart-shaped. It is self-fertile.
Autumn Glo Apricot  www.homestratosphere.com	Dwarf tree (10 to 12 feet), 4-in-1 and 3-in-1 planting techniques are used. Fruits are yellow in color.
Golden Kist Apricot  tucson.com	Earliest fruit to ripen, orange freestone skin fruit tastes excellent and are used to cook, can, and dry food.

Varieties	**Salient features**
Tomcot  www.pomonafruits.co.uk	Fruit is oval-shaped, medium- to large-sized fruit that is 5.4 cm in diameter with a pale orange tint skin. Mellow orange flesh has a sweet flavor. Canning is not an option.
Rival  growgreatfruit.com	Originated in France has a large size comes in yellow with rosy cheeks and is harvested in Mid to late Summer.
Autumn Royal  www.lecooke.com	Originated in southern California comes in medium to large and has a slightly yellow to orange blush harvested in the Late season of May to July.
Perfection  www.vanwell.net	Originated on the Russian Chinese Border and mainly harvested in the early mid-season of has a hint of light yellow to orange color.

Varieties	Salient features
Shipley early	Early variety, Fruits are medium-sized and creamy white appearance, the flesh is sweet and juicy with a bitter kernel and the seed is large and round with cling stone.
New Castle	Regular and prolific bearer with round, medium-sized yellow-colored fruits. Ripens after Shipley early. Leading variety for the mid hills.
Harcoat  minnetonkaorchards.com	New introduction for the mid hills. Fruits are round, light yellow, medium-sized, larger than New Castle and ripen almost with New Castle under Kandaghat conditions.
Turkey	Late variety for the mid-hills and ripens during first week to middle of June. Fruits are round, medium-sized and creamish white in appearance.
Kaisha	Prolific bearing variety that ripens in mid-June. Fruits are round medium in size with yellow skin. Flesh is orange yellow, sweet and tasty.
Blenheim 	Fruits are large in size, yellow and ripen in the middle of June. Flesh is yellow firm and juicy.
Moorpark  growgreatfruit.com	It is often a shy bearer variety, ripens during mid to last week of June.
St. Ambroise	It is a shy bearer under the mid hills. The fruits are oval large and orange yellow. The kernel is somewhat flat unlike other cultivars where it is round. The flesh is deep yellow, firm and sweet.

Varieties	Salient features
Nugget  anfic.com.au	Regular bearing variety that ripens in the middle of June. Fruit has a medium to large size and a round shape. Its flesh is characterized by its deep yellow color, firm texture and juicy. Additionally, the kernel of this variety has a sweet flavor.
Nari	An important variety of the high hills. Fruits are small to medium in size, slightly oblong. The seed is almost round with sweet kernel. The fruits start drying on the trees thus exposing the seeds. Fruits are harvested during the second week of September under Spiti conditions in Himachal Pradesh.
Kaisi	Fruits of Kaisi are slightly larger in size as compared to Nari. The flesh is light yellow in colour and very sweet in taste. The kernel is small and sweet.
Sharkarpara  Alajil et al., 2021	Sweet cultivar with high TSS of 26 per cent. The fruits ripen in June under high hills and during third week of September under Spiti conditions. If the chilling requirement is not fully met it becomes a shy bearer.
Halman  @halmanapricot	The important cultivar of Ladakh used as dessert and also dried on a commercial scale. Because of high sugar content, it is highly suitable for drying. Fruits are medium-sized and golden in colour.

Varieties	**Salient features**
Raktsey- Karpo  www.thestatesman.com	Apricots with white seed stone, is unique to Ladakh. It is the most preferred cultivar for fresh consumption. The phenotype - white seed stone is associated with sweet kernel, brightly coloured fruit with high with a high content of total soluble solids (TSS) and sorbitol (a sugar alcohol which is used as a sweetener in sugar-free sweets and chewing gum). Geographical Indication was given to Raktsey-Karpo apricot of Ladakh in the year 2022.
Wenatchee Moorpark  minnetonkaorchards.com	Introduction made from USA, these varieties are shy in bearing and ripen during mid-June coinciding with the onset of monsoon.
Tilton  minnetonkaorchards.com	Freestone Variety. Introduced from USA.
Tokpopa	Local selection from Ladakh. The fruits are large, round and orange yellow. It is a popular variety among the locals and is mostly used as a table variety. The fruits ripen from mid to last week of August.

Central Institute of Temperate Horticulture (CITH), Srinagar, Jammu and Kashmir, India has released three varieties of Apricot through clonal selection in the year 2009.

1. CITH Apricot-1  Sofi et al.2009	Self-fertile and mid-season blooming, this variety is well-suited for cultivation in the entire temperate region of the North Western Himalayas. Its fruits are of a larger size (50-60g), round in shape with an orange hue and a reddish tinge on one side (25-30%). This cultivar is a high yielder (15-20 t/ha) and boasts low acidity and high total soluble solids (14° Brix). Ideal for both fresh consumption and processing. Early maturing with good quality fruit. This variety is also tolerant to major pests and diseases.
2. CITH Apricot-2  Sofi et al.2009	Self-fertile and blooms during the early to mid-season, making it ideal for cultivation in the entire temperate region of the North Western Himalayas. The fruits are yellowish-orange in color, medium-sized (40-50g), oblate, asymmetrical, with a slightly pointed beak. They have low acidity, high total soluble solids (14° Brix) and are high-yielding (12-15 t/ha), with mature trees expected to produce 20-25 kg per tree. This variety is suitable for both fresh consumption and processing, maturing early and offering superior quality with widespread acceptance by both growers and consumers. Additionally, this variety is tolerant to leaf curl and stigmina blight.
3. CITH Apricot-3  Sofi et al.2009	Self-fertile and blooms during the early to mid-season, making it ideal for cultivation in the entire temperate region of the North Western Himalayas. Fruit is very attractive with a bright orange color covering 30-40% of its surface. It is medium-sized (40-45g), boasts low acidity, high total soluble solids (16° Brix), and is a high-yielding variety (10-12 t/ha), making it ideal for use as a dessert.

Three precocious and high-yielding varieties, namely KS-1, AS-1 and AS-2, have been developed and are gaining popularity in the region. These varieties have a total soluble solids (TSS) range of 13.1 to 14.6 °Brix (CITH, Annual report, 2015-16).

Hybrids

Three apricot hybrids of regular bearing with low chilling requirement and high yielding namely Chaubattia Alankar, Chaubattia Madhu and Chaubattia Kesri for commercial cultivation have been released.

Hybrids	Parentage	Ripening time
Chaubattia Alankar	Kaisha x Chargamz	End May to starting of June
Chaubattia Madhu	Turkey x Chargamz	Early ripening (First fortnight of June)
Chaubattia Kesri	St. Ambriose x Chargamz	Mid-season (Third week of June)

Propagation And Rootstocks

Commercial propagation

Apricot is commercially propagated by vegetative methods such as budding and grafting as it maintains the desirable genetic characteristics. Whip and tongue grafting / Chip budding (carried out during September has a high rate of success) and T- budding is usually practiced.

According to the finding of Dwivedi *et al.*, (2000a), T-budding when executed in mid-August determined to be the most effective method of propagation when compared to patch budding. Chip budding performed in the final week of February was found to produce over 70% of salable plants (Dwivedi and Singh, 2001). It is recommended to use tongue and cleft grafting performed in the first week of February as the preferred method for propagating apricots through grafting (Dwivedi *et al.*, 2000b). Apricot propagation can also be achieved through seeds, however, stratification at 4°C for 72 days is required. The germination process can be expedited by removing the kernel and treating it with GA_3 (500 ppm) or kinetin (5 ppm) before planting.

Rootstocks

Clonal Rootstock

Apricot seedlings are commonly propagated onto wild apricot, wild plum and peach. These seedlings are considered to be ideal for creating a strong graft union, leading to the production of vigorous and thriving trees. The common rootstock utilized for apricot primarily belongs to the same species. The popular Myrobalan seedling rootstock and its clones (Myrobalan B and Myrobalan 29C) are widely used to graft different apricot varieties. By pairing specific scion varieties with rootstocks like Mariana GF 8-1, Greengage CD-4 or Damas1869, it's possible to see increased yields in apricot production (Dimitrova, 2001). Mariana GF 8-1 has been scientifically shown to increase its longevity. Additionally, Pollizo prune (*Prunus institia* L.) – a Spanish rootstock in the Mediterranean region of Spain has demonstrated resistance to flooding (Domingo *et al.*, 2002). ACLSV (Apricot Chlorotic Leaf Spot Virus) is responsible for the serious viral disease of apricot in Spain known as Viruela. Scientific literature reports have shown that GF305 peach and Real Fino apricot seedlings are effective in resisting ACLSV, as they displayed no symptomology on the leaves (Ibarra *et al.*, 2010).

- Wild Apricot (Chulli): Graft union is good and produces more vigorous trees.
- Peach: Well suited for light soils and dry conditions, but may result in a rough or enlarged graft union.
- Myrobalan Plum: Well suited in heavy soils under excessive soil moisture.

Seedling rootstock

- Seeds are collected from fully ripe and wild apricot fruits
- To break the dormancy, seeds are stratified at 4 °C for 72 days.
- Stratified seeds are then sown in nursery beds, where they grow into a graftable size after a year
- Those seedlings that have reached the thickness of a pencil within a year are grafted using the tongue method in February, while those of a smaller girth are T-budded in May-June.

The few species which are used as seedling rootstocks are,

- *Prunus armeniaca*
- *Prunus mandshurica*
- *Prunus mume*
- *Prunus cerasifera*
- *Prunus domestica* (can be used as both seedling and clonal rootstock)
- *Prunus microcarpa*

Prunus microcarpa is a promising rootstock candidate for apricot. In a study by Ugur (2022), the effects of grafting apricot cultivars (Hacıhaliloğlu', Hasanbey', Kabaaşı', and Roxana) onto *Prunus microcarpa*, *Prunus armeniaca* L. and *P. institia* rootstocks were examined. *P. armeniaca* and *P. institia* served as control rootstocks. Results indicated that the cultivars grafted onto *P. microcarpa* exhibited early blooming, approximately 3-5 days ahead of the control rootstocks and displayed notable dwarf stature with an average height of 83.65 cm. It is a common occurrence for the yield per tree to suffer a reduction due to the dwarfing effect of grafted apricot varieties on *Prunus microcarpa*. However, as a result of *P. microcarpa's* high yield efficiency, higher yields per unit area can be anticipated through appropriate spacing.

In vitro propagation

Apricots exhibit high levels of heterozygosity and thus, require an effective protocol for micropropagation. Micropropagation presents a potential solution for the challenges posed by the clonal propagation of elite apricot varieties and lays the foundation for genetic transformation experiments against viruses.

Explant source

Shoot tips or Axillary shoots or Single node buds can be used as a source of explant for apricot *in vitro* culture.

Disinfectants used

Sodium hypochlorite (28%) + 70% ethanol for about 15 min. + 1 min. respectively and 100% double distilled water wash is done thrice (Khan *et al.*, 2021).

Inoculation of shoot tips (1 – 1.5 cm long) is done in MS media + 1 mg/l of BAP (6-Benzyl Adenine Phosphate) + 0.01 mg NAA (Napthalein acetic acid). The shoots should be moved to freshly prepared medium every three weeks. It should be maintained for 9 weeks. After nine weeks they are transferred to the rooting medium. IBA and NAA @ 6 mg/l and 2 mg/l respectively increase the rooting percentage and number of roots followed by hardening of the rooted *in vitro* plants. (Šiško *et al.*, 2022).

Fig. 1: Regeneration of shoots of apricot (P. armeniaca): (a) shoot developed after disinfection and induction of bud; (b) shoots used as an explant in experiment; (c) shoots obtained on medium supplemented with meta-topolin (Šiško *et al.*, 2022)

Cultivation

Planting and Orchard Establishment

A square system of planting with a distance of 6 x 6 metre is traditionally adopted. There is, however, a recent trend towards rectangular patterns because more plants can be placed in a single row. Planting should be done during the dormant season *i.e.,* at the end of December to February or mid of March. The Pit size of $1m^3$ is dug a month before planting and 1 kg SSP is added to each pit. In plains, square and triangular planting systems and in hill slopes, a contour system is followed.

High-Density Planting (HDP)

HDP aims to optimize land utilization for maximum early-season yield while facilitating manageable orchard operations. The implementation of the HDP system, in conjunction with appropriate canopy management techniques and a drip-fertigation system, is critical for enhancing productivity (Kumar, 2013). High and medium-density orcharding in apricot with suitable varieties have been standardized and are being demonstrated by CITH, Srinagar. Apricot varieties such as CITH-AP-3, CITH-AP-2, CITH-AP-1, Harcot and Rival on a standard rootstock were planted at 5 x 5 m spacing (625 trees/ha) and trained to a modified leader yielded 23.64 t/ha, 23.02 t/ha, 19.06 t/ha, 18.20 t/ha and 17.51 t/ha respectively. The above yield was recorded from a 10th-year-old tree (CITH, Annual report, 2014-15).

Training and pruning

The relationship between fruiting behavior and growth pattern in apricots is strongly intertwined. The chilling requirement of apricot vary depending on the characteristics of the bud (floral or vegetative) and the shoot (brindle, long brindle, sylleptic or spur) and may ultimately shape the growth habit and fruiting behavior (Guerriero and Viti, 1997). To effectively plan pruning methods, it is crucial to distinguish the shoot based on its unique rate of growth post-bud break. Therefore, it is essential to continuously monitor its physiological behavior and understand the pattern of the shoot growth.

Training

A minimal fraction of the fruit develops on one-year-old lateral shoots, while the majority of them on spurs have a lifespan of three years. It is imperative to train apricot trees to ensure a strong framework that can sustain a substantial crop load and facilitate the convenient use of harvesting tools and equipment. Regular pruning on an annual basis is necessary to achieve consistent and uniform yields and desirable fruit size. Open center and modified leader systems are commonly followed in apricot. To mitigate the risk of damage resulting from the weight of a bountiful harvest, many trees benefit from the utilization of ropes or wires secured around their scaffold branches. In recent times, new plantations have adopted high-density arrangements in the form of a perpendicular 'V'. Some of these feature training wires, while others stand independently (Bhat *et al.*, 2021).

Open center is followed in hilly regions. The one-year-old plant is headed back to 60-70 cm above the ground having 3 to 5 well-spaced primary branches in all directions. Undesirable shoots are pinched during summer.

In the case of a modified leader system, the main branch is headed back to 50-70 cm above the ground and 4-6 primary branches are retained.

Pruning

The apricot tree bears most of its crop on short spurs which have a life span of 3-4 years.

In the first dormant season, three to five primary branches are selected for growth. These branches should be positioned at a 45° angle, spaced 10-15 cm apart, and arranged in a spiral pattern around the tree trunk. The lowest branch should be positioned 40-45 cm above the ground. These primary branches are then headed back by half to encourage the growth of secondary branches.

During the second phase of dormant pruning, on each primary branch, 4-5 evenly spaced secondary branches are judiciously selected and the remaining ones are trimmed. By the end of the **third year,** pruning is limited to the removal of branches that overlap or encroach upon each other.

In the case of young-bearing trees, a corrective type of light pruning is done whereas severe pruning is employed on older trees to maintain the proper ratio of growth and fruiting.

Fruit thinning - Apricot tends to set heavily. Thus, to generate fruit of a commercially viable size, thinning is required. When the crop is abundant, spacing between fruits of 4-8 cm should be maintained, whilst the tree is less substantial, spacing may be reduced to 4-5 cm. Each spur should bear two to three fruits. The thinning process has to be performed within 40 days following full bloom. Fruit thinning leads to an increase in fruit size, and regular fruit production, reduces limb damage caused by excessive crop weight and preserves the tree's vitality. The most efficacious method of thinning is achieved through the foliar spray of NAA @ 25 to 50 ppm twenty days following the fruit set.

Use of Plant Growth Regulators

Auxins successfully improve the size of the fruits. NAA application @ 25 to 50 ppm is ideal when sprayed twenty days following the fruit set. NAA @ 50 ppm increases both thinning rate and fruit size. (Hajam *et al.*, 2018). IBA @ 2000 ppm increases the rooting percentage and quality of roots. Prohexadione-Ca (growth retardant) and NAA trigger a rise in fruit weight and yield (Mesa *et al.*, 2012).

As a result of the application of three synthetic auxin compounds at the inception of pit hardening, resulted in a significant increase in fruit size and overall yield (Stern *et al.*, 2007). The anatomical investigation depicted that the primary impact was the prompt instigation of cellular augmentation. The compounds 3,5,6-TPA, 2,4-DP, 2,4-D and NAA did not exhibit any adverse impact on the fruit quality at harvest and post-storage, nor did it affect the subsequent season's blooming. The highest increment in growth rate was observed with application of the synthetic auxin such as 2,4-DP-P resulting in a six-fold increase in the size of the fruit as compared to the control. Subsequently, after being stored, these fruits were subjected to severe internal discoloration. In addition, fruit splitting was reported with elevated concentrations. Thus, it is advisable to closely monitor the concentrations. The use of gibberellins is an alternate approach to the manual thinning of stone fruits (Southwick *et al.*, 1995).

Intercultural operations

Intercropping

It is recommended that during the first 3 to 4 years of the orchard's existence, while the plants are still in their infancy, intercropping with pea, bean, soybean, cowpea, as well as strawberry and tomato, is pursued. This practice not only enriches the soil but also provides economic benefits to the farmers or growers.

Orchard soil and weed management

The prevalent technique of floor management employed in apricot orchards is a combination of sod cultivation and mulching with black polyethylene and desiccated grass. Utilization of atrazine or diuron as pre-emergence at a rate of 4.0 kg/ha and the application of gramoxone (2 kg/L) or glyphosate (800 ml/ha) as a post-emergence were determined to be both efficacious and cost-effective in suppressing weed proliferation. The implementation of a 10 to 15 cm mulch layer of dry grass around the base of the trees also serves to mitigate weed growth.

Nutrient Management

Apricot trees heavily remove soil nutrients and need a combination of organic and chemical fertilizers to grow and produce fruit effectively. The optimum manurial requirements depend on factors such as the tree's age, soil type, regional climate, and cultural practices. It is recommended to apply farmyard manure in December to January, in addition to the full amount of phosphorous and potassium.

Nitrogen is administered in two split doses: (i) the first half in two to three weeks before flowering (ii) the remaining half one month following flowering. Under rain-fed conditions, application (broadcasting) of the latter portion of nitrogen should be completed by the onset of monsoon or through 2-3 foliar applications of urea after fruit set. In steep topography areas with high rainfall, it is preferable to apply nitrogen-based fertilizers through banding rather than broadcasting.

Recommended fertilizer schedule for apricot

Tree age (year)	FYM (kg)	CAN (g)	N (g)	SSP (g)	P_2O_5 (g)	MOP (g)	K_2O (g)
1	10	280	70	220	35	165	100
2	15	560	140	440	70	335	200
3	20	840	210	660	105	500	300
4	25	1120	280	880	140	670	400
5	30	1400	350	1100	175	835	500
6	35	1680	420	1320	210	1000	600
7 and above	40	2000	500	1560	250	1170	700

Source: Kasaka, 2006

Particulars	FYM (kg/tree)	N (g/tree)	P2O5(g/tree)	K2O (g/tree)
Himachal Pradesh	60	500	250	600
Punjab	25	500	120	500
Uttaranchal	10	300	300	500

Source: Shamrao, 2020

Irrigation

Apricots are resilient to arid conditions, but they still require irrigation, especially during crucial stages of fruit growth and development. The highest water demand occurs from late April to mid-June, which coincides with the fruit development stage. Irrigating the soil when 20% of moisture has been depleted from its field capacity enhances fruit size and yield. The irrigation frequency should be every 10 days in May and in June every 6-8 days. Sprinkler irrigation is better than surface irrigation in steep slopes.

The research was conducted to explore the feasibility of using deficit irrigation methods to conserve water without negatively impacting yield and quality in apricot trees. The results showed that fruit yield and quality did not increase proportionally as tree growth increased based on the amount of water applied. Thus, yields per unit of trunk cross-sectional area and yields per unit of crown volume were higher in the treatments that received less water compared to those that received more water. However, the use of deficit irrigation is not recommended for apricot production because it causes water stress during the critical period (immediately after harvest for 1 ½ months), which in the following year can result in a substantial decrease in fruit yield. (Torrecillas *et al.*, 2000).

Flowering, pollination and fruit drop

In apricot trees usually, three buds develop in the axillary region of each leaf node on a shoot or spur. The central bud is vegetative, while the two lateral buds are floral. The timing and flowering duration can vary based on the variety and prevailing weather conditions. In mid-hill regions, flowering typically occurs in March, while it takes place in late March and April in higher-hill regions. Most commercial apricot varieties are self-fertile, meaning they can produce fruit without the aid of a pollinator. However, certain cultivars, such as Charmagz and Perfection, are self-incompatible.

When grown in suitable climatic conditions, apricot cultivars tend to have a good fruit set rate, with a fruit set of 40-60% in commercially grown mid-hill cultivars. However, there is a significant fruit drop rate of 79% in these cultivars, primarily during the second week after the fruit set. Fruit falls off because most of the trees produce significantly more flowers than required. Not all the flowers are successfully pollinated.

Fruit drops can be managed in the following ways

- Thinning out the maturing apricots from the branches, leaving 5-10 cm between the remaining fruits, results in larger and fleshier fruits.
- Spraying of NAA @ 10 ppm at the commencement of pit hardening reduces pre-harvest fruit drop.

Physiological disorders and management

Apricots are very sensitive to fruit disorders, which show up as internal breakdown or disintegration and gel breakdown (translucent, gelatinous material) (flesh browning). Apricots are also sensitive to fungal species including *Monilinia fructicola, Botrytis cinerea, Penicillium expansum, and Rhizopus stolonifer* decomposing them.

1. Pit Burn

Causes

- Pit burn occurs between the period when the fruit is green and has reached sufficient coloration for harvest and the discoloration is due to heat injury and it increases with higher temperature and longer durations of exposure to sun.

Symptoms

- When apricots are exposed to higher temperatures (>38 °C) before harvest, the flesh around the stone softens and turns brown. Excessive heat causes the flesh surrounding the pit to mature faster than the rest, which cannot be observed from the exterior of the fruit.

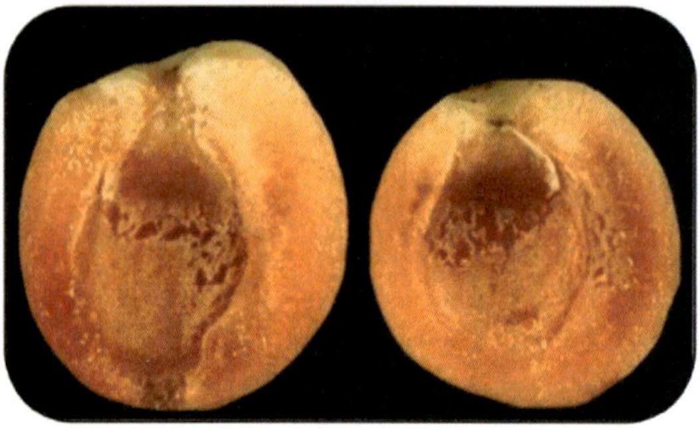

Management

- Using potassium-based fertilizer, adequate irrigation and aeration to cool the soil and reduction of nitrogen-based fertilizers.
- It can be controlled by avoiding heat injury and providing shades to the fruits.

2. Gel breakdown (GB) or chilling injury

Causes

- It occurs when the fruits are stored at a low temperature between 2.2 – 7.6 °C.
- Harvesting all or partial fruit at an advanced post-optimum ripeness is the most frequent cause of gel breakdown.

Symptoms

- It is characterized by the breakdown of tissue and in some cases, it may be accompanied by a spongy texture and the formation of a gel. In the early stage, water-soaked spots appear and then turn brown. Such type of fruits affected by chilling injury has short market life and lose flavour due to which it has a very less degree of customer acceptability and short market life.

Management

- Maintain the temperatures just above freezing. Being a progressive disorder, GB levels will increase with longer storage, hence the storage should be curtailed down to 28-38 days.

3. Internal Breakdown

Causes

- Internal breakdown normally appears when the fruit is stored at room temperature while undergoing ripening after being stored in cold temperatures.

Symptoms

- The visible symptoms of internal breakdown are browning of the flesh, a lack of juiciness due to a leathery or mealy texture, a black pit cavity, flesh translucency, an accumulation of red pigment (bleeding), and a decline in flavor.

Management

- Storing the fruit at a temperature < 0°C but above freezing is beneficial in delaying internal breakdown.
- The utilization of CA (Controlled atmospheric) in combination with a temperature nearly to 0 °C had been found beneficial.

4. Skin Discoloration

Causes

- Any kind of damage, such as picking the fruit, unloading it from bins, packing it, or even rubbing it against each other, can cause this discoloration.

Symptoms

- Brown and black spots or stripes appear and are limited to the skin.

Management

- Reduce fruit damage. Maintain the picking containers dirt-free, and harvesting equipment clean.

5. Surface Pitting and Bruising

Causes

- Bruises may occur from rough handling in the harvesting, packing process, and in transit.

Symptoms

- Small sunken areas on the surface are called pits while larger flattened areas are referred to as bruises. Bruises are defined as discoloration of the flesh that extends deeper than 1/8th of an inch (3.1 mm) or any bruise with a diameter exceeding 3/8th of an inch (9.5 mm) or an accumulation of smaller bruises that negatively affect the appearance, edibility or shipping quality of the apricot.

Management

- Pre-harvest application of chelated calcium at 0.05% calcium decreases pitting.
- Hydro-cooling with calcium may also decrease the pitting.

Major insect pests and their management

1. Flat Headed Borer: *Chrysobothris mali* Horn.

Biology

In Indian conditions, the adults generally emerge in June-August after completing their winter phase as larvae inside the tree. Further lay eggs on bark crevices and the larvae immediately after hatching start feeding sapwood after boring through the bark.

A 0.4-inch-long adult beetle has a dark bronze body color and coppery markings on the wing. Larva that is fully developed has a flat, noticeable body expansion right below the head and is light in colour and one generation every year.

Nature of damage

Damage caused by beetles is frequently observed on susceptible parts of a tree, such as weakened, young, injured or damaged limbs with sunburn, infestations from scale insects, bacterial canker or major pruning wounds. The beetles lay their eggs at these sites of injury. Once the eggs hatch, the larvae excavate large cavities just below the bark and burrow into the tree's core. They feed under the bark of limbs or trunk, causing the surface above their mines to appear rough and cracked. The gum oozes out from the attacked portions and the branches start drying and slowly the plant starts declining and the yield is affected.

Management

- To guard the tree trunk against sunburn and infestations from flat headed borers, cover or paint the trunk with white, water-based paint or whitewash above and 1 inch below the soil line at the time of planting.
- Proper training and pruning are important to protect the trunk from direct exposure to the sun as the borer often invades sunburned areas.

- Plant a sound, vigorous and healthy tree as it has a much better chance of overcoming the attack than a weak tree.
- Spraying the trunk with methyl parathion or fenitrothion (0.1%) at regular intervals protects the plants against this disease.

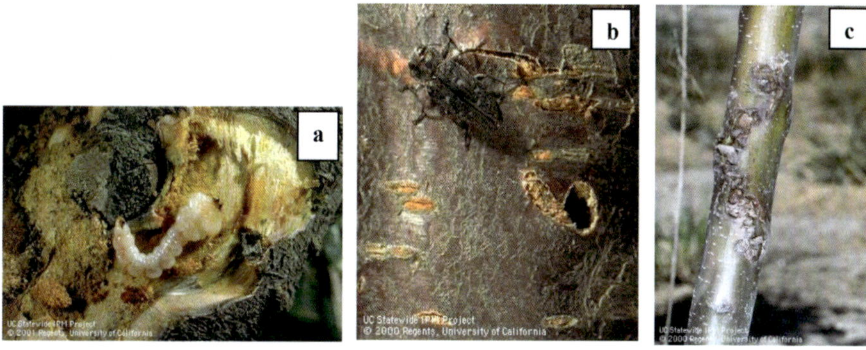

Fig 2: a. larva of Chrysobothrismali; **b.** adult Chrysobothrismali, next to its exit hole; **c.** cracked, rough bark caused by larvae of Chrysobothrismali, tunneling under bark

2. Codling moth: *Cydia pomonella*

Cydia pomonella is a serious pest affecting apricots and is prevalent in the fruit-growing regions of Ladakh. It is one of the most persistent, damaging and hard-to-manage pests of fruit crops in Ladakh.

Biology

Codling moth adults begin to emerge around the end of June. The female moth, after mating, lay eggs singly on the young foliage and occasionally on fruits, twigs and spurs. The eggs are small, circular and white or transparent and they hatch within 7 to 15 days after being laid. The newborn larvae generally start boring from the end of the blossom or calyx and feed on the exterior before boring inside the fruits after which they start feeding the flesh and developing seeds until they reach full maturity. Furthermore, the mature larvae exit by making a hole in the fruit and spinning a cocoon in the loose bark, fissures or detritus. They then pupate in September and transform into adult moths from May to June

Nature of damage

Apricot begins to ripen from July to mid of September in Ladakh condition. The larvae penetrate and tunnel to the core of the fruits and come out from an exit hole filling with reddish brown, crumbly droppings called frass. The larvae feed on the whole pulp of the infected fruits, causing them to shrivel, shrink in size, and develop a thin covering. The fruits also turn brown as the thin outer skin dries out, and eventually drop prematurely. The larvae migrate towards the healthy ripe fruits and starts feeding in them and often infest 40-90% of the fruit.

Management

- The pupae (cocoons) should be collected and destroyed from April to June.
- Remove all the dead wood and clear away loose bark and mosses through pruning

- Dispose of the infected fruits from the ground and kill all the larvae present inside the fruit.
- Restrict the crawling of larvae by banding the trunk with a sack or corrugated paper below the junction of branches and a few inches above the ground surface as the mature larvae will get trapped and prevent them from seeking a place to pupate.
- Keep the orchard clean by removing all the fallen leaves, twigs and infected fruits.
- Installation of pheromone traps on trees aids in monitoring and trapping male insects before mating, thereby further multiplication is restricted.
- Releasing the egg parasitoid, *Trichogramma* spp., at a rate of 2,000 adults per tree during June effectively aids in controlling the codling moth population.

Scientists at SKUAST-Leh have reported that Trichogramma platneri or Trichogramma embryophagum is highly effective in controlling pest infestation and eliminating it in its early stages.

Fig. 3: a. Cydiapomonellaa. adult moth, **b.** larvae and **c.** red-ringed entry hole on the exterior part of apricot Lamo, 2017

3. Hairy caterpillar: *Lymantria obfuscate* Walker

Biology

Eggs are the resting phase of this insect where they overwinter. Young and older caterpillars feed during the day and night respectively. During early summer (June to early July), this insect enters a pupal stage. The pupae resemble dark brown, hairy, shell-like structures that measure about two inches in length. It takes 10 to 14 days for adult moths to emerge from the pupae. The female moths, who are unable to fly, lay between 500 to 1,000 eggs underneath tree barks. The male moths, which are smaller in size, can be differentiated from the females by the unique V-shape marking on their wings. This moth species only has a single generation a year.

Nature of Damage

Caterpillars are gregarious and voracious feeders and feed exclusively during night hours. In cases of severe infestation, only the hard veins are spared and entire leaves are devoured. The complete defoliation of the host plant leads to the failure of fruit development.

Management

The viability of the eggs may be significantly reduced by using the mineral oil-based ESSO tree spray oil (3%), IOC tree spray oil (6%), Power oil WSE 7267 (9%) and Sandolin A, a dinitro or the cresol preparation, at a concentration of 1%, 2% or 3%.

Natural enemies: *Telenomus* sp, *Cotesiamelanoscela*, *Glyptapantelos* sp, Tachinid (*Pales* sp), *Brachymeria* sp.

Fig. 4: Lymantria obfuscate. **a**. adult male and **b.** adult female

4. Peach leaf curl aphid (*Brachycaudus helichrysi*)

Biology

It overwinters as an egg, emerges as a nymph in the spring and feeds on developing buds. Both winged and wingless types reproduce parthenogenetically. The duration of the nymphal stage is 7-9 days. Adults generate 8–22 nymphs per day and have a lifespan of 2–3 weeks. Life cycles last around 22 to 25 days. It has 12–14 generations annually, depending on the conditions.

Nature of damage

The aphid colonizes young shoots, buds and grafts causing serious damage. Adults and nymphs both ingest the sap found in leaves, shoots, and fruits. Curled leaves are the result of it and lead to wilting of terminal shoots and stunted tree growth.

Management

It can be controlled by the application of tar oil or DNC/petroleum washes during the dormant season.

Predator

Scymnus sp, *Chilomenes sexmaculatus*, *Chrysoperla zastrow* sillemi, Coccinellids, Predatory mite and parasitic wasps.

Fig. 5: The adult aptera of Brachycaudushelichrysi is variable in colour, ranging from yellow to green to pink to white, often shiny with a slight wax dusting

Major disease and their managemen

1. Bacterial gummosis

Causal Organism: *Pseudomonas syringae*

Symptoms

- Attacks all parts of the plant except the roots.
- Circular to elongated soaked gumming lesions arise on bark and fruits
- The fruits become unfit for marketing

Management

- Streptocyclin (1g/10L) before the onset of monsoon reduces the incidence of gummosis.
- Spray copper oxychloride (3%) after leaf fall.

Source: extension.usu.edu

2. Apoplexy

Causal Organism: Responsible for apoplexy or dieback can be broadly divided into two groups: (a) Physiological-rootstocks, soil and climatic factors, pruning and other cultural practices (b) Parasitic-bacteria, fungi, viruses and pest - *Pseudomonas syringae* pv. *Syringae* bacterium and *Valsaria insitiva* (syn. *Cytospora cincta*) fungus are the main causative agents, *Eutypa lata, Phytophthora* spp., *Monilia laxa* and mycoplasma (Paunovic, 1989 and Klement *et al.,* 1972)

Symptoms

- Wilting of an entire or part of the tree and sudden death
- Necrosis of phloem, cambium and xylem of branch and trunk
- Brown discoloration on the collar and roots

Management

- Spraying of Bordeaux mixture before and after pruning

3. Brown rot and blossom blight

Causal Organism: *Monilinia laxa*

Symptoms

The initiation phase of the blossom and twig blight is marked by the demise of tender flowers, spurs and foliage.

- Infection is transmissible and spreads from flowers to twigs forming canker.
- Small and circular spots on the fruits covered with ash color tufts.
- Wilting of flowers.
- In high humidity conditions, infected areas develop grayish-brown spore masses.

Source: Vikaspedia

Management

- Captan @ 0.3% - 3 weeks before harvest
- Carbendazim @ 0.1% at flowering stage at 7-10 days interval.

4. Powdery Mildew

Causal Organism: *Sphaerotheca pannosa*

Symptoms

- Leaves get curled as they expand
  A disease affecting fruit typically manifests as circular white spots that enlarge over time. The white mycelium sloughs off, revealing rusty round lesions, with some cases showing a ringspot pattern where the center tissue is unaffected.
- Leaves affected by the disease exhibit dry, brown patches coated with a white powdery substance, while infected new leaf tissue appears stunted, distorted and dwarf.
- Infected new leaf tissue is dwarfed, distorted and stunted shoot growth

Management

- 2 sprays – 2% wettable sulphur before flowering and after fruit set
- To enhance air circulation and decrease humidity within the canopy, it is advisable to selectively prune branches. Additionally, to prevent the spread of the disease, it is crucial to take certain measures, such as removing and properly disposing of all the fallen parts of the plant on susceptible hosts, ensuring adequate plant nutrition, and pruning the overlapping trees and shrubs to reduce shading.

5. Verticillium wilt

Causal Organism: *Verticillium dahliae*

Symptoms

(*Source*: NMSU-PDC)

- The disease typically becomes apparent in the early summer, characterized by a gradual loss of leaves from infected branches, starting from the base of each limb.
- The leaves may turn yellow and dull in appearance before falling off prematurely. In cases of severe infection, by the end of summer, only a small cluster of leaves may remain at the tip of the affected branch.
- Dieback and wilting of tree
- Stunted and unproductive for several years.

Management

- Sterilization of soil before planting as it is soil-borne.
- To minimize the spread of *V. dahliae*, it is important to only use planting material with high-quality and free of disease as its transmission through vegetative planting stock is quite common.
- Suppress weeds after planting.
- Important weed hosts of *Verticillium* belong to the families - Amaranthaceae (red-root amaranthus), Solanaceae (nightshade), and Chenopodiaceae (fat hen).

Harvesting and post-harvest management

Apricots usually reach maturity between early May and late June depending upon variety and location. Apricot trees begin producing fruit in their fifth year with their peak fruit production at 8 to 10 years after planting and continue to do so for up to 35 years. Apricot is a climacteric fruit with a moderate rate of respiration. The fruit is highly sensitive to shriveling. Therefore, for fresh consumption, it is harvested while the fruits are still firm. However, if the fruit is intended for drying, freezing or canning, it is harvested when fully ripe. Change of surface color from green to yellow, days from full bloom to harvested and TSS are considered as the best indices of maturity. Due care should be taken during and after harvesting as apricots are highly perishable. To prevent fruits from being directly exposed to sunlight during the process of grading and packing, harvesting should be carried out in the morning hours. After harvesting the fruits are graded based on size (Special – 42mm and above, Grade-I – 36-42 mm and Grade-II - <36 mm) and packed in corrugated fiberboard. Though it can also be packed in wooden boxes CFB is mostly preferred as it fetches a better price because of minimal bruising damage as compared to wooden boxes. Apricot fruits are usually marketed fresh, however, being perishable they are

processed into various value-added products such as squash, nectar and jam. However, the fruit's shelf life can be extended up to 1-2 weeks, if stored at 0°C temperature and 90-95% relative humidity. Sun-drying apricots are the oldest known preservation technique used by Ladakhis, and it remains widely popular even today (Hussain *et al.*, 2014).

Yield

The crop yield gets highly influenced by production practices and the variety grown. A yield of 50 to 85 kg per tree or 15-25 tonnes per hectare can be obtained.

Modern technologies

Microsatellites also referred to as simple sequence repeats (SSRs) exhibit high levels of polymorphism evenly distributed throughout the genome. Their ability to inherit according to Mendelian principles and to be transported across species make microsatellites well-suited for studies of phylogenetics and population diversity. They have been utilized in various *Prunus* species, including apricot, for exploring the crop's evolutionary history and for characterizing traditional and local cultivars from countries such as Spain, Turkey, Tunisia, and Iran.

The ancestry of most recent apricot cultivars remains largely unknown. To address this knowledge gap, 202 apricot accessions, consisting of both landraces and releases from breeding programs across various countries, were analyzed using SSRs. This characterization aimed (i) to evaluate the current genetic diversity (ii) to determine the relationships between cultivars and (iii) to calculate the extent of population structure among the predominant cultivars grown globally.

The presence of 1% PPE (Pomegranate Peel Extract) in Nanochitosan significantly curtailed the growth of psychrophilic bacteria, yeasts and molds during storage, thereby enhancing the shelf life and quality of apricots. The use of nano chitosan coatings enriched with PPE significantly retained DPPH radical scavenging activity (Gull *et al.*, 2021)

RNA-Seq and real-time qPCR were used to investigate the changes in gene expression. The findings suggested that genes involved in the biosynthetic pathways with differential expression, such as those involved in phenylpropanoids, flavonoids, starch, sucrose and carotenoids, showed differences in expression levels and could be used as markers for determining fruit quality traits such as colour and soluble solid content. Candidate genes such as carotenoid cleavage dioxygenase and the gene sucrose synthase act as a potential marker for the ripening process and quality traits of fruits specifically white skin ground color, flesh color and high soluble sugar content. These genes are involved in the metabolism of carotenoids and starch/sucrose, respectively. The candidate genes found in the study could be utilized in marker-assisted selection for apricot breeding to improve the early identification of fruit genotypes with both high quality and nutraceutical value.

Starch and sucrose metabolism, plant-pathogen interactions, and plant hormone signal transduction pathways are concentrated in the DEGs, according to the Kyoto Encyclopedia of Genes and Genomes (KEGG) research. In addition, 11 drought-related DEGs were employed in quantitative real-time PCR (qRT-PCR) to support the RNA-seq findings. Second, the physiological study of apricot leaves under continuous drought stress revealed that the interior microstructure of the leaves had altered to tolerate the stress. A larger degree of membrane damage was also visible in plants under long-term drought stress, which inhibited photosynthesis in the affected leaves. This finding advances our knowledge of the molecular and physiological mechanisms behind the drought response in the apricot fruit crop and improves our understanding of how this crop adapts to drought.

Future prospects

Conserve the rich genetic resources of the Ladakh region. In this area, apricots are cultivated naturally without the use of pesticides or chemical fertilizers. Ladakh apricots ought to be certified as organic and promoted as such. The focus is on enhancing the cultivation of premium quality apricots, specifically the Halman and Raktsey Karpo varieties, and the utilization of solar drying methods for the fruits.

It is crucial to grow well-known cultivars on a big scale to achieve homogeneity in goods, both fresh and dried. The classic grafting method is hard and time-consuming for mass replication of a chosen mother tree. Results will be seen quickly if tissue culture technology is used to multiply a few chosen mother trees on a wide scale which should be standardized before taking up in a commercial scale.

At the moment, branches are shaken or beaten to remove the apricot fruit, which results in a high rate of waste. Even though SKUAST-K Leh has created a low-cost apricot harvest net, mechanical harvesting interventions are required to increase harvesting efficiency and preserve the quality of gathered fruits.

The absence of coordinated postharvest management for apricots necessitates its development and also the development of a marketing infrastructure is required for fresh and processed apricots, *i.e.,* there is a need to develop a systematic fruit collection, processing, drying and marketing in the region.

Integration of nutraceutical and pharmaceutical industries is necessary for the advancement of nature-based supplements.

References

Alajil O, Sagar VR, Kaur C, Rudra SG, Sharma RR, Kaushik R, Verma MK, Tomar M, Kumar M and Mekhemar M. (2021). Nutritional and phytochemical traits of apricots (*Prunus armeniaca* L.) for application in the nutraceutical and health industry. *Foods*. 10(6):1344.

Angmo P, Angmo S, Upadhyay SS, Targais K, Kumar B and Stobdan T. (2017). Apricots (*Prunus armeniaca* L.) of trans-Himalayan Ladakh: Potential candidate for fruit quality breeding programs. *Scientia Horticulturae*. 218:187-192.

Anonymous. (2019). *Lymantria obfuscata* (Indian gypsy moth). CABI Compendium. CABI International. doi: 10.1079/cabicompendium.31813.

Batu A. Kirmacı B and Akbulut E. (2007). Apricot molasses production technique. *Electronic Journal of Food Technologies*. 2:53-57

Bhat R. Bhat KM, Hussain S. Mir MM. Iqbal U and Bashir M. (2021). Canopy Management in Stone Fruits. *Production Technology of Stone Fruits*.103-129.

De Candolle A. (1886). Origin of cultivated plants. D. Appleton and Company, 1, 3 and 5 bond street, New York, USA. 468.

Dimitrova M. (2001). Evaluation of some plum rootstocks as rootstock for apricot in the orchard. VII International Symposium on Plum and Prune Genetics, Breeding and Pomology 577: 311-314.

Domingo R. Pérez-Pastor A and Ruiz-Sánchez MC. (2002). Physiological responses of apricot plants grafted on two different rootstocks to flooding conditions. *Journal of Plant Physiology*. 159(7): 725-732.

Dwivedi SK and Singh B. 2001. Studies on chip budding in apricot (*Prunus armeniaca* L.) in cold arid conditions of Ladakh. *Indian Journal of Horticulture.* 58:215-217.

Dwivedi SK, Attrey DP and Paljor E. 2000a. Standardization of budding in apricot (*Prunus armeniaca* L.) in cold arid conditions of Ladakh. *Progressive Horticulture*, 32:117-120.

Dwivedi SK, Singh B and Paljor E. (2000b). Studies on vegetative propagation of apricot (*Prunus armeniaca* L.) through grafting in Ladakh. *Indian Journal of Horticulture.* 57:39-41.

Ercisli S. (2009). Apricot culture in Turkey. *Scientific Research and Essays*. 4 (8):715-719.

FAO (http://www.fao.org)

FAOSTAT (http://www.fao.org/home/en)

García-Gómez BE, Ruiz D, Salazar JA, Rubio M, Martínez-García PJ and Martínez-Gómez P. (2020). Analysis of metabolites and gene expression changes relative to apricot (*Prunus armeniaca* L.) fruit quality during development and ripening. *Frontiers in plant science.* 11:1269.

Geuna F. Toschi M and Bassi D. (2003). The use of AFLP markers for cultivars identification in apricot. *Plant Breeding*. 122: 526–531.

Groppi A, Liu S, Cornille A, Decroocq S, Bui QT, Tricon D. ..... and Decroocq V. (2021). Population genomics of apricots unravels domestication history and adaptive events. *Nature Communications*. 12(1):1-16.

Guerriero R and Viti R. (1997). Problemi relativi alla biologia fiorale e di fruttificazione dell'albicocco. *Italus Hortus*. 4(2):29-38.

Gull A, Bhat N, Wani SM, Masoodi FA, Amin T and Ganai SA. (2021). Shelf-life extension of apricot fruit by application of nanochitosan emulsion coatings containing pomegranate peel extract. *Food chemistry*. 349:129-149.

Hajam MA. Hassan GI. Parray EA. Wani MA. Shabir A. Khan IF. Wani AW. Bhat TA and Masoodi L. (2018). Transforming fruit production by plant growth regulators. *Journal of Pharmacognosy and Phytochemistry.* 7(1):1613-1617.

Hormaza JI, Yamane H and Rodrigo J. (2007). Apricot. In: Genome Mapping and Molecular Breeding in Plants, Vol 4: Fruits and Nuts. C Kole (ed.). *Springer*-Verlag Berlin Heidelberg. 171-187.

http://nmsuplantclinic.blogspot.com/2015/06/apricot-fruit-with-powdery-mildew-photo.html

https://extension.usu.edu/pests/research/backyard-apricot-pests

https://influentialpoints.com/Gallery/Brachycaudus_helichrysi_Leaf-curling_plum_aphid.htm

https://ipm.ucanr.edu/PMG/V/D-AM-VDAH-FO.001.html

https://vikaspedia.in/agriculture/crop-production/integrated-pest-managment/ipm-for-fruit-crops/ipm-strategies-for-apricot/apricot-insect-pests

https://www.ipm.ucanr.edu/agriculture/apricot/pacific-flatheaded-borer

https://www.itis.gov/servlet/SingleRpt/ SingleRpt?search_topic= TSN&search_value =24769#null

Hummer KE and Janick J. (2009) Rosaceae: taxonomy, economic importance, genomics. In: Folta KM, Gardiner SE (eds) Genetics and genomics of rosaceae. *Plant genetics and genomics: Crops and Models*. *Springer*. 6.

Hussain A. Dawa S and Akbar PI. (2014). Solar apricot dryers and drying processes in the high altitude cold-arid Ladakh region of India. *International Journal of Ambient Energy*. 35: 180-185.

Hussain SZ. Naseer B. Qadri T. Fatima T. and Bhat TA. (2021). Apricots (*Prunus armeniaca*) - Morphology, Taxonomy, Composition and Health Benefits. In *Fruits Grown in Highland Regions of the Himalayas: Nutritional and Health Benefits*. Cham: *Springer International Publishing*. 91-102

Ibarra AG. Rubio M. Dicenta F and Gomez PM. (2010). Evaluation of Resistance to Apple Chlorotic Leaf Spot Virus (ACLSV) in Controlled Greenhouse Conditions in Apricot Breeding Programme of CEBASCSIC in Murcia (Spain). *Acta Horticulturae*. 862: 487-490.

Karatas F and Kamisli F. (2007). Variations of Vitamins (A, C and E) and MDA in apricots dried in IR and microwave. *Journal of Food Engineering*. 78: 662-668.

Kasaka N. (2006). XIIIth IS on Apricot Breeding & Culture, *Acta Hort*, 717, *ISHS*.

Khan FA. Afzal A. Erum S. Irum S. Khan H. Hussain Z. Sher H. Bostan N and Rabi F. (2021). Optimization protocol for *in vitro* establishment of selected apricot germplasm. *Fresenius Environmental Bulletin*. 30(9):10899-10907.

Kiralan M. Özkan G. Kucukoner E. Ozcelik MM. (2019). Apricot (*Prunus armeniaca* L.) Oil. In Fruit Oils: Chemistry and Functionality; *Springer International Publishing*: New York, USA. 505–519.

Klement Z. Rozsnyay ZS and Martos A. (1972). A téli fagy és a lombhullás utáni állapot szerepe a baktériumos kajszi gutaütésben. *Növényvédelem*. 8: 209-214.

Kumar N. (2013). Drip-fertigation to increase the productivity of mango cv. Alphonso raised under Ultra-High-Density Planting (UHDP) system. Paper presented at the XXII International Plant Nutrition Colloquium meeting (IPNC) held at Istanbul, Turkey from 19th to 22nd August 2013.

Kumar N. (2019). High Density Planting in Mango-Prospects and Problems. *Advanced Agricultural Research and Technology Journal*. 3:47-53.

Lamo K. (2017). Codling Moth a Notorious Pest of Apricot in Sham conditions of Ladakh Region. *Biotecharticles*.

Layne REC, Bailey CH and Hough LF. (1996). Apricots. In: Fruit Breeding, Vol 1: Tree and Tropical Fruits. J Janick, JN Moore (ed.), *John Wiley & Sons*, New York, USA, pp 79–111.

Mesa K. Reginato G. Contador L and Infante R. (2012). Prohexadione Calcium and Naphthalene Acetic Acid Sprays Improve Fruit Size and Maintain Fruit Quality of Castlebrite' Apricot. *European Journal of Horticultural Science*. 77(3):115- 121.

Miller NF. (1999). Agricultural development in western Central Asia in the Chalcolithic and Bronze Ages. *Vegetation History and Archaeobotany* 8:13-19.

Moustafa K and Cross J. (2019). Production, Pomological and Nutraceutical Properties of Apricot. *Journal of Food Science and Technology*. 56:12–23.

Naryal A, Angmo S, Angmo P, Kant A, Chaurasia OP and Stobdan T. (2019a). Sensory attributes and consumer appreciation of fresh apricots with white seed coats. *Horticulture, Environment and Biotechnology*. 60: 603-610.

Naryal A, Bhardwaj P, Kant A, Chaurasia OP and Stobdan T. (2019b). Altitude and seed phenotypic effect on amygdalin content in apricot (*Prunus armeniaca* L.) kernel. *Pharmacognosy Journal*. 11:332-337.

Naryal A, Dolkar D, Bhardwaj AK, Kant A, Chaurasia OP and Stobdan T. (2020). Effect of altitude on the phenology and fruit quality attributes of apricot (*Prunus armeniaca* L.) fruits. *Defence Life Science Journal*. 5:18-24.

Oomah B and Mazza G. (2000). Functional foods. Wiley Encyclopedia of Food Science and Technology. 2:1176-1182.

Paunovic SA. (1989). Apricot problems yesterday, today and tomorow. In IX International Symposium on Apricot Culture 293:21-24.

Potter D. Eriksson T. Evans RC. Oh S. Smedmark J. Morgan DR. Kerr M. Robertson KR. Arsenault M and Dickinson TA. (2007). Phylogeny and classification of Rosaceae. *Plant Systematics and Evolution. Springer.* 266(1/2):5-43

Roussos PA, Denaxa NK, Tsafouros A, Efstathios N, Intidhar B. (2016). Apricot (*Prunus armeniaca* L.). In Nutritional composition of fruit cultivars. *Academic press.* 19-48.

Shamrao BS. (2020). Production technology of peach, plum and apricot in India. In Prunus. *Intechopen.* 96-101.

Sharma JB. Chauhan N. Rana K and Bakshi M. (2020). Evaluation of rootstocks for temperate fruit crops-A Review. *International Journal of Current Microbiology and Applied Sciences*. 9(11):3533-3539. doi: https://doi.org/10.20546/ijcmas.2020.911.422.

Shi S. Li J. Sun J. Yu J and Zhou S. (2013). Phylogeny and classification of *Prunus sensu lato* (Rosaceae). *Journal of Integritive Plant Biology*. 55(11): 1069–1079.

Singh DB. Sharma OC. Mir JI. Sharma A. Kumawat KL and Rai KM. (2016). ICAR-Central Institute of Temperate Horticulture (CITH) - Annual Report 2015-16.

Singh, A. (2016). Fruit physiology and production. 5th Ed. Kalyani Publishers.

Sinskaya EN. (1969) Historical geography of cultivated floras (at the dawn of agriculture). Kolos, Leningrad, USSR (in Russian).

Šiško M, Ternjak T and Grobelnik-Mlakar S. (2022). The Effects of Different Cytokinin Types and their Concentration on *in vitro* Growth of Apricot (*Prunus armeniaca* L.) Shoots. *Agricultura*. 19(1): 1-6.

Sofi AA, Verma MK, Ahmad N, Verma RK, Dinesh K, ... and Dar GA. (2009). New apricot varieties released. *Indian Journal of Horticulture*. 66:1-3.

Southwick SM. Weis KG. Yeager JT and Zhou H. (1995). Controlling cropping in 'Loadel' cling peach using gibberellin: Effects on flower density, fruit distribution, fruit firmness, fruit thinning, and yield. *Journal of the American Society for Horticultural Science.* 120:1087–1095.

Stern RA. Flaishman M and Ben-Arie R. (2007). The effect of synthetic auxins on fruit development, quality and fnal fruit size in 'Canino' apricot (*Prunus armeniaca* L.). *Journal of Horticultural Science and Biotechnology*. 82:335–340.

Stobdan T, Namgial D, Chaurasia OP, Wani M, Phunchok T and Zaffar M. (2021). Apricot (*Prunus armeniaca* L.) In Trans-Himalayan Ladakh, India: Current Status and Future Directions. *Journal of Food and Agriculture Research*. 1:86-105.

Torrecillas A, Domingo R, Galego R and Ruiz-Sanchez MC. (2000). Apricot tree response withholding irrigation at different phenological periods. *Scientia Horticulturae.* 85(3): 201-215.

Tsering S, Deldan N, Chaurasia OP, Mushtaq W, Tsewang P and Mohmad Z. (2021). Apricot (*Prunus armeniaca* L.) in Trans Himalayan Ladakh, India: Current Status and Future Directions. *Journal of Food and Agriculture Research.* 1(1):86-105.

Ugur R. (2022). *Prunus microcarpa*: a potential rootstock for apricots. *Turkish Journal of Agriculture and Forestry.* 46(1):49-58.

6

Advances Production Technology of Cherry

Abdullah Zaid[1], Shivam Chaurasia[2] and Paramanand Prajapati[2]

[1]M.Sc. Research Scholar, Department of Fruit Science, College of Horticulture, Banda University of Agriculture and Technology Banda Uttar Pradesh

[2]Ph.D. Research Scholar, Department of Fruit Science, College of Horticulture, Banda University of Agriculture and Technology Banda Uttar Pradesh

Introduction

Cherry is a fleshy fruit of stone, it occupies a prominent position among the temperate fruit grown commercially all over the world. Cherry belongs to the genus *prunus* and the family *Rosaceae*. Other stone fruits such as plum, peach, almond, apricot also come under the genus prunus. The cultivated cherries are divided into two main group sweet cherries (*Prunus avium*) are mainly cultivated for table purpose and are consumed as fresh fruits (Fig. 1) , sour cherry (*Prunus cerasus*), is mainly cultivated for processing purpose which is a cross of *Prunus avium* × *Prunus fruticosa*. Somatic chromosome number of sweet cherry is 2n=16 and somatic chromosome number of sour cheery is 2n=32. Cherry requires highest chilling hours among the temperate fruits with 2000 to 2700 hours. In India cherry is commercially grown in Kashmir valley and other adjoining regions with an area of around 1,110 ha. With and annual production of around 605 MT (Bal, 1997). In these areas the span of winter is wide and very much pronounced and according to Food and Agriculture Organization Corporate Statistical Database (Anon. 2022) Turkey is largest producer of cherry in the world, around 42% of world production originated in Turkey, and around 14% production in the United States. Among the temperate fruit cherry comes first in the market. The primary center of origin is supposed to be the Caspian and Black sea region from

Fig. 1.1: Cherries

where it has been spread throughout the world (Oz., 2013). The earliest records indicate that it was first domesticated in Greece around 300 BC. It spread to Italy from Greece where it was established as a fruit crop by 37 BC. In India, it was introduced by Britishers in Kashmir, Kullu and Shimla hills during pre-independence era. At present European countries are growing cherries on large scales for table purpose and also for processing.

Importance, composition and uses

There are some cherry that are grown for their esthetic and decoration purposes and not for consumption and some cherry trees are prized for their wood. Cherries hold a high nutritional and medicinal significance. Cherry fruits consist of high content of anthocyanin which is responsible for red pigments (Jin, 2016). Principle coloring compound in cherry is keracyanin chloride and flavour is due to Methyl anthranilate and methyl salicylate. Cherry decreases pain and inflammation and act as antioxidant and anti-inflammatory and also help in heart related disease and diabetes (Cook, 2019). Cherry contains more calories than apple and it is also rich in protein, sugars, carotene and folic acid. The sweetness of the cherry is due to its fructose and glucose content, followed by arabinose and sorbitol (Pacifico *et al.*, 2014) As for its acidity, it is thought to be explained by the presence of organic acids, mainly malic acid, followed by ascorbic, tartaric and shikmic acid (Legua, 2017). The six compounds responsible for cherry aroma were reported to be benzaldehyde (22.1%), benzyl alcohol (20.8%), eugenol (9.5%), nonanal (6.9%), trans-2-hexenal (6.8%) and isoamyl butyrate (5.3%) (Legua, 2017). Cherries contain carotenoids, most notably β-carotene and lutein, zeaxanthin is also present in a minute quantity (Ferretti, 2010). sweet cherry contains water around 80–83% (Serradilla *et al.*, 2016) and sour cherry fruits, contains water around 81–88% (Filimon *et al.*, 2011). Fruits are good source of following substances such as fructose 5 to 7.6g in sweet cherry and from 3.5 to 4.9 g in sour cherry (Papp *et al.*, 2010; Ballistreri *et al.*, 2013). 6 to 10 g glucose in both sweet and sour cherries (Papp *et al.*, 2010; Ballistreri *et al.*, 2013), 260mg potassium in sweet cherry and 200mg in sour cherry (McCune *et al.*, 2011), 30mg calcium, 12 mg magnesium and 18 mg vitamin C per 100 g of edible fruit (Mitic *et al.*, 2012; USDA ARS, 2016). Sweet cherry is also used in canned fruits cocktails, ice cream, juice making, bakery products and fruits salads whereas kirschwasser wine is made from sour cherry is very popular in European countries.

Botany

Sweet cherry

It is a tall tree with upright branches and huge, thin leaves that are serrated and hairy underneath. The long petioles have two or more enlarged glands. Drupe fruit with edible mesocarp and epicarp, as well as a double sigmoid growth curve is found. The fruit has a deep cavity, a cordate form, and a rounded or pointed apex. Only lateral, basic flower buds which are borne on 2-year-old shoots or at the base of 1-year-old shoots contain floral buds. Flowers have 5 petals, several stamens, a single style, an ovary with a single carpel that contains two ovules, and are white in colour. Cherry's closest botanical relatives are plums, not peaches or apricots. The colour of cherry fruit varies, depending upon the variety and can be yellow, red, or purplish black. The anthocyanin pigment gives cherries their red colour, and methyl anthranilate and methyl salicylate give them their flavour. The flesh has a soft or firm texture, tastes delicious, and is either yellow, crimson, or dark purple in colour. Sweet cherries have been divided into two pomological categories based on their flesh and shape: the Heart Group and the Bigarreau Group. Heart-shaped fruits are having soft and tender flesh and are in heart shape. Fruits can range in colour from dark with reddish juice to light with clear juice. The cherries in the Bigarreau Group are typically rounded.

Tart (sour) cherry

Prunus cerasus, is called as "red cherry" or "sour cherry" "And it's thought to have originated from a *Prunus avium* pollen grain that had been crossed with a *Prunus fruiticosa* pollen grain. The tree is often tiny (15–20 feet tall), moderately spreading, and lacks a sturdy trunk. The tree looks bush-like, and have comparatively weak limbs which frequently droop. Although they are naturally cold hardy, trees live less time than sweet cherries. About 40% of the flowers emerge from axillary buds on one-year-old wood, while slightly more than half of the blooms start as buds on spurs. The fruit is smaller than a sweet cherry and has a stronger acid and lower sugar content. Tart cherries can be processed, preserved, made into jam, and used as pie filling.

Taxonomy

The main criteria of assessment of the genetic diversity of fruit varieties is typically based on phenotypic and morphological analysis of characters. In most of cases it is very difficult or even impossible to distinguish between many cultivars because they are very similar morphologically (Anna Lisek *et al.*, 2006). Cherry belongs to the genus *Prunus*, sub genus *Cerasus* and family *Rosaceae*. According to (Kolesnikova, 1966) the sub genus *Cerasus* comprises of around 150 species but only few of them produces edible fruit suitable for consumption. Whereas among them sour cherry was referred to as *Prunus cerasus* which is a hybrid of *P. avium* × *P. fruticosa* and sweet cherry was referred to as *Prunus avium* (Hummer & Janick., 2009) these two species are of commercial importance throughout the world. Duke cherry (*P. gondouinii)* also produces edible fruit and it is a interspecific hybrid of *P. avium* × *P. cerasus*. Tree and fruit characteristics of duke cherries are intermediate of sweet and sour cherries. Various other popular species are bird cherry (*P. pennsylvania),* dwarf cherry (*P. pumila),* European dwarf cherry (*P. fruiticosa),* wild black cherry (*P. serotina),* wild Himalayan cherry *(P. cerasoides). P. avium* is the oldest species with a basic chromosome number of x=8, but it is usually diploid (2n=16) but triploid and tetraploid (2n=24 and 32) forms are also found. The wild forms of sweet cherry occur throughout the temperate region of mainland Europe with the greatest concentration being between south of the Caspian and Black sea.

Soil and climatic requirement

Soil

A well-drained sandy loam soil is considered best for cultivation with a pH 6.5-7.0, which can hold, moisture during summer are most suitable. The cherry plant is very prone to water logging conditions so good drainage facility is very important and therefore heavy soil should be avoided. While sour cherries performs best on relatively heavier soil, morphologically sour cherries withstand much wider range of soils as compared to sweet cherry.

Climate

Cherry cultivation needs a chilly climate. It can be effectively grown in regions that are 1,600 to 2,700 meters above mean sea level, although it needs 2000 to 2700 hours of chilling time throughout the winter. When opposed to sour cherries, the sweet cherry is more susceptible to both heat and cold. Cherry blossoms are extremely susceptible to cold, so locations free from frost such as valleys and hill slopes with good drainage should be utilized. It is not advisable to choose southern or south-western faces for cultivation at lower altitudes since they are warmer. In general, cherry cultivation is best suited to North-East or North-West aspects. 100 to 120 cm annual rainfall that is evenly distributed throughout the year is ideal for cultivation, although profuse blossoming is caused by high rainfall during the flowering season.

Varieties

On the basis of the texture of flesh, sweet cherries have been divided into two pomological group heart group and bigarreau group.

1. **Heart Group:** Fruit has a heart-shaped form and sensitive, velvety flesh. Fruit can range in colour from dark red juice to light with clear juice. Important varieties of this group include Red Heart and Black Heart.
2. **Bigarreau group:** Fruits has a rounded form, and their juice ranges in colour from dark to light crimson. The new, promising types Sam, Summit, Sue, Sunburst, Lapins, Compact Stella, and hybrid (13-17-40)

Important cultivars with their features

Bing: it is one of the most important cultivar of cherry throughout the world, the fruits produced is of the best quality but it is susceptible to cracking during rainy season

CITH Cherry-01: This is new released variety and has developed through clonal selection from the variety 'Double (Bigarreau Napoleon)'. Regular bearing in nature, precocious variety and its Yield is higher as compare to double. It is suitable for cultivation in Kashmir valley and also recommended for other Himalayan Regions. It has excellent consumer acceptability due to its attractive bright and shiny red colour, taste and overall appearance.

CITH Cherry-02: It was created by clonal selection from the variety "Mishri (Bigarreau Noir Grossa)" and is a newly released variety. It bears regularly in young age, and produces more (3.85 kg/tree) than "Mishri." This early to mid-maturing cultivar has a more appealing red skin coloration and superior quality.

Governor Wood: This variety yields large sized fruits with heart shaped, it ripens after early rivers and they need to be plucked green, regular bearing habit with high yielding capacity, it is susceptible to brown rot.

Early purple black heart: This is one of the earliest cultivar ripens during the third week of April, fruits is sweet and juicy with poor shelf life

Compact lambert: It is a mutant variety developed through radiation of lambert at Summerland in the year 1964. Late maturing but high yielding in nature, fruit is black in colour with good size

Compact Stella: it is also a mutant variety developed through radiation of Stella at Summerland during the year 1973, early maturing and heavy yielding crop with semi dwarf growing habit, heart to oval shaped fruit with black in colour.

Bedford prolific: its skin is black in colour, early maturing ripens by the middle of May month, flesh has good taste and attractive aroma.

Early rivers: this is the finest early cherry cultivar and it ripens in the beginning of the May month, fruits are large, black, and oval with a small stone and flesh is pink in colour which is highly flavoured and it is also resistant to cracking.

Cavalier: this is a new black colour cultivar that matures early, probably around June 3. Fruit is medium to large, dark red, and has dry, firm flesh with good flavor, and is resistant to cracking.

Napoleon: This cultivar is also known as Napolean Wax or Royal Anne. The fruit has brilliant red cheeks and with a pale yellow tint. The fruit has a good fresh eating quality, medium in size, firm, sweet, and juicy and also it is fairly cracking resistance.

Stella: The Canadian Department of Agriculture Research Station at Summerland, British Columbia, created this cultivar in 1968. It was created from a cross between Lambert and John Innes Seeding. This cherry is massive, black, and heart-shaped. It has pleasant aroma, medium-firm flesh, although it is susceptible to cracking.

Lapins: This cultivar was developed from a 'Van' and 'Stella' hybrid and released in 1983 by the Agriculture Canada Research Station in Summerland, British Columbia. The fruit ripens about June 15 and is quite huge, resistant to cracking, and flavorful.

Sour Cherries

Montmorency: It was believed to be originated in Montmorency valley in France, during the 17 century it is one of the oldest known cultivar of cherry, trees are hardy and it is the only sour cherry cultivar which is used for commercial cultivation, this cultivar occupies about 90% of area among the tart cherry production. Fruit are large with bright red colour and clear juice

North star: It is originated as cross of English Morello × Serbian pie, tress is small in branches with thin branches, medium sized fruits with round in shape and dark red in colour, at some place it is also used as dwarfing interstock for sweet cherries, it is suitable for growing in home garden.

Meteor: It is originated as a cross of Montmorency × vladimir in the year 1952. Plant is vigorous with dense canopy, fruit is medium in size with oval in shape with bright red colour skin but juice is colourless.

Danube: It ripens a few days earlier than "Montmorency" and is a new tart cherry for fresh consumption. The fruit is larger than most tart cherry, dark red, and sweeter. In Europe, this cultivar is commonly cultivated.

Balaton: originated as a regional variety in Hungary. The fruit is bigger, firmer, sweeter, and redder than "Montmorency," and it has flesh that is luscious. These cultivar produce somewhat larger than average pits and is causes a challenge for processing industry.

Others : Early richmond, Hebros, English morellos, Ferracida and Black Tartarian

Propagation

Like many of the other fruits, sweet cherry doesn't produce uniform seedlings. Thus, vegetative propagation method is adopted to get desired fruit and tree characteristics. Clones can be maintained by budding and grafting, by stopping natural mutations or sports. Cherry is usually propagated by whip or tongue grafting on seedling of wild cherry such as paja and Tongue grafting is recommended during month of February-march and common method of clonal rootstock multiplication is mound layering, the grafted seedling becomes mature for transplanting after two years.

Rootstocks

Cherries typically do not reproduce from seed. However, growing it from a cutting is also challenging. Rootstocks are budded or grafted with the appropriate variety on a commercial scale. Sweet cherries are propagated through grafting on seedlings as well as on clonal rootstocks. Sweet cherries were traditionally propagated mostly on seedlings of Mahaleb (*Prunus mahaleb*) or Mazzard (*Prunus avium*).

Seedling rootstock

Paja *(Prunus cerasoides*), bird cherry (*Prunus paddum*), mahaleb, and mazzard are popular and mostly used seedling rootstocks.

Propagation of seedling rootstocks: Paja seeds don't need to be chilled in order to emerge from dormancy. Although the majority of the older plantations relied entirely on paja rootstock, now it has been discouraged as a rootstock due to delayed incompatibility. Prior to planting, mahaleb and mazzard seeds need to be stratified. To break seed dormancy, seeds are stratified by being placed between layers of sand in a cold location at 2 to 4^{0}C for 80 to 120 days for mahaleb and 120 to 150 days for mazzard. Seeds are first soaked in 500 ppm GA_3 for around 24 hours. The medium is kept wet throughout stratification. These are transplanted 6 cm deep and 10 to 15 cm apart in rows separated at 20 to 25 cm when the embryonic root emerges from the seed coat.

Clonal rootstocks

Commercially, clonal rootstocks are advised for growing its plants because trees on paja exhibit delayed incompatibility. Colt, Giesela, Charger, SL64, and Mazzard F 12/1 are clonal rootstocks. Colt is a semi-dwarf tree that has strong anchorage, is compatible with practically all sweet cherry varieties, is tolerant to gummosis and crown rot, and is only partially resistant to the stem-pitting virus and bacterial canker. However, it is sensitive to the oak-root fungus. The rootstock Mazzard F 12/1 is moderately vigorous and challenging to root rot.

Propagation of clonal rootstock: Mound layering or trench layering is the common method of clonal rootstock multiplication. The stool beds are established during December by planting healthy mother plants at 30-45 cm apart in row spaced at 60-70 cm. Before new growth starts, the mother plants are cut back to 2.5 cm above the ground level. New shoots develop on the stub in the spring. When the shoots are 25-30 cm long, their bases are covered with a mound of soil or saw-dust, building the mound to a height of 20-25 cm as the shoots grow during spring. The suckers are ringed at the base and then covered with soil to encourage rooting. Mazzard F12/1 faces difficulty in rooting IBA (7,500 ppm) is applied to the ringed portion of the shoots during summer. The shoots are separated in winter and then lined out in the nursery beds. If the suckers are well-rooted and more than 0.8 cm in diameter, they are grafted in the spring, otherwise they are kept in the bed for a year to produce strong plants for grafting. Colt rootstock is easy to root and can also be multiplied through cuttings. Hardwood cuttings of 30-45 cm length and of pencil thickness are taken in February. Cuttings are treated with IBA (2,500ppm) for 10 seconds and planted in nursery beds for rooting. Rooted cuttings are lined out in December and grafted with scion variety in March

Important rootstock and their features

Mahaleb: *Prunus mahaleb* L. has a wild seedling known as mahaleb. It is 20% smaller than a mazzard, so it is somewhat dwarfing. Trees are more productive at a young age, but they frequently live shorter lives because they are less tolerant of moist soils and less tolerant of cold weather. It is resistant to certain nematodes, crown gall, and bacterial canker. It shouldn't be grown on soils with inadequate drainage because it is highly vulnerable to collar rot. 'Van' is one variety that does not work well with mahaleb.

Mazzard: Larger trees are produced by it, and it also induces fruiting at a later age and is more tolerant of damp soils. It is also compatible with all sweet cherry cultivars. It is winter hardy and very vigorous rootstock.

Paja: In India, it is the most widely used rootstock and requires the least amount of chilling.

Colt: (*P. avium* x *P. pseudocerasus*) Colt was introduced as a semi-dwarfing rootstock by the research facility at East Malling, England, in the 1970s. However, it produces a robust tree with a similarly low precocity to "Mazzard" in the irrigated orchards of the Pacific Northwest. 'Colt' is also susceptible to dry soils and chilly winter temperatures. Due to its resistance to cherry stem pitting, a crippling virus disease that is common in California, "Colt" has been widely planted there. Additionally, it has demonstrated resilience to gopher damage, Phytophthora root rot, and bacterial canker, but is vulnerable to crown gall. 'Colt' does well in PNW replant conditions where cherries follow cherries on unfumigated lands. (Long, 1995)

Mahaleb x Mazzard (MxM) clones: These were chosen from a population of open pollinated mahaleb seed in an effort to find dwarfing rootstocks that will resist a variety of soils and diseases. The common ones are MxM 2 and MxM 60.

Gisela: The Gisela rootstocks were developed in the early 1960s, at Justus Liebig University in Giessen, Germany.

Gisela 1: It is a hybrid of *Prunus fructicosa* 'Klon 64' and *Prunus avium*. It promotes early and heavy production, creates a tree that is about 25% the size of Mazzard, and is triploid in nature. Cherries, both sweet and tart, can be used with it. Prunus necrotic ringspot virus and prune dwarf virus are two pollen-borne viruses that can affect it.

Gisela 5: It is a hybrid of *Prunus cerasus* 'Shattenmorelle' and *Prunus canescens*. It is a triploid that can be used for both tart and sweet cherries and grows trees that are roughly 45% the size of Mazzard's. Gisela 5 results in a heavy and early production. It can withstand heavy soils and cherry virus infections.

Dwarfing rootstocks: CAB 6 P, CAB 11E, W-10, W-11, W-13, M X M-14, M X 39, M X M-46, M X M-27 Colt.

Cold hardy rootstocks: CAB 6P, CAB 11E, W-10, @-11, W-13.

Crown and root rot-resistant rootstock: M x M-2, Colt, M x M 39, M x M-97, W-10, W-11, and W-13.

Resistant to crown gall: F 12/1 s

Resistant to gummosis and canker: F 12/1, M x M-14, 39, 97 and 60

Planting

Cherry farming in India is generally limited to steep or muddy terrain, hence the planting is done on a terrace or contour system. For the creation of an orchard, a square scheme of layout is suggested in valley locations. The planting distance is influenced by the rootstock type and soil fertility. In Himachal Pradesh 6m × 6m spacing is advised for plants grown on seedling rootstock. Spacing can be lowered to 4.5 x 4.5 m for semi-dwarfing rootstock like colt. One month before to planting, 1 m x 1 m x 1 m pits are dug and filled with a mixture of 35–40 kg farmyard manure and 50–60 kg super phosphate. Finally, the pits are filled to a height of 15 cm.

Orchard management

Before establishing the orchard, soil should be tested by soil scientist to determine soil pH, soil fertility status, and slope of land and, organic matter content of soil. Apply lime if soil is too acidic because mountainous soil are generally acidic in nature and also apply fertilizers and farm yard manure to maintain the proper fertility status of soil as recommended by the scientist. Sweet cherry trees on vigorous rootstocks should be planted at a spacing of 4 to 5 m within the row and 6 to 7 m between columns. Sweet cherry trees on dwarfing rootstock can be planted 3 to 4 m within the row and 5 to 6 m between columns. Trees planted in holes dug with a spade or auger. Make a hole that is 18 inches broad and 18 inches deep. To prevent the roots from drying out, place the root system in the hole, fill it with soil to about halfway, and tamp it down. If required, raise the tree so that the graft union is about 3 inches above the soil surface. Then, fill the hole with dirt and tamp it down. To get rid of any air pockets in the soil, water the tree with at least 5 litres per tree. Unless at least 1 inch of rain has fallen since the last watering, water the tree every 14 days. To get rid of grass and weeds within two feet of the trunk, use herbicides or do mechanical weeding to protect the orchard from weeds.

High density planting

Sweet cherries are the choicest fruit among stone fruits. It has been successfully growing in selected pockets of Kashmir valley. Bing and Summit grafted on Gisela-6 rootstock planted at closer spacing found very productive. New Cherry clones selected by CITH, Srinagar also showed promising results under high density. Trees trained with vertical as well as open center canopies are have positive effect on yield and fruit quality

Training

Cherry trees are mostly trained using a modified leader method. When plants are planted, they are headed back at 60 to 80 cm. In the first dormant pruning, 3 to 5 wide, angled branches spaced 20 to 25 cm apart are chosen while maintaining the core leader system. 40 to 60 cm above the ground should be the height of the lowest branch. Only one-fourth of the growth is clipped from the selected scaffold branches, which are being brought back to their minimum length. In the second dormant pruning, 3 to 4 well-spaced main branches are chosen, with the fourth growth being clipped out, and 3 to 4 well-spaced subsidiary branches are chosen for each main scaffold.

Pruning

Pruning is done on sweet cherry trees to preserve the canopy, size, and shape of the trees, to allow good light to penetrate inside the canopy, and to maintain balance between vegetative and reproductive growth. Cherry plants need more corrective pruning than excessive branch heading back. Pruning is necessary to keep the center of bearing trees open. In order to promote fresh growth, the top is kept relatively low. Pruning should only be done to remove dead, sick, and entangled branches. Fruits develop laterally on spurs of one-year-old shoots, where the majority of the fruit is developed. Fruiting spurs have an average productive life of 10 to 12 years.

Manuring

Cherry requires all necessary nutrients for greater growth and higher-quality fruits since it responds well to nutrients and fertilizers. It has a high need for mineral nutrients since fruit development and vegetative growth occur at the same time. The age or size of the tree, the type and fertility of the soil, cultural practices, and the expected fruit output are taken into consideration when determining how much manure and fertilizer should be administered. In addition to a complete application of super phosphate and muriate of potash in December, farmyard manure should also be used. About 30 cm out from the tree trunk, fertilizer is spread in a tree basin. Application of FYM or sheep dung is done per plant wise. Nitrogen plays an essential role in improving yield and quality of the fruits. Nitrate fertilizers around 165g per tree per year is recommended in initial day of planting and later its dose will increase as per plants ages.

Water management

Cherry trees perform well with less water as compared to other temperate fruits and most of the cherry orchards are not irrigated in north western Himalayan region. Due to undulated land and less availability of irrigation water, in our country, cherries are primarily cultivated in rain-fed environments. The plantations should be often irrigated because the distribution of rainfall is quite variable and unequal throughout the year due to reduced rainfall in April and May. For greater fruit size and quality, plants should receive weekly irrigation during the period of fruit growth and development. Supplemental irrigation also helps young trees to develop and established trees, produce more, especially in times of drought or in areas with poor soils that cannot hold water.

Weed management

Weed is a plant growing out of their natural space. Weeds compete for water, minerals and nutrients and cherry trees do not perform well in grass and weeds effected orchards. The size of the weed-free region around a tree closely relates to how much growth it produces. A higher yield of cherries is obtained by using herbicides or mechanical methods to keep a weed-free area of 3 to 4 feet on both sides of the tree. If using an organic mulch, such as wood chips or straws, it must be removed by November to prevent wildlife from congregating around the tree.

Pollination and fruit setting

Since the majority of cherry varieties are inherently incompatible with one another, variety compatibility is necessary for cross-pollination. For proper fruit set, the orchard should be planted with a variety of cross-compatible and cross-incompatible groups and types. It is important to take care to ensure that they do not share the same sterility alleles and that their flowering periods do not overlap. If their flowering periods coincide, the universal donor cultivars like Stella, Vista, Seneca, Vic, and Vega are suitable cherries to plant with any type to obtain good fruit set. Plants of one variety should be next to those of the other in the planting pattern for effective pollination. For a proper fruit set, cherry trees need a specific number of chill hours. In India, where the environment is ideal for cherry growth, cherries are mostly farmed in the northern regions of Himachal Pradesh, Jammu and Kashmir, and also in Uttarakhand. In India, the cherry fruit setting process typically begins in March and lasts till May. The pollination process begins to occur during this time as the cherry tree flowers begin to bloom. Honeybees, which usually live in cherry orchards throughout the blooming season, pollinates cherry. According to a research in the Journal of Apicultural Research, honeybees are India's main cherry pollinators. (Bhat *et al.*2017 & Kumar *et al.*2019).

Major insect pest and their management

Black cherry aphid (*Myzus cerasi)*

Aphid attack the leaves as soon as the bud burst and they continue to feed as long as the growth remain succulent. Infested leaves of sweet cherry curl and cluster then they turn yellow and die. The aphid also feed on the fruit.

It can be controlled by spraying 1 ml rogor or metasystox per litre of water before or at bud swell stage

Flat headed borer : *(Sphecnoptera lafertei)*

Pest bores inside the trunk of the tree, insecticides such as aldrin, dieldrin, and parathion was found effective against freshly hatched larva.

Major disease and their management

Bacterial canker: *Pseudomonas syringae* pv. *morsprunorum* and *Pseudomonas syringae* pv. *syringae* are the principle organism that cause it. It causes canker when it enters through openings like stomata, blooms, and lenticels, as well as through wounds, pruning cuts, and places of winter harm. Cankers sites exude amber-colored gum and serve as borers' entry points. Controlling bacterial canker is quite challenging. It works to apply the Bordeaux mixture twice before bud break. In the spring and summer, use an antibiotic, such as streptomycin, right before debudding or pruning.

Crown Gall: It is caused by bacteria *Agrobacterium tumefaciens*, this bacterium is very common and found extensively in most soils. An abnormal vegetative growth, or round gall (1/4 inches to more than a foot in diameter), occurs largely below the soil's surface on the tree's trunk and roots. Use sterile rootstocks to prevent damage to the tree's roots and trunk.

Brown Rot: It is one of the most common diseases of cherries, it is caused by fungus *Monilinia fructicola* and also *Monilinia laxa.* Wind and water splashes spread this disease rapidly. Blossoms suddenly wither and turn brown. Dark spots expand rapidly which results in rotting of entire fruits. Tuft of gray to buff colour spores cover the surface of collapsed fruit. Fruits mummify and may remain on tree.

Cherry leaf spot: it is caused by *Cercospora circumscissa*, it causes severe defoliation in cherries. Round, with light to dark brown necrotic spots and brownish red edges appear on the leaves which enlarge and later drop out leaving shot holes.

Four applications of benomyl or thiabendazole from mid-April at monthly interval have been found effective in controlling the disease.

Yield

Yield of sweet cherry is highly variable every year and poor yield is of frequent occurrence the main cause of poor fruit yield is inadequate fruit setting due to improper pollination or in adequate pollen transfer. Sweet cherry tree grows larger than sour cherry and also bears heavily each tree produce around 12 to 15 kg of fruits annually and one hectare orchard yields around 8 to 10 tons of fruits

Harvesting and post-harvest management

Cherry start bearing from fifth year after planting and reaches full bearing stage after ten years, cherry has long production cycle of up to 50 years of age. The yield and quality of cherry is affected by the stage of maturity at which it is harvested. Early picking of fruits causes flat fruits with less yield as cherries mostly develop rapidly during last few days before reaching maturity. Harvesting of over ripe fruits causes loss in weight, volume and quality. Colour development, TSS and flavour is evaluated for judging the time of harvesting. Cherries are harvested in late spring or summer after the full maturity of fruits is attained and fruits starts to ripe Hand picking is generally practiced and it requires maximum care to protect the fruit quality so this process becomes labour consuming. Fresh fruits are plucked with stem when the surface colour changes from green to red. The fruits are packed in boxes lined with paper. Generally, 1 kg cardboard boxes are used for packing now a days. Sweet cherries can be stored at room temperature 17^{0}C to 23^{0}C for four days, eight days at 8^{0}C twelve day at 2^{0}C and fourteen days at 0^{0}C and 85% relative humidity (Muthu Krishnan and Palaniswamy, 1966)

Future prospects

Cherry is an important fruit crop that is gaining popularity in India due to its high demand and economic value. However, cherry cultivation in India is still in its nascent stage, and there is a significant scope for crop improvement to enhance its production and quality. Crop improvement programs can play a crucial role in developing new cherry varieties that are well-adapted to the Indian climatic conditions. The current cherry cultivars available in India are mostly imported from other countries, and they may not be suitable for the local environment. Therefore, researchers are working on developing new cultivars through traditional breeding methods and biotechnology tools such as genetic engineering, genomics, and marker-assisted breeding. The focus is on developing cherry varieties with desirable traits such as high yield, better fruit quality, disease resistance, and adaptability to different agro-climatic regions. Another area of research in cherry crop improvement is the development of early-ripening cherry varieties. Currently, the cherry harvesting season in India starts from May and continues until June, which overlaps with the harvesting season of other fruits such as mango and litchi. Therefore, developing early-ripening cherry varieties can help farmers to take advantage of the market demand for fresh cherries during the offseason.

In addition to developing new cultivars, the adoption of modern techniques and technologies can also contribute to the improvement of cherry cultivation in India. For instance, high-density planting, drip irrigation, and integrated pest management are some of the practices that can be used to enhance the productivity and quality of cherry crops. Furthermore, the use of precision agriculture techniques such as remote sensing, GIS, and GPS can help farmers to optimize their production practices and reduce input costs.

The future prospects of cherry cultivation in India are promising, given the growing demand for high-quality fruit in the domestic and international markets. The increasing popularity of cherries as a health food and the rising demand for processed cherry products also present new opportunities for the crop's growth. However, to fully realize the potential of cherry cultivation, there is a need for concerted efforts by farmers, researchers, and policymakers. In conclusion, crop improvement programs can play a vital role in enhancing the productivity, quality, and sustainability of cherry cultivation in India. Developing new cherry varieties with desirable traits, adopting modern techniques and technologies, and exploring new markets are some of the ways to tap the potential of this crop. With proper support and guidance, cherry cultivation can become a profitable venture for Indian farmers and contribute to the country's economy.

Refrences

Bal, J.S. (1977) Fruit Growing. *Kalyani Publishers*, Ludhiana, India, pp, 414-422.

Ballistreri, G., Continella, A., Gentile, A., Amenta, M., Fabroni, S. and Rapisarda, P. (2013) Fruit quality and bioactive compounds relevant to human health of sweet cherry (*Prunus avium* L.) cultivars grown in Italy. *Food Chemistry* **140**, 630–638.

Bhat, S. A., Sharma, R. K., & Gani, M. (2017). Cherry production in India: Status and prospects. *International Journal of Agriculture, Environment and Biotechnology*, **10**(2), 247-253.

Cook MD, Willems MET. Dietary Anthocyanins: A Review of the Exercise Performance Effects and Related Physiological Responses. *Int J Sport Nutr Exerc Metab*. 2019; **29**(3): 322–30.

FAO. 2022. FAO Production Statistics. http://faostat.fao.org

Ferretti G, Bacchetti T, Belleggia A, Neri D. Cherry antioxidants: from farm to table. *Molecules*. 2010 **12**; 15:6993-7005.

Filimon, R.V., Beceanu, D., Niculaua, M. and Arion, C. (2011) Study on the anthocyanin content of some sour cherry varieties grown in Iasi area, Romania. *Cercetari Agronomice in Moldova* **1**, 81–91.

Hummer, K. and J, Janick. (2009). *Rosaceae*: taxonomy, economic importance, genomics. *Springer.*

Jin W, Wang H, Wang J, Yang Y, Zhang X and Yan G. (2016). The R2R3 MYB transcription factor PavMYB10.1 involves in anthocyanin biosynthesis and determines fruit skin colour in sweet cherry (*Prunus avium* L.) PubMed -NCBI. *Plant Biotechnol J.* **14**(11): 2120–33.

Kolesnikova, V. (1966) *Fruit Biology*. *Mir Publishers*, Moscow.

Kumar, R., Sharma, H. K., & Kumar, A. (2019). Honeybees as pollinators of cherries: a review. *Journal of Apicultural Research*, **58**(1), 45-54.

Legua P, Domenech A, Martínez JJ, Sánchez-Rodríguez. L, Hernández F, Carbonell-Barrachina AA. (2017). Bioactive and Volatile Compounds in Sweet Cherry Cultivars. *J Food Nutr Res.* **5**(11): 844–51.

Lisek A, Korbin M, and Rozpara E (2006). Using simply generated RAPD markers to distinguish between sweet cherry (*Prunus avium* l.) cultivars. J. Fruit *Ornam. Plant*

McCune, L.M., Kubota, C., Stendell-Hollins, N.R. and Thomson, C.A. (2011) Cherries and health: a review. *Critical Reviews in Food Science and Nutrition* **51**, 1–12.

Mitic, M.N., Obradovic, M.V., Kostic, D.A., Micic, R.J. and Pecev, E.T. (2012) Polyphenol content and antioxidant activity of sour cherries from Serbia. *Chemical Industry and Chemical Engineering Quarterly* **18**, 53–62

Muthukrishnan, C.R. and P.K. Palaniswamy, (1966*). S. Indian Hort.* **14**(1-4) 60-62.

Oz MH, Vurgun H, Bakir M, Buyuk İ, Yuksel C, unlu HM, Cukadar K, Karadogan B, Kose O, Ergül A. Molecular analysis of East Anatolian traditional plum and cherry accessions using SSR markers. *Genet Mol Res.* 2013 Nov 7; **12**(4):5310-20. doi: 10.4238/2013. November.7.6. PMID: 24301792.

Pacifico, S., Di Maro, A., Petriccione, M., Galasso, S., Piccolella, S., Di Giuseppe, A.M.A., Scortichini, M. and Monaco, P. (2014) Chemical composition, nutritional value and antioxidant properties of autochthonous *Prunus avium* cultivars from Campania Region. *Food Research International* **64**, 188–199.

Papp, N., Szilvassy, B., Abranko, L., Szabo, T., Pfeiffer, P., Szabo, Z., Nyeki, J., Ercisli, S., Stefanovits-Banyai, E. and Hegedus, A. (2010) Main quality attributes and antioxidants in Hungarian sour cherries: identification of genotypes with enhanced functional properties. *International Journal of Food Science and Technology* 45, 395-402

Serradilla, M.J., Hernandez, A., Lopez-Corrales, M., Ruiz-Moyano, S., Cordoba, M.G. and Martin, A. (2016) Composition of the cherry (*Prunus avium* L. and *Prunus cerasus* L.; Rosaceae). In: Simmonds, M.S.J. and Preedy, V.R. (eds) Nutritional Composition of Fruit Cultivars. *Academic Press,* London, pp. 127-147

7

Advances Production Technology of Almond

***Satyarath Sonkar*[1*], *Anushi*[1], *Rajat Rajput*[2], *Nitin Kumar Chouhan*[1] *and Vikas Patel*[3]**

[1]*Department of Fruit Science, Chandra Shekhar Azad University of Agriculture & Technology, Kanpur, Uttar Pradesh (208002), India*
[2] *Department of Horticulture (Vegetable Science), School of Agricultural Sciences and Rural Development, Nagaland University Medziphema-(797106), Nagaland*
[3]*Department of Vegetable Science, Banda University of Agriculture & Technology, Banda, Uttar Pradesh (210001), India*
[1*]Corresponding Author E-mail: satyarthag242@gmail.com

Introduction

A significant nut that originated in Central Asia, the almond (*Prunus communis*) is now grown all over the world in hot, dry Mediterranean climatic zones (Monteiro *et al.*, 2001). The Rosaceae family it belongs to (Martinez-Gómez *et al.*, 2007). Number of chromosomes is 2n = 16. Today, the USA produces the most almonds, followed by Australia and Spain. Due to genetic, ecological, and processing-related variables, different cultivars of almond exhibit varied chemical profiles. Regular nut consumption has been linked to health benefits, particularly in the fight against cardiometabolic disorders (Massantini *et al.*, 2022). Consuming nuts has been shown to have beneficial benefits against a variety of illnesses, including obesity, hypertension, diabetes mellitus, and metabolic syndrome, according to epidemiological research and clinical trials (Chen *et al.*, 2006). Regular nuts consumers also exhibit better metabolic profiles and smaller waist circumferences. The edible portion of an almond, known as the almond kernel, is made up of two enormous cotyledons that are protected by an exterior

Fig. 1.1: Almond

hull and an inside shell (Aydin *et al.*, 2003). When the seed reaches maturity, the hull opens and the seed readily separates. Approximately 50% of almonds are lipids, 25% are proteins, and 20% are carbs. Almonds also have a low moisture content and a variety of minor beneficial components (Ahrens *et al.*, 2005). Almonds' macro- and micronutrient composition is related to their consumption's positive effects on health (Esfahlan *et al.*, 2010). The lipid profile can be highlighted as one of the components with beneficial health properties. According to Roncero *et al.*(2020), monounsaturated fatty acids (MUFA) make up 60% of the lipid profile, followed by polyunsaturated fatty acids (PUFA), fibre, vitamins, minerals, phytosterols, and polyphenols. The wide variety of cultivars, cultivation practises, and meteorological factors are all to blame for the startling variations in the chemical composition of almonds. Almonds are frequently eaten raw or roasted, which alters their constitution. The roasting procedure alters the lipid content chemically and microstructurally, favouring its oxidation and changing antioxidant components (Amarowicz *et al.*, 2005). Additionally, prebiotics are being used more and more to encourage gut microbiota modification towards a profile that promotes health. Almonds' fibre and polyphenol content may act as a substrate for microbial fermentation in the gut, helping to control the makeup of the microbiota (Bolling *et al.*, 2017).

Importance, uses and nutrient composition

Almonds hold great importance in India and have been widely used for various purposes for centuries. From culinary applications to health benefits, almonds have become an integral part of Indian culture and traditional practices. The Indian subcontinent has a rich history of incorporating almonds into its culinary traditions, and they are used in a variety of dishes. Almonds are known for their delightful flavor and texture, and they add a distinct nutty taste to both sweet and savory preparations. In Indian cuisine, almonds are commonly used in traditional sweets like kheer, halwa, and barfi. They are also added to rice dishes like biryani and pulao, providing a crunchy element and enhancing the overall taste of the dish. Almonds are often ground into a fine powder and used as a base for gravies and curries, adding richness and thickness to the preparations. Apart from their culinary significance, almonds offer numerous health benefits. They are highly nutritious, packed with essential nutrients such as healthy fats, protein, fiber, vitamin E, magnesium, and antioxidants. Consuming almonds regularly can have a positive impact on overall well-being. Research studies have shown that almonds contribute to heart health by reducing LDL cholesterol levels and promoting cardiovascular function (Sabaté *et al.*, 2003). Incorporating almonds into a heart-healthy diet has been found to result in significant improvements in lipid profiles and a reduced risk of heart disease.

Furthermore, almonds are known for their role in weight management. Despite being calorie-dense, almonds have been shown to promote satiety and can aid in weight loss when consumed as part of a balanced diet. A study published in the European Journal of Clinical Nutrition demonstrated that including almonds in a weight loss program led to greater reductions in body weight and waist circumference compared to a low-calorie diet without almonds (Foster *et al.*, 2012). The fiber and protein content of almonds contribute to their satiating effect, helping individuals feel fuller for longer.

Almonds also find application in traditional Indian medicine systems like Ayurveda. Almond oil, derived from almonds, is highly valued for its therapeutic properties. It is commonly used in skincare and haircare products due to its moisturizing and nourishing effects. Almond oil is rich in vitamin E, an antioxidant that helps protect the skin from damage caused by free radicals and environmental factors (Korać and Khambholja, 2011). It is believed to improve skin complexion, reduce dark circles, and promote healthy hair growth. A new American Dietetic Association research found that eating almonds increases the amount of vitamin E in your red blood cells and lowers your chance of developing high cholesterol. Increased blood levels of vitamin E produce antioxidants that stop your cells from clogging with growing cholesterol (Wijeratne *et al.*, 2006). So, by eating a handful of almonds every day, you may increase the amount of vitamin E in your blood and reduce your chance of acquiring cholesterol. It is thought that eating almonds and other nuts together is excellent for your heart.

Nutrient composition

According to the USDA, the nutritive value of an apple (100g of edible portion) is as follows:

Nutritive value of almond fruit/100 g edible portion:

Contents	Amount with Unit	Contents	Amount with Unit
Energy	579 kcal	Niacin (Vitamin B3)	3.385 mg
Carbohydrates	21.69 g	Folate (Vitamin B9)	50 µg
Protein	21.15 g	Calcium	264 mg
Fat	49.93	Iron	3.72 mg
Fiber	12.5 g	Magnesium	270 mg
Sugars	3.9 g	Phosphorus	481 mg
Vitamin E	25.63 mg	Potassium	705 mg
Riboflavin (Vitamin B2)	1.014 mg	Zinc	3.08 mg

Source: USDA Nutrient Database: g = grams, kcal = kilocalorie, µg = micrograms; mg milligrams

Area and production

Almond trees cover 10 thousand hectares in India, producing 11 thousand MT in total (Anon, 2020). With the proper farming tools, techniques, and agricultural resources, an area and yield 10 hectare and 11 MT (2021-22) of almonds. Jammu and Kashmir - The region produces an average of 12.93 tonnes of almonds annually, satisfying 91.26% of India's need (thousands of tonnes). With an average production of 0.76, Himachal Pradesh is the second-largest producer of almonds in India (000 tones).

The third-largest producer of almonds is Maharashtra because Indian badam can thrive in higher temperatures. In India, Karnataka ranks as the fourth-largest producer of almonds. The most productive areas are Bangalore, Hubli, Belgaum, Davanagere, and Mangalore, which are also the most well-known. Andhra Pradesh - Andhra Pradesh offers a sizable market for suppliers, wholesalers, distributors, and exporters in India, making almond cultivation there possible. Uttar Pradesh - After determining that it is possible to cultivate almond trees, Uttar Pradesh is putting together plans and resources to help increase India's output of almonds. Kerala - Kerala sees potential in almond tree growing for improved yield and revenues as a result of discovering the health advantages of almonds in India.

Soil

In India, the PH of the soil should be between 7.0 and 8.5 for almond production. Insufficiently draining soils are unsuitable for almond trees. Deep, loamy soil that can withstand drought is required. Proper farmyard manure should be added to the soil, along with regular watering. To grow, almond trees require full light and loamy soil that drains well. Although the almond tree is drought-resistant and may live for many years on dry soils, its yield is much diminished in such circumstances. The importance of low soil moisture at the conclusion of the growing season is lower than it is at the start.

The soil must be ploughed to a depth of 20-27 inches prior to planting the young almond trees (50-70 cm). Tillage seeks to eradicate perennial weeds and fluff the soil, which is essential for the growth of the root system. Before ploughing, almond producers frequently collect soil samples and send them to a lab to assess if any remedial activities are required. To promote soil fertility and enhance soil texture, many almond producers apply 20-30 tonnes of manure per hectare well before planting the young trees.

Climatic requirement

The ideal summer temperature for almond growing is often between 300°C and 350°C. For a shorter amount of time, blossoms can endure cold temperatures between -2.2°C and -3.3°C. However, if this temperature persists, the flowers are more susceptible to injury.

Flowers that haven't opened can survive down to -2.2°C. While flowers in the petal stage might suffer harm if the temperature drops between 0.50 and 1.10 degrees Celsius.

Important Rootstock

Prunus dulcis M., Viking, Atlas, *Prunus besseyi* x Myrobalan plum, Pollizo" plums,

P. besseyi x peach

Krymsk86: *Prunus persica* x prunus spp

Kryansk 1: *Prunus tomentosa* x *Prunus cerasifera*

Commercial Varieties

There are several varieties of almonds grown around the world. Here are some popular almond varieties:

California Paper Shell: This variety of almond is known for its thin and easily crackable shell. It is favored for its convenience in shelling and is commonly used for snacking or as an ingredient in various culinary applications.

Shalimar: The Shalimar almond variety is known for its sweet flavor and large kernel size. It is often used in confectionery and baking due to its desirable taste and appearance.

Nonpareil: Nonpareil is one of the most popular and widely grown almond varieties. It is highly regarded for its mild and delicate flavor, making it a preferred choice for both raw consumption and processing into almond products.

Merced: Merced almonds are characterized by their medium-sized kernels and good flavor. They are commonly used in baking, cooking, and as a topping for salads, cereals, and desserts.

Pranyaj: This almond variety is specifically cultivated in India. It is known for its excellent taste and is often consumed as a snack or used in traditional Indian sweets and dishes.

IXL-Drake-Ne Plus: This almond variety is a combination of three cultivars, namely IXL, Drake, and Ne Plus. It is favored for its sweet and nutty taste and is commonly used in both raw and processed forms.

Ne Plus Ultra: The Ne Plus Ultra almond variety is known for its high oil content and rich flavor. It is often used in almond oil production and is also enjoyed as a snack or used in various culinary preparations.

Waris: The Waris almond variety is specifically grown in India. It is prized for its large kernel size and excellent taste, making it a popular choice for both raw consumption and culinary purposes.

Peerless: Peerless almonds are characterized by their medium-sized kernels and good flavor. They are often used in baking, cooking, and as a topping for various dishes.

Monterey: Monterey almonds are known for their versatility and adaptability to different growing conditions. They have a good flavor profile and are widely used in both raw and processed forms.

Sonora: Sonora almonds are recognized for their sweet and delicate flavor. They are commonly used in baking, cooking, and as a topping for desserts and salads.

Fritz: The Fritz almond variety is known for its large kernels and rich flavor. It is commonly used in almond butter production and is also enjoyed as a snack.

J K 55: J K 55 is an almond variety known for being the earliest ripening cultivar. It is favored by growers because it allows for an early harvest, providing an advantage in regions with shorter growing seasons

H-15: H-15 is a variety resulting from a cross between the Thin-shelled and Sloh almond cultivars. It inherits traits from both parent varieties, potentially combining characteristics such as thin shells and desirable kernel qualities.

AH 258: AH 258 is a variety derived from a cross between the Petrick's Wonder and Sloh almond cultivars. It is known for its specific traits inherited from its parent varieties, and it may exhibit a combination of characteristics like kernel size, flavor, or shell thickness.

Primorskiji: Primorskiji is an almond variety that originated in the Primorsky region of Russia. It is specifically adapted to the local growing conditions and climate of that region.

Prabhat: Prabhat is an almond variety with specific characteristics that make it distinct. The specific traits and qualities of Prabhat almonds may vary depending on the region and growing conditions in which they are cultivated

Mukhdoom: Mukhdoom is an almond variety that is known for its large kernel size and rich flavor. It is particularly popular in India and is used in various traditional Indian sweets and dishes.

Mission: Mission is a widely grown and popular variety of almonds. It is known for its medium-sized kernels and sweet flavor. Mission almonds are often used in both raw and processed forms, including snacking, baking, and making almond products.

Thompson: Thompson almonds are recognized for their medium-sized kernels and good flavor. They are commonly used in a variety of culinary applications, including baking, cooking, and snacking.

Supernova: Supernova is a mutant variety of almonds. Mutant varieties arise from natural genetic mutations, resulting in unique characteristics. The specific traits of Supernova almonds may differ from the parent variety.

Sloh: Sloh almonds are derived from a cross between a peach and an almond. This hybridization results in a unique combination of flavors and characteristics, potentially including attributes from both parent species.

Merced: Merced almonds are known for their medium-sized kernels and good flavor. They are widely used in baking, cooking, and as a topping for various dishes.

Jordonalo: Jordonalo is a variety of almonds with specific traits that distinguish it. The specific characteristics of Jordonalo almonds may vary depending on the region and growing conditions in which they are cultivated.

Eureka: Eureka is an almond variety with specific qualities that make it unique. The precise attributes and qualities of Eureka almonds can vary depending on factors such as location and cultivation practices.

Propagation

They are commonly propagated through different methods to ensure their successful growth and development. Almonds can be propagated through seeds, grafting, budding, or tissue culture. The various ways almonds can be propagated.

Seeds: The most common method of propagating almonds is through seeds. Almonds are the seeds of the almond tree (Prunus dulcis), and they can be directly planted in well-draining soil. It's essential to choose healthy, viable seeds and soak them in water overnight before planting. The seeds should be planted about 2 inches deep and spaced adequately. Germination can take anywhere from 3 to 6 weeks, and the resulting seedlings can be transplanted to their permanent locations once they are sturdy enough.

Budding: Budding is also popular methods for propagating almond trees. Budding, considered the most popular method (Mir *et al.*, 2016), involves collecting dead wood during the winter dormancy period and storing it until spring. During spring, T-budding is performed by attaching a bud from a desired variety onto the rootstock of another tree. This allows the desired traits to be transferred, and the bud develops into a new almond tree (Uematsu *et al.*, 1987).

Fig. 1.2: Budding

A technique for propagation known as "soft wood cutting" involves removing immature branches from a woody plant and forcing them to take root. Choose a few young, malleable offshoots from the almond tree in the spring after it has leaflet out and formed new shoots for softwood cuttings. Check to see if there are fresh shoots emerging from the tree's graft union rather than suckers coming from the graft's base. Prepare a seeding tray or tiny pots with a quality compost or potting media before cutting off the shoots for softwood cuttings. Use a pencil or dowel to make holes in the potting material for the cuttings. A rooting hormone should also be kept on hand. Cut the juvenile offshoots you choose for almond tree propagation right below a leaf node using a clean, sharp knife. The chosen shoots should be 3 to 4 inches (8 to 10 cm) length. Any leaf buds or leaves on the lower part of the cutting should be removed. T-budding involves making a T-shaped cut in the rootstock, inserting an almond bud beneath the bark of the cut, and holding it there using grafting tape or a thick rubber band. Shield or chip budding involves removing a shield-shaped chip from the rootstock and replacing it with a chip that fits perfectly and has an almond bud on it. Grafting tape is then used to fix this chip bud into position. Almond buds are grafted onto the chosen rootstock using one of two techniques i.e., T-budding or chip/shield budding by making precise incisions with a grafting knife.

Fig. 1.3: T-budding

Grafting: Another popular method of almond propagation is grafting. This technique involves joining a desired almond variety (scion) onto a rootstock of a closely related almond species or hybrid. Grafting allows for the propagation of specific almond traits, such as disease resistance or improved yield. It also ensures that the resulting tree will be genetically identical to the parent tree. Grafting should be done during the dormant season when the trees are not actively growing.

Tissue Culture: In recent years, tissue culture has gained popularity as a method of propagating almonds. It involves growing almond plantlets from small tissue samples in a laboratory setting. This technique allows for rapid multiplication of genetically identical plants and can help overcome limitations such as disease susceptibility or slow growth rates. Tissue culture propagation is particularly useful for almond varieties that are difficult to propagate through traditional methods.

Planting

Almond plants are extremely accessible in any large nursery from July to September. The Graft plant begins producing fruit after the third year of germination whereas the almond plant produces fruit once each year. The almond tree will begin to bloom with lovely pink blossoms in January. April is when March turns into fruit, and ripening begins in July. At time of blooming stop the irrigation. To give the roots a good start in life and ensure that they are properly saturated before being inserted into the ground, they should get a thorough dose of water prior to planting. The tap root should get extra attention to prevent it from being twisted out of shape, and the hole should be excavated broad and deep enough to accommodate the entire root system. Almonds, like many other nut trees, are extremely sensitive to damage to their tap root, thus they should not be cut or forced into a hole that is not large enough for them. To avoid matting, the remaining roots should also be treated delicately and dispersed gently. The same depth at which they were cultivated in the nursery should be used for planting. This is true for both potted trees and bare root plants. As you fill the hole, the soil should be tightly packed around the roots. Giving your newborn tree two buckets of water after you've filled the hole can help it adapt well to its new surroundings. You may also fertilize your tree at this time to give it a little push, but it is better to wait until spring if you planted in the autumn.

Nutrition management

Nutrition management plays a crucial role in maintaining the health and productivity of almond orchards. Adequate nutrient supply, including macronutrients and micronutrients, is essential for the growth, development, and optimal yield of almond trees. Here is some information on the recommended nutrient management practices for almond orchards, including the application rates of various nutrients, along with a reference for further exploration.

Organic Matter: Adding farmyard manure (FYM) or compost to the soil can improve its structure, water-holding capacity, and nutrient availability. The recommended quantity of FYM for almond orchards is around 10-15 kilograms per tree per year.

Macronutrients

Nitrogen (N): Nitrogen is essential for vegetative growth, leaf development, and overall tree vigor. The recommended nitrogen application rate for almond orchards is approximately 200-300 grams per tree per year.

Phosphorus (P): Phosphorus promotes root development, flowering, and fruit formation. Almond trees require around 100-150 grams of phosphorus per tree per year.

Potassium (K): Potassium is crucial for improving fruit quality, water regulation, and overall tree health. Almond orchards typically require approximately 250-350 grams of potassium per tree per year.

Micronutrients

Zinc (Zn): Zinc is important for enzymatic activities, hormone synthesis, and overall plant growth. Almond trees may need about 100-150 grams of zinc sulfate per tree per year.

Boron (B): Boron is essential for flowering, fruit set, and proper cell division. The recommended boron application rate for almond orchards is around 50-100 grams of borax per tree per year.

It is crucial to note that nutrient requirements may vary depending on soil conditions, tree age, and local factors. Regular soil testing and consultation with agricultural experts can help determine the specific nutrient needs of your almond orchard.

Water management

Water management is a crucial component of successful almond cultivation in India, especially considering the limited availability of water resources in many regions. Almond trees have specific water requirements that must be met throughout their growth stages to ensure optimal development, flowering, and fruit production. In India, where water scarcity is a pressing concern, efficient water management practices are essential to maximize water utilization and minimize wastage. One of the key aspects of water management in almond cultivation is irrigation. Almond trees require a consistent and well-regulated water supply, particularly during critical periods such as fruit set and nut development. The recommended irrigation depth for almond orchards in India typically ranges from 450 to 600 millimeters (mm) per growing season, depending on factors such as climate, soil type, and tree age.

To optimize water usage, growers are increasingly adopting advanced irrigation techniques such as drip irrigation. This method delivers water directly to the root zone, reducing evaporation and ensuring targeted water application. By utilizing drip irrigation systems, growers can achieve higher water-use efficiency compared to traditional flood or sprinkler irrigation. Moreover, precision irrigation technologies enable growers to monitor soil moisture levels and adjust irrigation schedules accordingly, ensuring that water is provided when and where it is most needed by the almond trees. In addition to efficient irrigation, other water management strategies can further enhance water conservation in almond orchards. Mulching the soil surface with organic materials, such as straw or wood chips, helps minimize evaporation and retain soil moisture. This practice reduces the need for frequent irrigation and helps maintain optimal soil moisture conditions for almond tree roots. Regular monitoring of soil moisture levels using sensors or probes can provide valuable insights into the water status of the orchard, enabling growers to make informed decisions about irrigation scheduling and duration.

Water management in almond cultivation goes beyond irrigation practices. It also involves implementing water-saving measures, such as optimizing orchard layout and design to minimize runoff and utilizing cover crops or intercropping to enhance soil water-holding capacity. Additionally, incorporating techniques like deficit irrigation during less critical growth stages can help conserve water without compromising almond tree health or productivity.

Orchard floor management

Orchard floor management plays a vital role in almond cultivation in India. Proper management of the area beneath almond trees helps optimize soil health, moisture retention, and weed control. One common practice is the use of mulch, which helps conserve soil moisture, suppress weed growth, and regulate soil temperature. Organic mulch materials such as wood chips or straw can be applied to the orchard floor. Additionally, regular weed control through manual removal, mowing, or herbicide application is essential to minimize competition for water and nutrients. Maintaining a well-managed orchard floor promotes a healthy growing environment for almond trees, ultimately enhancing their productivity and overall orchard performance.

Inter-cropping

Inter cropping or the practice of growing additional crops alongside almond trees, can offer several benefits in almond cultivation in India. Suitable inter crops such as legumes, grasses, or cover crops can help improve soil health, increase biodiversity, and maximize land use efficiency. These inter crops can provide additional income for farmers, suppress weed growth, enhance nutrient cycling, and improve soil moisture retention. However, it is important to select inter crops that do not compete significantly with almond trees for resources and consider factors such as tree spacing, crop compatibility, and overall orchard management. Careful planning and proper management are crucial to ensure successful inter cropping in almond orchards and achieve the desired outcomes.

Training and pruning

For training in almonds, a modified central leader method must be chosen. Reduced trimming on younger trees can produce higher, earlier harvests, according to research. This is mostly caused by an expanded canopy that can generate crop. According to some studies, letting young trees unpruned initially provides advantages, but over time, these branches will split as the crop weight rises, reducing production. In order to solve the tight branch angles and close closeness of scaffolds, which result in included wood and weak branches, some structure of first and second leaf trees is advised. There is no one perfect number, and a farm-level selection should take tree vigour, spacing, and management approaches into consideration. If the trees are headed higher at planting, there will be more option to choose scaffolds, but trimming will take longer. Heading the trees too short might cause problems with shaker head placement when they are fully grown, which would result in further bark damage. After the first year of development, choose scaffolds that are spaced apart rather than ones that are the same height as the tree trunk and on the same plane. The expanding limbs will be more securely bonded, resulting in less breaking, if the scaffolds are spaced vertically and all around the tree. Eliminate limbs with steep angles as they are frequently weaker. Ideal limbs have an angle of 45 to 60 degrees. Avoid making harsh heading cuts since doing so might encourage lanky, vertical regrowth the following spring. According to my experience, pruning trees to a wood diameter of 12" to 34" usually results in the optimal growth response. Finally, keep in mind that trees with less vigour will require less trimming than trees with higher vigour. There isn't much trimming required after the first year's scaffold selection. After the second growth season, 1-2 cuts are often required per tree to remove limbs that are vertical or crossing the tree. Additionally, limbs that are excessively low and in the way of orchard activities should be cut off. Pruning should concentrate on eliminating limbs that are in the way of operations after the third growth season. In mature orchards, hedging or making arbitrary heading cuts in the top canopy is typically not advised. Although this procedure is helpful for boosting light on the orchard floor, it frequently results in the removal of sizable portions of the fruitful canopy. Additionally, these heading cuttings produce a strange pattern of regeneration in which new limbs develop vertically. Consider cutting chosen limbs thinner instead to aid boost light penetration. It's possible that this kind of maintenance trimming won't be required every year, which would let the expense to be spread out over three to four years. In order to prevent wound infections by pathogen fungi that disseminate spores during rain events, pruning should only be done when rain is not predicted. When it comes to developing, young trees, this is especially true. Larger wounds take the slowest to heal and are sensitive for up to two weeks following pruning. Pruning paints are not advised despite this healing time since they might hinder the healing process. Though still being evaluated, ongoing research with biological agents have showed some promise. Pruning after leaf emergence won't affect a tree's growth.

Flowering and fruiting

Mid-February to mid-March sees the blossoming of almond tree buds into stunning white and light-pink blossoms. Beekeepers move hives into the orchards at this time to aid in pollinating the frui (Traynor *et al.* 2017). Although it can take an almond tree 5-12 years to begin bearing fruit, a mature tree can normally continue to provide fruit for up to 25 years. Almond trees can only be planted in particular regions across the world, comparable to the Mediterranean, in order to yield fruit. Although a severe frost might harm the blossoms, the tree buds must experience a period of cold weather before they can develop flowers. The first trees that bees often see after the winter are almond orchards. The bees may regain their vitality during the almond bloom because the almond pollen is very nourishing. During the spring and summer, the bees rely heavily on the pollen and nectar of almonds that were stored in the hive during the bloom. Almond trees are the first to bloom in the spring and produce blossoms on their current-season growth.

Fig. 1.4: Flowering in almond orchards

Physiological disorders and their management

Physiological disorders can pose significant challenges to almond cultivation in India, affecting tree health, yield, and overall orchard productivity. Two common physiological disorders observed in almond orchards are Muletail or Crazy top and Gummosis. Effective management strategies are crucial to minimize the impact of these disorders on almond trees.

Muletail or Crazy top

Muletail or Crazy top is a disorder characterized by abnormal shoot growth, where the terminal bud fails to develop properly, resulting in a cluster of twisted or stunted shoots resembling a "crazy top." This disorder is often associated with environmental factors such as cold temperatures during bud development or nutrient imbalances. To manage Muletail or Crazy top, it is important to ensure proper nutrition, especially balanced levels of nitrogen, phosphorus, and potassium, and provide adequate protection against extreme cold temperatures during bud formation.

Gummosis

Gummosis is a physiological disorder that leads to the formation of gum-like exudates on the trunk and branches of almond trees. It is caused by a combination of factors, including fungal or bacterial infections, physical injuries, or environmental stressors such as extreme temperatures or waterlogging. Effective management of Gummosis involves implementing preventive measures, such as practicing proper sanitation, pruning infected branches, and providing good orchard drainage. In severe cases, the application of appropriate fungicides or bactericides may be necessary to control the spread of the disease.

Major Diseases and their management

(1) Almond brown line and decline: Cause: Phytoplasma

The signs are tree development is slowed, the leaves are drooping or withering, and the bark has brown necrotic spots. mostly seen on young trees.

- Management: plant trees from stock that is free of pathogens. Stunted trees ought to be taken out and replaced, and only certified pathogen-free trees should be planted.

(2.) Almond kernel shrivel: Cause: Phytoplasma

The signs are late flowering, stunted new growth, darker, smaller leaves, and harvested nut kernels that had shrunk. Where peach rootstock has been utilised for grafting, it is most typical.

- Management: To manage, cut down diseased trees. Plant only approved trees; remove sick ones.

(3) Almond leaf scorch: Cause: *Xylella fastidiosa*

Its signs include Chlorotic leaf margins, necrotic leaf margins that start at the leaf tip and progress to the base, and areas of necrotic tissue with chlorotic edges are all examples of this. (Sisterson and coworkers, 2008) Each year, more trees will become infected; the bacteria may also infect rye, blackberry, and nettle; if these plants are close, they may serve as reservoirs; and it is spread by leafhoppers and spittlebugs.

- Management: Early detection (when the illness only affects one branch) and cutting the major scaffold 5 to 10 feet below the symptoms will eliminate the disease; later infections may call for the removal and replacement of the tree.

(4) Crown gall: Cause: *Agrobacterium tumefaciens*

The signs are Galls of different sizes on roots and the root crown below the soil line; occasionally, galls will appear on the trunk; these are initially light-colored bumps that grow larger and darken; they may be soft and spongy or hard; if galling is severe and girdles the trunk, young trees will become weaker due to constricted vascular tissue; they may become stunted and rarely die. The bacteria infects host plants through wounds and causes plant cells to multiply and become undifferentiated, which results in the development of galls.

- Management: Plant trees in soils that drain well, use disease-free nursery stock, and minimise plant trauma. A biocontrol agent can be used to treat recently opened wounds (*Agrobacterium tumefaciens* K84).

(5) Alternaria leaf spot: Cause: *Alternaria alternata*

The signs are Light yellow necrosis that dries out and turns tan in the middle of the leaves can occur. Infected leaves can develop light brown lesions that grow to create circular lesions on the leaf blade or semi-circular lesions on the border.

- Management: If signs of Alternaria are apparent, treatment with the right fungicide should be done in late spring.

(6) Anthracnose

Blossom blight, limb dieback, leaf death with connected leaves, and nuts with orange lesions are all symptoms. All cultivars are vulnerable; it happens more frequently in warm, humid weather (Rubio *et al.*, 2003).

- Cause = *Colletotrichum acutatum*
- Management = Disease control requires fungicide treatment and cultural norms. A low angle nozzle should be used in orchards using spray irrigation to minimise wetness of the leaves. Orchards with a history of anthracnose infections should be treated at 5-10% bloom and treatments should be repeated every 10 to 14 days. Dead diseased branches should also be removed.

(7) Brown Rot Blossom Blight: Cause: *Monolinia laxa*

Blighted blossoms, browning and necrotic stigma and anthers, browning and collapsing blossoms, light brown powdery fungal masses that may be seen on infected flowers, gummy exudate around flower bases, and cankers developing on twigs near the blossoms are all signs of the disease.

- Management: Fungicide application at 5-10% bloom and full bloom to protect flowers; one application at full bloom usually sufficient if there is no rainfall; two or three applications should be made if bloom is accompanied by rainfall.

(8) Hull rot: Cause = *Rhizopus stolonifera*

Tan lesions on the hulls that expand and cause the fruit to shrivel are symptoms, as are dark grey spore masses visible between the hull and shell, leaves that are adjacent to the infected fruit may curl and wither, and leaf death occurs on the side of the stalk that is closest to the fruit.

- Management: Management should be used while managing irrigation. Demethylation inhibitor and quinone outside inhibitor fungicide may be administered in addition to irrigation management at the hull split.

(9) Shot hole: Cause = *Wilsonomyces carpophilus*

The signs are Circular purple patches on leaves that get larger, turn chlorotic, then turn tan; when lesions dry, a missle of the lesion falls out of the leaf, creating tiny holes. Water-borne spore transmission makes illness more prevalent in moist environments.

- Management: A fall treatment with fungicide is necessary if fungal fruiting structures are present in the autumn (visible under a hand lens as little black dots in the centre of lesions); fungicide should be sprayed before wet periods to safeguard the tree.

(10) Verticillium wilt (Blackheart): Cause = *Verticillium dahlia*

The signs are A tree's leaves on one side are yellowing and drooping in the early seasons. A difficulty arises if an orchard is interplanted with other vulnerable plants, such as cotton, tomatoes, or melons. The fungus overwinters on the soil and returns every year.

- Management: Intercropping vulnerable plants in orchards, such as cotton, tomatoes, or melons, is not recommended. Fungi in the soil can be killed by solarizing or fumiganting the soil before planting.

Major insect pests and their management:

1. Pavement ant (Southern fire ant)

Tetramorium caespitum, Solenopsis xyloni, S. *molesta* and various hemipteran pests also attack on orchard of almonds (Rijal *et al.*, 2021). Hollowed-out nuts on the ground are a symptom. It is common in orchards that use spray or drip irrigation.

- Management: In April and May, check the orchard for ants; before harvest, use ant baits to control large ant populations; as soon as possible, remove nuts from the orchard floor (Zacheo *et al.*, 2000).

2. Navel Orangeworm (*Amyelois transitella*)

The navel orangeworm is another significant pest that attacks almonds, particularly during the fruiting stage. The larvae of this moth bore into the almond kernels, causing direct damage and facilitating secondary infections. To control navel orangeworm infestations, insecticides such as Chlorpyrifos or Carbaryl can be used. Chlorpyrifos is commonly applied at a rate of 1 to 1.5 ml per liter of water, while Carbaryl is applied at a rate of 2 g per liter of water (Nahar *et al.*, 2019).

3. Almond Aphids (*Brachycaudus amygdalinus*)

Almond aphids are small, sap-sucking insects that infest almond trees, causing stunted growth and reduced vigor. To manage almond aphids, systemic insecticides like Imidacloprid or Acetamiprid can be used. Imidacloprid is typically applied as a soil drench at a rate of 0.5 to 1 g per liter of water, while Acetamiprid is applied as a foliar spray at a rate of 0.5 to 1 ml per liter of water (Prasad *et al.*, 2017).

Maturity indices

Maturity indices are important indicators that can help determine the optimal time for harvesting almonds. The following are some of the commonly used maturity indices for almonds:

Hull split: The hull is the outer covering of the almond fruit, and its split is an important indicator of maturity. When the hull begins to split, it is a sign that the almond kernel inside has reached full maturity.

Nut size: The size of the nut is also an important indicator of maturity. Fully mature almonds are typically larger and heavier than immature ones.

Kernel moisture content: The moisture content of the almond kernel is another important indicator of maturity. Fully mature almonds typically have a lower moisture content than immature ones.

Internal color: The color of the almond kernel can also indicate maturity. Fully mature almonds typically have a uniform light beige color, while immature almonds may have a greenish tint.

Sugar content: The sugar content of the almond kernel can also indicate maturity. Fully mature almonds typically have a higher sugar content than immature ones.

Overall, a combination of these maturity indices should be used to determine the optimal time for harvesting almonds. This can help ensure that the almonds are of the highest quality and have the best flavor and nutritional content.

Harvesting and Post-harvest handling

Within three years of transplanting, almond trees start to yield fruit. And the tree begins producing high-quality nuts after it is 6-7 years old. And the almonds may be picked after 8 months of blossoming. Make sure the pods are green to yellow to determine whether the fruit is ready to be harvested. Almond nuts might be collected using poles. Additionally, the nuts must be manually dried in dry air or exposed to the sun until the moisture content reaches 5-7%.

Post-harvest handling of almonds is an important process that involves various activities aimed at preserving the quality and safety of the harvested nuts. The following are some of the key steps involved in post-harvest handling of almonds:

Harvesting: Almonds are typically harvested mechanically or manually. The harvested nuts should be handled carefully to avoid damage that can lead to spoilage or contamination.

Cleaning and sorting: After harvesting, the nuts should be cleaned to remove any debris, leaves, or twigs. They should also be sorted to remove any damaged or diseased nuts.

Drying: Almonds should be dried to reduce their moisture content to a safe level. This can be done using natural air drying or mechanical drying methods.

Storage: Proper storage is essential to maintain the quality of the almonds. They should be stored in a cool, dry, and well-ventilated area to prevent moisture buildup and pest infestation.

Processing: Almonds can be processed into various forms such as roasted nuts, almond butter, almond milk, and almond flour. The processing should be done under hygienic conditions to prevent contamination.

Packaging: The almonds should be packaged in airtight containers to prevent moisture buildup and to protect them from pests and contamination.

Overall, post-harvest handling is critical for maintaining the quality and safety of almonds. Proper handling and storage can help to reduce spoilage, improve shelf life, and maintain the nutritional value of the nuts.

Yield

The yield is 3-6 kg of nuts per tree or 1-2 tonnes per acre. But in Kashmir, it was possible to produce 10-20 kg of nuts per tree. From July through September, fruits are primarily collected. With a crack, the nut's colour changes from green to yellow. Before the nut falls, the husk breaks (Pascuzzi *et al.*, 2017), and the drupe is extracted.

Future Prospect

Almond is an important nut crop with great economic and nutritional value. The almond industry is growing rapidly worldwide, with an increasing demand for almonds in various forms such as whole nuts, milk, oil, and butter. Advances in production technology of almond are therefore essential for improving crop yields, enhancing product quality, reducing production costs, and meeting the increasing demand.

One of the most important aspects of production technology for almonds is the development of new varieties that are adapted to different climatic and soil conditions. This is particularly important in regions where water resources are limited and climate change is affecting crop productivity. Advances in genetic engineering, genomics, and biotechnology have enabled the creation of new varieties with improved traits such as drought tolerance, disease resistance, and higher yields. Another important area of research in almond production technology is the optimization of irrigation and fertilization practices. Water and nutrient management are crucial for almond trees, as they require large amounts of water during their growth and fruiting stages. The use of precision irrigation techniques such as drip irrigation, combined with the application of controlled-release fertilizers, can help to optimize water and nutrient use efficiency, reduce waste and pollution, and improve yields.

The adoption of new technologies for pest and disease management is also critical for almond production. Integrated pest management (IPM) strategies that combine biological, cultural, and chemical methods have proven to be effective in controlling pests and diseases, while reducing the use of harmful chemicals. New biological control agents such as microbial pesticides, pheromone traps, and biocontrol agents can also be used to control pests and diseases more sustainably and cost-effectively. Finally, advances in post-harvest technology are essential for maintaining product quality and reducing waste. Proper storage, handling, and processing techniques can help to minimize losses due to spoilage, insect damage, and other factors. Technologies such as vacuum cooling, controlled atmosphere storage, and value-added processing can also help to enhance product quality, extend shelf life, and increase market value. In conclusion, the advances in production technology of almond are of great importance for the future of this industry. The development of new varieties, optimization of irrigation and fertilization practices, adoption of IPM strategies, and advances in post-harvest technology are all essential for improving crop yields, enhancing product quality, reducing production costs, and meeting the increasing demand for almonds in the global market.

References

Ahrens, S., Venkatachalam, M., Mistry, A. M., Lapsley, K., and Sathe, S. K. (2005). Almond (*Prunus dulcis* L.) protein quality. *Plant Foods for Human Nutrition*, 60, 123-128.

Amarowicz, R., Troszyńska, A. and Shahidi, F. (2005). Antioxidant activity of almond seed extract and its fractions. *Journal of Food Lipids*, 12(4), 344-358.

Anonymous, (2020). National Horticulture Board, Indian Horticulture Database. Ministry of Agriculture & Farmers Welfare, Government of India. www.agricoop.nic.in

Aydin, C. (2003). Physical properties of almond nut and kernel. *Journal of food engineering*, 60(3), 315-320.

Bolling, B. W. (2017). Almond polyphenols: Methods of analysis, contribution to food quality, and health promotion. *Comprehensive Reviews in Food Science and Food Safety*, 16(3), 346-368.

Chadha, K.L. (2019). Hand Book of Horticulture. Second Edition, Volume 1, Indian Council of Agricultural Research, New Delhi. pp. 231-234.

Chen, C. Y., Lapsley, K., & Blumberg, J. (2006). A nutrition and health perspective on almonds. *Journal of the Science of Food and Agriculture*, 86(14), 2245-2250.

Esfahlan, A. J., Jamei, R., & Esfahlan, R. J. (2010). The importance of almond (*Prunus amygdalus* L.) and its by-products. *Food chemistry*, 120(2), 349-360.

Foster, G. D., Shantz, K. L., Vander Veur, S. S., Oliver, T. L., Lent, M. R., Virus, A. and Szapary, P. O. (2012). A randomized trial of the effects of an almond-enriched, hypocaloric diet in the treatment of obesity. The American Journal of Clinical Nutrition, 96(2):249-254.

Garcia-Perez, P., Xiao, J., Munekata, P. E., Lorenzo, J. M., Barba, F. J., Rajoka, M. S. R., and Simal-Gandara, J. (2021). Revalorization of almond by-products for the design of novel functional foods: An updated review. *Foods*, 10(8), 1823.

Imtiyaz A. Wani, Rayees A. Ahanger, Hilal A. Bhat, Abid A. Lone, Tauseef A. Bhat, Imtiyaz A. Malik and G.I. Hassan. Division of Fruit science SKUAST-K Shalimar, Srinagar-191121

Korać, R. R. and Khambholja, K. M. (2011). Potential of herbs in skin protection from ultraviolet radiation. Pharmacognosy Reviews, 5(10):164-173.

Ladizinsky, G. (1999). On the origin of almond. *Genetic Resources and Crop Evolution*, 46, 143-147.

Li, S., Chu, S., Lu, J., Wang, P., & Ma, M. (2018). Molecular and structural properties of three major protein components from almond kernel. *Journal of Food Processing and Preservation*, 42(3), e13536.

Martínez-Gómez, P., Sánchez-Pérez, R., Dicenta, F., Howad, W., Arús, P., & Gradziel, T. M. (2007). Almond. *Fruits and nuts*, 229-242.

Massantini, R., & Frangipane, M. T. (2022). Progress in almond quality and sensory assessment: An overview. *Agriculture*, 12(5), 710.

17. Mir, M. A., Shameem, R., Pandit, A. H., Bhat, K. M., Wani, W. M., Jan, A., and Hussain, S. (2016). Propagation of almond through budding under temperate conditions of north western Himalaya. *Applied Biological Research*, 18(3), 299-304.

Monteiro, A., & Cordeiro, V. (2001, May). Almond growing in Tras-os-Monte's region (Portugal). In *III International Symposium on Pistachios and Almonds* 591 (161-165).

Nahar, N., Shams-Ud-Din, M., & Ali, M. M. (2019). Management of navel orangeworm (Amyelois transitella Walker) infestation in almond (Prunus dulcis) orchards. *International Journal of Agriculture and Biology*, 21(2), 291-296.

Nanos, G. D., Kazantzis, I., Kefalas, P., Petrakis, C., & Stavroulakis, G. G. (2002). Irrigation and harvest time affect almond kernel quality and composition. *Scientia Horticulturae*, 96(1-4), 249-256.

Pascuzzi, S., & Santoro, F. (2017). Analysis of the almond harvesting and hulling mechanization process: A case study. *Agriculture, 7(*12), 100.

Prasad, A., Shukla, N., Kumar, V., & Singh, S. (2017). Integrated management of aphid, Brachycaudus amygdalinus (Homoptera: Aphididae) on almond (Prunus dulcis) in Jammu. Journal of Entomology and Zoology Studies, 5(4), 733-736.

Remón, J., Latorre-Viu, J., Matharu, A. S., Pinilla, J. L., & Suelves, I. (2021). Analysis and optimisation of a novel 'almond-refinery' concept: Simultaneous production of biofuels and value-added chemicals by hydrothermal treatment of almond hulls. *Science of The Total Environment*, 765, 142671.

Rijal, J. P., Joyce, A. L., & Gyawaly, S. (2021). Biology, ecology, and management of hemipteran pests in almond orchards in the United States. *Journal of Integrated Pest Management*, 12(1), 24.

Roncero, J. M., Álvarez-Ortí, M., Pardo-Giménez, A., Rabadán, A., & Pardo, J. E. (2020). Review about non-lipid components and minor fat-soluble bioactive compounds of almond kernel. *Foods*, 9(11), 1646.

Rubio, M., Martínez-Gómez, P., Dicenta, F., & Weber, W. E. (2003). Resistance of almond cultivars to Plum pox virus (sharka). *Plant Breeding*, 122(5), 462-464.

Sabaté, J., Fraser, G. E., Burke, K., Knutsen, S. F., Bennett, H. and Lindsted, K. D. (2003). Effects of walnuts on serum lipid levels and blood pressure in normal men. The New England Journal of Medicine, 328(9):603-607.

Siriwardhana, S. S., & Shahidi, F. (2002). Antiradical activity of extracts of almond and its by-products. *Journal of the American Oil Chemists' Society*, 79, 903-908.

Sisterson, M. S., Chen, J., Viveros, M. A., Civerolo, E. L., Ledbetter, C., & Groves, R. L. (2008). Effects of almond leaf scorch disease on almond yield: implications for management. *Plant Disease*, 92(3), 409-414.

Takei, Y., & Yamanishi, T. (1974). Flavor components of roasted almond. *Agricultural and Biological Chemistry*, 38(12), 2329-2336.

The Ministry of Agriculture & Farmers Welfare, formerly the Ministry of Agriculture, national horticulture board.s

Traynor, J. (2017). A history of almond pollination in California. *Bee World*, 94(3), 69-79.

Uematsu, C., Akihama, T., & Inokuma, C. (1987). Propagation of almond by shoot tip culture. *Japanese Journal of Breeding*, 37(4), 464-468.

Wijeratne, S. S., Abou-Zaid, M. M., & Shahidi, F. (2006). Antioxidant polyphenols in almond and its coproducts. *Journal of Agricultural and Food Chemistry*, 54(2), 312-318.

Zacheo, G., Cappello, M. S., Gallo, A., Santino, A., & Cappello, A. R. (2000). Changes associated with post-harvest ageing in almond seeds. *LWT-Food Science and Technology*, 33(6), 415-423.

8

Advances Production Technology of Walnut

Amit Kumar*[1] and Sajeel Ahamad[2], Satvaan Singh[3]
Khursheed Alam[4] and Imran Ali[5]

[1]Ph.D. Research Scholar, Department of Fruit Science,
[3]Department of Floriculture and Landscape Architecture
[4]Ph.D. Research Scholar, Department of Vegetable Science,
[5]M.Sc. Fruit Science, Department of Fruit Science College of Horticulture, Sardar Vallabhbhai Patel University of Agriculture and Technology, Meerut-250110, Uttar Pradesh, India.
[2]Ph.D. Research Scholar, Division of Food Science & Post-harvest Technology, Indian Agricultural Research Institute, New Delhi-110012, India
**Email:* amitworld701@gmail.com

Introduction

Walnut (*Juglans regia* L.) is an important temperate nut crop grown in various parts of the world. It belongs to the family Juglandaceae and is commonly known as English walnut, Persian walnut, or common walnut. It has a chromosome number of 2n=32 (Kalia *et al.*, 2016). Walnuts are a rich source of omega-3 fatty acids (Husain *et al.*, 2013). The walnut fruit is a drupe that contains a nut surrounded by a hard shell. The nut is enclosed in a fleshy green or brownish-yellow hull, which splits open when the fruit is ripe. Walnut is an important nut crop due to its high nutritional value and medicinal properties. The kernel of the walnut is rich in protein, fat, and minerals such as calcium, phosphorus, and magnesium. It also contains vitamins B1, B2, and E. In addition to being used as food, walnut has been used in traditional medicine for the treatment of various ailments such as diabetes, cardiovascular diseases, and cancer (Liu *et al.*, 2017). Walnut oil is rich in linoleic acid and is used in the production of cosmetics and soap. The leaves and bark of the tree are used in traditional medicine to treat various ailments such as diabetes, fever, and inflammation.

Fig. 1.1: Walnut

Importance and uses

Walnut is a valuable nut crop that is grown in several regions of the world, including India. The nut is a rich source of several nutrients, including protein, fiber, omega-3 fatty acids, and antioxidants. Walnuts have several health benefits, including reducing the risk of heart disease, improving brain function, and reducing inflammation. They are also used in several industries, including food, pharmaceuticals, and cosmetics.

Walnuts have been an important food source for humans for thousands of years. In ancient times, they were believed to have medicinal properties and were used to treat several ailments (Martinez *et al.*, 2010). Today, they are used in the food industry to produce confectionery, bakery products, and as an ingredient in several cuisines. They are also used in the pharmaceutical and cosmetic industries due to their anti-inflammatory and anti-aging properties (Alasalvar *et al.*, 2010; Bolling *et al.*, 2011).

Walnuts are a rich source of antioxidants, which are important for maintaining good health. Antioxidants help to neutralize free radicals in the body, which can cause cell damage and lead to several chronic diseases, including cancer and heart disease (Alasalvar *et al.*, 2010). In addition, walnuts are an excellent source of omega-3 fatty acids, which are important for brain health and reducing inflammation in the body (Bolling *et al.*, 2011). Walnuts also contain fiber, which is important for maintaining healthy digestion and reducing the risk of colon cancer (Martinez *et al.*, 2010).

Walnuts are also used in the cosmetic industry due to their anti-aging properties. The nut contains several compounds that help to protect the skin from damage caused by UV rays and reduce the appearance of wrinkles and fine lines (Alasalvar *et al.*, 2010). In addition, walnuts are used in the pharmaceutical industry to produce medicines that help to reduce inflammation and treat several chronic diseases, including arthritis and asthma (Bolling *et al.*, 2011).

Nutrient composition: According to the USDA National Nutrient Database, 100 grams of walnuts contain:

Table 1: Nutritional value per serving of 100 g

Contents	Amount with unit	Contents	Amount with unit
Energy	654 kcal	Zinc	3.09 mg
Protein	15.2 g	Vitamin C	1.3 mg
Fat	65.2 g	Thiamin	0.341 mg
Carbohydrates	13.7 g	Riboflavin	0.13 mg
Calcium	98 mg	Niacin	0.479 mg
Iron	2.91 mg	Vitamin B6	0.583 mg
Magnesium	158 mg	Folate	98 μg
Phosphorus	346 mg	Vitamin E	19.08 mg
Potassium	441 mg	Vitamin K	2.7 μg
Sodium	2 mg		

Source: USDA National Nutrient Database

Origin and Distribution

Walnut is native to the mountainous regions of Central Asia, including the areas now known as Kazakhstan, Kyrgyzstan, Uzbekistan, and Turkmenistan. It was introduced to Europe by the Greeks and Romans and eventually spread to other parts of the world. Today, walnut is grown in various parts of the world, including Europe, Asia, Africa, and the Americas. In India, walnut is mainly grown in the temperate regions of Jammu and Kashmir, Himachal Pradesh, and Uttarakhand. The walnut trees are usually found in hilly regions at an altitude of 1500-3000 meters above sea level. The crop is grown on a commercial scale in these regions and provides livelihoods to thousands of farmers.

Area and production

In India, walnut is mainly grown in the temperate regions of Jammu and Kashmir, Himachal Pradesh, and Uttarakhand. The crop is of great economic and ecological importance in these regions. The walnut industry provides employment to thousands of people, and it also helps in the conservation of the fragile ecosystem of the Himalayas. According to the National Horticulture Board (NHB), the area under walnut cultivation in India was 109 thousand hectares in 2020-21. The production of walnuts during the same period was around 281 thousand metric tonnes.

Botany

Walnut is a deciduous tree that can reach a height of up to 30 meters. The leaves are alternate, compound, and pinnate, with 5 to 9 leaflets. The flowers are monoecious, with the male flowers borne in catkins and the female flowers in spikes. The fruit is a large, round, and hard-shelled nut enclosed in a green, fleshy husk (Kalia *et al.*, 2016). The leaves of the walnut tree are alternate, compound, and pinnate. The leaflets are oval-shaped and serrated at the margins. The flowers of the walnut tree are monoecious, which means that both male and female flowers are produced on the same tree. The male flowers are yellow-green in color, while the female flowers are reddish-green. The male flowers are produced in hanging catkins, while the female flowers are produced in clusters at the end of the branches. The walnut nut is a large, single-seeded nut that is encased in a hard, woody shell. The fruit of the tree is a drupe that contains a nut surrounded by a hard shell. The walnut shell is divided into two halves, which are separated by a partition.

Soil and climatic requirements

Walnut trees require a deep, well-drained soil that is rich in organic matter and has a pH of 6.0-7.5. They are sensitive to waterlogging and therefore require a soil with good drainage. In addition, walnut trees are sensitive to salinity, and therefore the soil should have a low salt content. In India, the most suitable soils for walnut cultivation are loamy, alluvial soils with good drainage and moderate fertility. In terms of climate, walnut trees require a temperate climate with cool winters and mild summers. They are adapted to a wide range of temperatures, with the optimal temperature range for growth being between 18°C and 24°C. However, temperatures below -20°C can damage the trees, and therefore frost protection measures may be necessary in areas where temperatures regularly fall below this threshold.

The rainfall requirement for walnut trees is moderate, with an optimal range of 700-1000 mm per year. In India, the most suitable regions for walnut cultivation are those with a humid subtropical or a temperate climate, such as the regions of Jammu and Kashmir, Himachal Pradesh, and Uttarakhand. In these regions, the monsoon season provides the necessary moisture for walnut growth, and supplementary irrigation is generally not required. Apart from soil and climate, other factors that influence walnut production include nutrient management, pest and disease control, and pruning. Nutrient management is particularly important for ensuring optimal growth and yield, and studies have shown that the application of nitrogen, phosphorus, and potassium can significantly increase nut yield and quality (Gholami *et al.*, 2013).

Important commercial varieties

Most walnut plantations consist of seedling-origin plants, each being a unique genotype due to the crop's cross-pollination. The world's walnut varieties are generally developed through seedling selection. Numerous varieties, including exotic cultivars such as Lake English, Colby Blackmore, Turtle 16, Turtle 31, Nielson, Orth, Serr, and Natra Attiro, have been introduced or selected from indigenous material in India and are thriving under the conditions of Himachal Pradesh (Rathore, 1986).

In India, a few seedlings with high shelling percentages, better kernel quality, and soft shells have been identified and released for cultivation, such as Gobind, Roopa, Ratna, Pratap, and Sholding in Himachal Pradesh, and Hamdan, Sulaiman, CITH-Walnut-1, 2, 3, 4, 5, 6, 7, 8, 9, and 10 in J & K, as well as Charata No- 2, 4, 6, 13, and 14 in Uttarakhand (Rathore, 1986), which are showing promise. Commercial walnut cultivars with good qualitative traits are widely grown in countries like the USA, France, and Spain. Among these, only a few exotic cultivars, such as Opex Culchery, Tutle, Nugget, and Chenovo, have been introduced and planted in the field gene bank of CITH, Srinagar, to observe their comparative performance. The most commonly grown varieties of walnut in India are Chandler, Howard, Serr, Tulare, and Vina. These varieties are known for their good nut quality and high yield potential. Recently released varieties include Lara, Fernette, Ivanhoe, and Pedro.

India has several commercial walnut varieties with unique features that make them suitable for cultivation in different regions of the country. Here are some of the most important commercial varieties of walnut in India (Singh & Chauhan 2017; Singh & Bhandari 2006; Prasad & Singh 2004)

Pusa Khor

The Pusa Khor walnut is a variety developed by the Indian Agricultural Research Institute (IARI) in New Delhi. It is a late-maturing variety and is known for its good yield and quality. This variety is primarily grown in the northern Indian states of Jammu and Kashmir, Himachal Pradesh, and Uttarakhand.

Chandler: Chandler is a popular variety of walnut grown in India. It is a large-sized nut with a thin shell and high-quality kernel. Chandler has a high yield potential and is known for its excellent taste and quality. It is a late-maturing variety, and the nut ripens in mid to late September.

Howard: Howard is a medium to large-sized nut with a good-quality kernel. The tree is vigorous and has a spreading habit. The nuts are borne in clusters of 2-5. The tree is self-pollinating and requires less chilling hours. Howard is an early-maturing variety, and the nut ripens in early September.

Tulameen: Tulameen is a medium to large-sized nut with a good-quality kernel. The nut has a thin shell, making it easy to crack. The tree is productive and has an upright habit. The nuts are borne in clusters of 2-3. Tulameen is a late-maturing variety, and the nut ripens in mid to late September.

Lake: Lake is a medium to large-sized nut with a good-quality kernel. The nut has a thin shell, making it easy to crack. The tree is productive and has an upright habit. The nuts are borne in clusters of 2-3. Lake is an early-maturing variety, and the nut ripens in early September.

Pedro: Pedro is a medium to large-sized nut with a good-quality kernel. The nut has a thick shell, making it difficult to crack. The tree is productive and has an upright habit. The nuts are borne in clusters of 2-3. Pedro is a late-maturing variety, and the nut ripens in mid to late September.

Gobind: The Gobind walnut is a variety developed by the Indian Council of Agricultural Research (ICAR) in the 1980s. It is a mid-season variety and is known for its good yield and quality. This variety is primarily grown in the state of Himachal Pradesh.

Tehama: The Tehama walnut is a variety that is native to California, USA, but has been introduced to India. It is a mid-season variety and is known for its good yield and quality. This variety is primarily grown in the state of Jammu and Kashmir.

Chakrata: The Chakrata walnut is a variety that is native to the Chakrata region in the northern Indian state of Uttarakhand. It is a mid-season variety and is known for its good yield and quality.

Roopa: The Roopa walnut is a variety of walnuts that is commonly grown in the northern Indian state of Jammu and Kashmir. These walnuts are oval in shape and have a hard shell. They are known for their sweet and nutty flavor, which makes them popular for use in desserts and baked goods.

Karan: The Karan walnut is another variety of walnuts that is also grown in Jammu and Kashmir. These walnuts are smaller in size compared to the Roopa walnut and have a softer shell. They have a slightly bitter taste and are often used for making walnut oil.

CITH Walnut-1

CITH Walnut-1 is a promising variety suitable for both export and domestic markets. It has medium tree vigor, light kernel color, bold nuts weighing 28 grams, and large kernel size of 14.21 grams. It has good kernel recovery at 50.75% and light shell color with intermediate shell seal and strength. The nuts are trapezoidal in shape and easy to remove kernel halves.

CITH Walnut-2

CITH Walnut-2 is another promising genotype with vigorous plant growth, medium-sized nuts and kernels weighing 16.0 grams and 8.0 grams, respectively. It has soft shells with a shelling percentage of 50%, light kernel color, large ovate nuts with medium shell texture, and intermediate shell color, seal, and strength. The shell has complete integrity, and the kernel has satisfactory flavor and is well-filled and plumy with easy-to-remove halves.

CITH Walnut-3

CITH Walnut-3 has medium tree vigor and produces large nuts that are long trapezoidal in shape with rough shell texture, medium shell color, strong shell seal and strength, and complete shell integrity. The kernel is light in color with satisfactory flavor and well-filled and plumy, but difficult to remove halves. The nut weight is 19.3 grams, and kernel weight is 9.8 grams with a kernel recovery of 50.7%.

CITH Walnut-4

CITH Walnut-4 bears heavily with high tree vigor and produces large ovate nuts with rough shell texture, light shell color, strong shell seal, intermediate shell strength, complete shell integrity, and thin, satisfactory flavored kernels that are well-filled and moderately plumy with very easy-to-remove halves. The kernel is light in color, and the nut weighs 19.08 grams, while the kernel weighs 10.09 grams with a kernel recovery of 50.9%.

CITH Walnut-5

CITH Walnut-5 is a high-yield variety with strong tree vigour, making it suitable for export markets. It has extra-light kernel colour, bigger nuts (19.5 g) and kernels (10.16 g), good kernel recovery (52.1%), light shell colour, ovate shape, and moderate ease of removing the full kernel halves.

CITH Walnut-6

CITH Walnut-6 is a desirable variety that produces superior quality nuts and has intermediate tree vigour. It has a broad ovate shape, light shell colour, intermediate shell seal, weak shell strength, satisfactory kernel flavour, well-filled kernel, moderate plumy, and easy-to-remove kernel halves. It yields in-shell nuts weighing 24 g, kernels weighing 12.2 g, and kernel recovery of 50.8%.

CITH Walnut-7

CITH Walnut-7 has high tree vigour and produces large nuts with elliptic shape, rough shell texture, medium shell colour, strong shell seal, strong shell strength, complete shell integrity, satisfactory kernel flavour, well-filled kernel, plumy, easy-to-remove kernel halves, and light amber kernel colour. It yields in-shell nuts weighing 23.3 g, kernels weighing 11.7 g, and kernel recovery of 50.2%.

CITH Walnut-8

CITH Walnut-8 has intermediate tree vigour and light kernel colour. It produces nuts weighing 20.4 g, kernels weighing 11.01 g, good kernel recovery (54%), light shell colour, long trapezoidal shape, very easy-to-remove kernel halves, rough shell texture, strong shell seal, and strong shell strength.

CITH Walnut-9

CITH Walnut-9 has high tree vigour, medium-sized nuts with round shape, rough shell texture, light-coloured shell, intermediate shell seal, intermediate shell strength, satisfactory kernel flavour, well-filled kernel, plumy, and easy-to-remove kernel halves. It yields in-shell nuts weighing 21.2 g, kernels weighing 10.7 g, and kernel recovery of 50.5%.

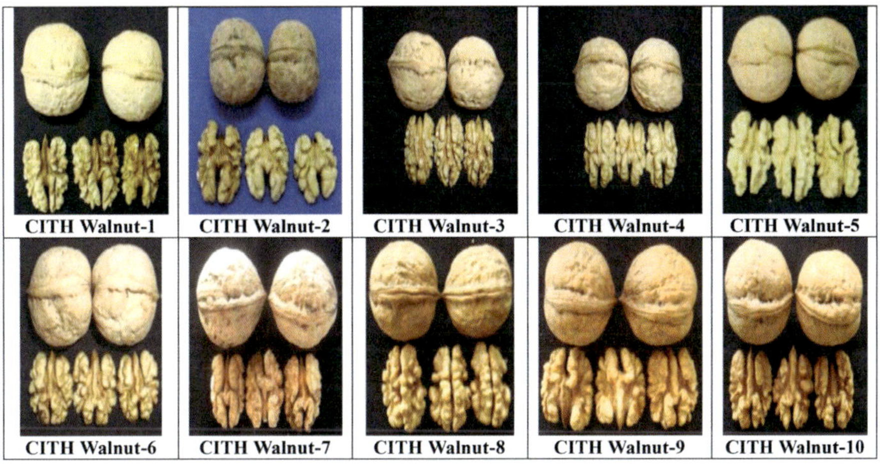

Fig. 1.2: Walnut Varieties

CITH Walnut-10

CITH Walnut-10 has high tree vigour, large size nuts with long trapezoidal shape, light colour, strong shell seal, intermediate shell strength, well-filled kernel, plumy, and moderate ease of removing the kernel halves. It yields in-shell nuts weighing 19.95 g, kernels weighing 11.08 g, and kernel recovery of 55.5%.

SR-11

It is a variety of walnut that has been developed and released by the Yashwant Singh Parmar University of Horticulture and Forestry (YSPU & F), Solan in India. It is a high-yielding variety that is known for its large-sized nuts with good kernel quality. The tree is also resistant to various diseases and pests. SR-11 has become a popular variety among walnut growers in India due to its desirable characteristics and is widely cultivated in the region.

Hamdan: Hamdan is a precocious and dwarfing variety with smooth shell, larger than the width of the nut. It produces nuts weighing 14 g and has a shelling percentage of 54 with good quality kernels. It has been released by SKUAST-K.

Sulaiman: Sulaiman is a precocious and semi-dwarfing variety with an average nut weight of 20 g, smooth shell surface, and somewhat round nut shape. It has a shelling percentage of 50 with good kernel characters and has been released by SKUAST-K.

Shimla Delicious

Shimla Delicious is a popular variety of walnut that is grown in the Indian state of Himachal Pradesh. It is known for its large and thin-shelled nuts with a good taste. The tree is vigorous and bears fruit in the late season, typically in the months of October and November. The variety is well adapted to the hilly terrain and cool climate of the region and is grown extensively in the districts of Shimla, Kinnaur, and Kullu.

The promising exotic walnut cultivars have distinct characteristics (Rathore DS 1986).

Hartley: Hartley is a medium to large-sized tree that is the most popular cultivar in California. The kernels are light in color and the tree produces a good yield. It is tolerant to codling moth and blight. The nuts have a broad, flat base and a pointed tip. The shell is light in color, thin, and well-sealed.

Payne: Payne is a leading cultivar in California that produces a heavy yield due to the fruitfulness of its lateral buds. The tree is round, vigorous, and requires heavy pruning to maintain its vigor. However, it is susceptible to codling moth and blight.

Ashley: Ashley is a cultivar from the breeding program of the pomology department at the University of California, Davis. Approximately 90 percent of its lateral buds produce female (pistillate) flowers, making it potentially nut-bearing. The tree matures its nuts relatively early because it comes into leaf early, which means it is not suitable for areas with heavy spring frosts. It is also extremely sensitive to walnut blight (*Xanthomonas juglandii*) if it is rainy in spring and is, therefore, not ideal for humid areas. The kernel weight averages 5.3 g, and the recovery average is 50 percent.

Chico: Chico is a cultivar that has many desirable features for home gardeners. It is upright, small, and 90 percent of its lateral buds are potentially nut-bearing, making it possible to prune without losing the crop. It is one of the heaviest-bearing varieties available, and the nuts mature early in the season. However, like many early leafing Californian selections, it is extremely sensitive to walnut blight if it is rainy in spring, and therefore not suitable for humid areas. Chico comes into leaf early, so it is not suitable for areas with heavy spring frosts. The nuts are small, with a kernel average of 5 grams, and the crack-out is 47 percent. But the quality of the kernels is good. Chico is pollenized by Payne, Serr, Vina, and Sunland.

Eureka: Eureka is a very large and spreading tree that is a 100 percent terminal bearer, so it is not ideal for home gardens.

Franquette: Franquette is an old French walnut variety that leafs out late, making it a good choice for regions with late spring frosts. However, its late flowering habit means that it often misses out on other varieties' pollen shedding, leading to poor nut set and yield. To improve pollination, Mayrick or Rex, both late flowering varieties, can be used as a pollenizer. While most Californian walnut varieties are susceptible to walnut blight and not suitable for humid areas, Franquette appears to have some blight resistance. On the downside, Franquette is a vigorous, slow-to-bear tree that spreads widely and produces nuts only at the tips of its branches. Nevertheless, its large and attractive nuts have an average kernel weight of 5.2 grams, with a crack-out rate of around 46%, although this may be lower in certain conditions. In addition, Franquette is reputed to produce high-quality timber.

Serr: Serr is a cultivar of walnut that was introduced in the late 1960s in California. This vigorous tree requires light pruning every year and produces large, thin but strong-shelled nuts on both terminal and lateral buds. It has a fair crop load and is an early-leafing variety that shows some resistance to bacterial blight. However, Serr can be difficult to control on very fertile, well-drained soils due to its excessive growth rate. It starts bearing fruit quickly after planting and has an average 50-50% recovery with larger nuts weighing about 7.6 grams. It requires pollination from Chico and Tehama varieties.

Propagation and Rootstocks

In India, walnut propagation is commonly done through seed, grafting, budding, and tissue culture techniques (Gajbhiye & Sharma, 2015). Seed propagation, while common, results in high variability in growth, yield, and nut quality. Grafting and budding techniques ensure uniformity in growth and fruiting, with grafting being the more preferred technique due to its higher efficiency and success rate. The choice of rootstock is crucial to walnut production, with different rootstocks used based on soil type, climate, and scion cultivar (Jain & Kumar, 2016). In India, rootstocks such as *Juglans regia*, Paradox, and Northern California Black are commonly used. Recent advances in biotechnology have led to the development of tissue culture techniques for walnut propagation, offering advantages such as disease-free plants, uniform growth, and rapid multiplication (Singh *et al.*, 2017). However, tissue-cultured plants may exhibit reduced vigor and delayed maturity compared to conventionally propagated plants. The most commonly used rootstocks in walnut are *Juglans nigra*, *J. regia* x *J. nigra* hybrids, and *J. hindsii*. Studies have focused on developing new rootstocks that can improve walnut production, such as the hybrid rootstock (*J. regia* x *J. nigra*) that exhibited better growth, yield, and nut quality than conventional rootstocks in India (Kumar *et al.*, 2018) and new rootstock (RX1) that showed better resistance to soil-borne diseases and improved yield compared to commonly used rootstocks in California (Martínez-Gómez *et al.*, 2016).

Planting and orchard establishment

The best time for planting walnut saplings is during the winter months, from December to February. During this time, the soil is cool and moist, which is ideal for the establishment of new trees. The pit size for planting walnut saplings should be at least 60 cm deep and 60 cm wide, and filled with a mixture of topsoil, compost, and well-decomposed cow dung. The planting depth should be such that the graft union is at least 5 cm above the soil level. After planting, the sapling should be watered thoroughly, and a 2-3 inch layer of mulch should be applied around the base of the tree to conserve soil moisture and control weeds.

In terms of spacing, the ideal spacing for walnut trees is 10-12 m between rows and 6-8 m within rows, resulting in a plant population of about 80-120 trees per hectare. However, the spacing can vary depending on the soil fertility, irrigation facilities, and training and pruning methods used. It is also important to maintain adequate space between the trees for proper air circulation and sunlight penetration, which can help reduce the incidence of pests and diseases.

After planting, the first year is critical for the survival and growth of the new walnut trees. Proper care and management practices must be followed to ensure the establishment of healthy and vigorous trees. One of the most important practices during the first year is irrigation, as young trees require consistent moisture to establish a strong root system. Drip irrigation is the most efficient method for water application in walnut orchards. Orchard establishment involves various activities such as site selection, land preparation, planting, irrigation, fertilization, and pest and disease management. In recent years, there have been significant advances in technology for orchard establishment, including the use of precision farming techniques, such as soil and plant analysis, drip irrigation, and micro-nutrient application, which can significantly improve the productivity and profitability of walnut orchards (Singh *et al.*, 2020).

Flowering, pollination, and fruit set

Walnut trees usually flower in the months of April and May. Walnut trees are monoecious, with separate male and female flowers borne on the same tree. The male flowers, known as catkins, appear in early spring, while the female flowers, which are located in the current season's growth, appear later in the season. The pollination period for walnut is usually in late April to early May in India (Rahman, 2017). The walnut tree is a cross-pollinated crop, which means that it requires pollen from another tree to produce viable seeds. Pollination in walnuts is anemophilous, which means that it is mainly carried out by wind. However, the efficiency of wind pollination is low, and bee pollination has been found to be more effective in increasing fruit set and yield. The female flowers are receptive to pollen for about 3-4 days after their emergence. The fruit set in walnut is usually high, and the crop load needs to be managed carefully to ensure optimum fruit size and quality. Various studies have reported that bee pollination can increase the number of fruits per tree, reduce the number of blank nuts, and improve nut quality parameters such as kernel weight and oil content. In India, honeybees such as *Apis mellifera* and *Apis cerana* are the most commonly used for pollination, and their effectiveness has been confirmed by several studies. For example, a study conducted by Singh *et al.*(2016) found that bee pollination increased fruit set by 24% compared to wind pollination.

Apart from bee pollination, other factors that influence fruit set in walnuts include temperature, humidity, and nutrition. Studies have shown that a warm and dry climate during flowering and early fruit set can increase the number of fruits per tree. In addition, adequate nutrition, particularly the availability of zinc and boron, is important for optimal fruit set and nut development (Chandra *et al.*, 2012).

Fig. 1.3: Walnut inflorescence, male and female flower

Use of growth regulators

The use of growth regulators is one of the important advances in the production technology of walnuts. Growth regulators are synthetic chemicals that regulate plant growth and development by modifying the activity of plant hormones. They can be used to promote or inhibit plant growth, depending on the type of regulator and its concentration. In walnuts, growth regulators are used to improve fruit quality, increase yield, and enhance stress tolerance.

One of the most commonly used growth regulators in walnut production is gibberellic acid (GA3). GA3 is used to promote shoot growth, increase fruit size, and improve fruit quality. For example, in a study conducted by Sharma *et al.*(2019), the application of GA3 at 40 ppm resulted in a significant increase in the number of nuts per cluster and kernel weight in walnuts. Another study by Akram *et al.*(2021) reported that the application of GA3 at 20 ppm increased the yield and quality of walnuts by improving nut size, kernel weight, and kernel percentage. Apart from GA3, other growth regulators such as cytokinins, auxins, and abscisic acid (ABA) have also been used in walnut production. Cytokinins are used to promote cell division and shoot growth, while auxins are used to stimulate root growth and development. ABA is used to regulate plant responses to stress, such as drought and salinity.

Auxins are plant hormones that promote cell elongation and differentiation, and are involved in many aspects of plant growth and development, including root formation, shoot growth, and fruit development. In walnut, the application of auxins can be used to promote root growth, improve nutrient uptake, and increase fruit size. The application of IAA (indole-3-acetic acid) has been found to improve root growth and nutrient uptake in young walnut trees (Nayyar *et al.*, 2016). However, excessive use of auxins can lead to excessive vegetative growth and reduced nut yield.

Ethylene is a plant hormone that is involved in many processes, including fruit ripening, senescence, and abscission. In walnut, the application of ethylene can be used to promote early fruit ripening, improve nut quality, and reduce pre-harvest fruit drop. The application of ethephon, a synthetic ethylene-releasing compound, has been found to improve nut quality by increasing kernel percentage and reducing shell thickness (Ghasemi *et al.*, 2018). However, excessive use of ethylene can lead to premature fruit drop and reduced yield.

The use of growth regulators is an important advance in the production technology of walnuts. Gibberellic acid is the most commonly used growth regulator in walnut production, with a concentration of 40 ppm being effective in improving yield and quality. Other growth regulators such as cytokinins, auxins, and ABA have also been used to enhance the growth and stress tolerance of walnuts. Further research is needed to optimize the application of growth regulators in walnut production in India.

Training and Pruning

Training and pruning are critical management practices for walnut orchards, as they help shape the tree into a strong, well-branched framework that will support the fruit load in later years. Training and pruning should begin during the first year after planting, and continue throughout the life of the tree.

The first step in training a walnut tree is to select a central leader, which will be the main trunk of the tree. The central leader should be straight and free of defects and should be allowed to grow to its full height without interference. Scaffold branches should be selected from the main trunk to provide support for the fruit load and should be spaced evenly around the tree to ensure good light penetration and air circulation. After the first year, walnut trees are commonly trained using the "Modified Central Leader System".

Pruning is necessary to remove any damaged, diseased, or crossing branches, as well as to maintain the shape and size of the tree. Pruning should be done during the dormant season, from December to February, to avoid interfering with the tree's growth and development during the growing season. The goal of pruning is to maintain a balance between vegetative growth and fruit production, while also ensuring that the tree is strong and healthy (Kumar *et al.*2017).

In addition to traditional pruning methods, new technologies such as mechanical pruning and hedging have been developed to improve the efficiency of pruning in walnut orchards. Mechanical pruning involves the use of machinery to prune the trees, which can be faster and more efficient than manual pruning. Hedging involves cutting the trees back to a certain height to promote branching, and can also be used to control the size of the tree (Sharma *et al.*2020).

Nutrition Management

Nutrition management for walnut orchards involves the application of fertilizers to provide the necessary nutrients for optimal growth and development of the trees.

In India, urea, diammonium phosphate (DAP), and potassium chloride (KCl) are commonly used fertilizers for walnut orchards (Kumar *et al.*, 2017). The recommended application rates of these fertilizers depend on the age of the trees, soil fertility status, and yield expectations. For young trees, the recommended application rate is 100-150 g urea, 150-200 g DAP, and 100-150 g KCl per tree per year. For mature trees, the recommended application rate is 400-500 g urea, 600-700 g DAP, and 300-400 g KCl per tree per year (Kumar *et al.*, 2017). These fertilizers should be applied in two to three equal splits during the growing season, starting in late winter or early spring and continuing through the end of summer.

In addition to traditional fertilizers, organic fertilizers such as farmyard manure (FYM) and vermicompost can also be used to improve soil fertility and nutrient availability (Singh & Kaur, 2018). The recommended application rates for FYM and vermicompost are 25-30 kg and 5-6 kg per tree per year, respectively (Kumar *et al.*, 2017).

Micronutrients are essential for the growth and development of walnut trees, and their deficiency can lead to reduced productivity and poor quality nuts. Among the important micronutrients for walnut cultivation in India are calcium, magnesium, iron, zinc, and boron. Calcium is required for the development of strong cell walls and helps in the movement of nutrients within the plant. In walnut cultivation, calcium can be applied in the form of calcium nitrate or calcium chloride (Kumar *et al.*, 2017). Magnesium is a component of chlorophyll, the green pigment that plays a crucial role in photosynthesis. Magnesium deficiency can lead to yellowing of leaves and poor growth. In walnut orchards, magnesium can be applied as magnesium sulfate or magnesium oxide (Kumar *et al.*, 2017). Iron is essential for the synthesis of chlorophyll and other important compounds in plants. Iron deficiency can lead to yellowing of leaves and stunted growth. In walnut cultivation, iron can be applied as iron sulfate or chelated iron (Kumar *et al.*, 2017).

Zinc is important for enzyme activity and plays a crucial role in the development of leaves, stems, and fruits. Zinc deficiency can lead to reduced leaf size, leaf chlorosis, and poor fruit development. In walnut orchards, zinc can be applied as zinc sulfate or zinc chelate (Kumar *et al.*, 2017). Boron is required for cell wall formation, pollination, and seed development. Boron deficiency can lead to hollow nuts and poor yields. In walnut cultivation, boron can be applied as borax (Sharma *et al.*, 2020).

It is important to note that excessive use of fertilizers can lead to nutrient imbalances and environmental pollution. Therefore, it is recommended to conduct regular soil and leaf analysis to determine the nutrient status of the trees and adjust the fertilizer application rates accordingly (Kumar *et al.*, 2017). Proper nutrition management practices, including the use of both traditional and organic fertilizers, are essential for the success of a walnut orchard in India (Singh & Kaur, 2018). These practices can help ensure that the trees are healthy and productive, leading to higher yields and better quality nuts.

Water Management

Water management is a crucial component of walnut production in India, particularly in areas with limited water resources. Proper irrigation practices are necessary to maintain healthy trees and ensure optimal nut production.

In general, walnuts require moderate to high amounts of water, with the highest water demand occurring during the period of rapid shoot growth and nut development. Water stress during these critical periods can result in reduced yield and poor nut quality. Therefore, it is important to provide adequate water to the trees during these stages.

The most common irrigation methods used in walnut orchards in India are surface irrigation, sprinkler irrigation, and drip irrigation. Surface irrigation is the most traditional method and involves flooding the field with water. Sprinkler irrigation involves applying water through overhead sprinklers, and drip irrigation involves providing water directly to the root zone of the tree through a drip system.

Drip irrigation is becoming increasingly popular in India due to its efficiency in water use and potential for higher yields compared to other methods. Research has shown that drip irrigation can result in water savings of up to 60% while maintaining or increasing nut yield and quality (Saini *et al.*, 2020).

In addition to irrigation, other water management practices such as mulching and soil management can also help conserve water and improve soil moisture retention. Mulching involves covering the soil around the tree with a layer of organic material such as straw, leaves, or wood chips, which can reduce water loss from the soil and improve soil moisture retention. Proper soil management practices such as tillage, soil amendments, and cover cropping can also improve soil water retention and reduce water loss through evaporation. Overall, proper water management practices are essential for the success of a walnut orchard in India. By providing adequate water to the trees through efficient irrigation methods and implementing other water conservation practices, growers can maintain healthy trees and optimize nut production.

Intercultural operation

Intercultural operations such as weed control, mulching, and soil cultivation are essential for maintaining good orchard health and productivity. Weed control is important for minimizing competition for water and nutrients and for reducing pest and disease incidence. Mulching can help in conserving soil moisture, regulating soil temperature, and suppressing weed growth. Soil cultivation is important for maintaining good soil structure, improving aeration and water infiltration, and controlling soil-borne diseases.

Orchard floor management

Orchard floor management plays a crucial role in the overall health and productivity of walnut trees. Proper management practices help control weeds, conserve soil moisture, prevent erosion, and create a favorable environment for walnut tree growth.

Weed management

Weed management is essential for maintaining good orchard health and productivity. Weeds compete with the trees for water and nutrients and can also harbor pests and diseases. Some of the commonly used methods of weed control in walnut orchards include manual weeding, herbicide application, and mulching. Herbicides such as Glyphosate, Oxyfluorfen, and Pendimethalin are recommended for pre-emergence and post-emergence weed control in walnut cultivation (Kaur *et al.*, 2018). The application rates and timings may vary depending on the specific herbicide and weed species present in the orchard. It is important to follow the manufacturer's instructions and adhere to recommended safety precautions when using herbicides.

Mulching

Mulching is an effective technique for weed suppression, moisture conservation, and soil temperature moderation. Organic mulches such as straw, wood chips, or compost can be applied around the base of walnut trees to create a weed-free zone. Mulches also improve soil structure, nutrient availability, and microbial activity, contributing to the overall health of walnut trees. Mulch should be applied in a layer of about 5-10 cm thick, avoiding direct contact with the tree trunk to prevent rotting.

Cover cropping

Cover cropping involves planting specific plant species or mixes of plants between walnut tree rows to provide additional benefits such as weed suppression, soil erosion control, and nitrogen fixation. Leguminous cover crops like clover or vetch help improve soil fertility by fixing atmospheric nitrogen, reducing the need for synthetic nitrogen fertilizers (Rashid *et al.*, 2020). The choice of cover crop species should be based on the specific requirements of the walnut orchard and local climatic conditions.

Top working

The most recommended method for establishing a walnut orchard is to plant the seedlings directly in the production field without grafting in a nursery, and then top-work the trees two years later. To prevent sap bleeding during the top-working process, it is advisable to either head back the stock two weeks before the operation or make a few slanting cuts through the bark into the wood just below the grafting point. This way, any bleeding that occurs later will happen through these cuts.

For the top-working process, it is important to select disease-free young trees and to cut the dormant scion wood from the parent tree ahead of time. The scion should be waxed, packed appropriately, and stored in a refrigerator at 4°C until needed. Bark grafting is recommended for top-working in late spring (end of March to April) or when new growth has already taken place. The graft union should be wrapped with a thick plastic sheet, and the stem portion should be painted with whitewash. Additionally, it is essential to remove any shoots other than the scion growing on the plant (Singh & Singh 2014).

Physiological disorders and their management

Physiological disorders are non-infectious diseases caused by unfavorable environmental conditions, nutritional imbalances, or cultural practices. Physiological disorders can have a significant impact on the yield and quality of the walnut crop. In this section, we will discuss the major physiological disorders that affect walnut production in India and their management.

Sunburn: Sunburn is a physiological disorder that occurs due to exposure to excessive sunlight. The symptoms of sunburn include discoloration and scorching of leaves, bark, and fruits. Sunburn can lead to reduced fruit size and quality, and in severe cases, it can cause fruit drop. To manage sunburn, trees should be planted in areas with partial shade. The use of reflective mulches, such as white plastic or aluminum foil, can also help to reduce the incidence of sunburn.

Brown line: Brownline is a physiological disorder that affects the bark of walnut trees. It is characterized by the formation of a brown line on the trunk, which can eventually girdle the tree and cause death. Brownline is caused by excessive exposure to sunlight and high temperatures. To manage brownline, trees should be planted in areas with partial shade, and the use of reflective mulches can be helpful. The application of a whitewash solution to the trunk of the tree can also help to reduce the incidence of brownline.

Tip-burn: Tipburn is a physiological disorder that affects the leaves of walnut trees. It is characterized by the browning and death of the tips of the leaves. Tipburn is caused by water stress, nutrient deficiencies, and excessive exposure to sunlight. To manage tipburn, trees should be irrigated regularly and fertilized with a balanced fertilizer. The use of reflective mulches and shading can also be helpful.

Nut drop: Nut drop is a physiological disorder that occurs when the nuts fail to develop properly and drop from the tree prematurely. Nut drop can be caused by a lack of pollination, water stress, nutrient deficiencies, and high temperatures. To manage nut drop, trees should be irrigated regularly and fertilized with a balanced fertilizer. The use of growth regulators, such as gibberellic acid, can also be helpful in reducing the incidence of nut drop.

Oil spot: Oil spot is a physiological disorder that affects the fruits of walnut trees. It is characterized by the formation of dark spots on the surface of the fruit. Oil spot is caused by the accumulation of tannins and other compounds in the fruit. To manage oil spot, trees should be irrigated regularly and fertilized with a balanced fertilizer. The use of growth regulators, such as gibberellic acid, can also be helpful in reducing the incidence of oil spot.

Blackline: Blackline is a physiological disorder that affects the shoots of walnut trees. It is characterized by the formation of a black line on the stem, which can eventually girdle the shoot and cause death. Blackline is caused by water stress and high temperatures. To manage blackline, trees should be irrigated regularly and planted in areas with good drainage. The use of reflective mulches and shading can also be helpful.

Physiological disorders can have a significant impact on the yield and quality of the walnut crop. The management of physiological disorders requires a holistic approach that considers the environmental, nutritional, and cultural factors that contribute to their development. Regular irrigation, balanced fertilization, and the use of growth regulators and reflective mulches can all help to reduce the incidence of physiological disorders in walnut trees.

Major insect-pests and their management

Walnut is susceptible to various insect pests, which can significantly reduce yield and quality of the nut. The damage caused by these pests may also make the nuts more susceptible to diseases. Here are some of the major insect pests of walnut in India and their management practices.

Codling Moth (*Cydia pomonella*)

Codling moth is one of the most destructive pests of walnut. The larvae of this moth tunnel into the nut and feed on the kernel, making the nut unmarketable. The damage caused by this pest is often difficult to detect until the nut is opened.

Management: To control codling moth, several insecticides can be used, including synthetic pyrethroids, organophosphates, and spinosad. However, the use of chemical insecticides should be done judiciously and only when necessary to avoid the development of resistance. Non-chemical management practices such as the use of pheromone traps to disrupt the mating of the moths can also be effective in reducing the population of this pest.

Walnut Husk Fly (*Rhagoletis completa*)

The walnut husk fly is a serious pest that infests the nut during the early stage of development. The female fly lays eggs on the developing nut, and the larvae tunnel into the husk and feed on the nut. The damage caused by this pest leads to premature nut drop and low-quality nuts.

Management: Several control measures can be employed to manage the walnut husk fly, including cultural practices such as timely harvesting and removal of dropped nuts, and the use of insecticides. The use of pheromone traps can also be effective in monitoring the population of the pest.

Walnut Aphid (*Chromaphis juglandicola*)

Walnut aphid is a common pest of walnut in India. The aphids suck the sap from the leaves and cause them to curl, turn yellow, and eventually drop off. The infestation of this pest can reduce the growth and yield of the tree.

Management: To control walnut aphids, various insecticides can be used, including neem oil, imidacloprid, and acetamiprid. The use of beneficial insects such as lady beetles can also be effective in reducing the population of the aphids.

Walnut Blister Mite (*Eriophyes erinea*)

Walnut blister mite is a tiny pest that feeds on the upper surface of the leaves, causing blister-like galls to form. The damage caused by this pest can reduce the growth and yield of the tree.

Management: To control walnut blister mites, various miticides such as abamectin, fenpyroximate, and spirodiclofen can be used. The use of dormant oil spray during the winter months can also help in reducing the population of the mite.

Walnut Twig Beetle (*Pityophthorus juglandis*)

Walnut twig beetle is a serious pest that is responsible for the spread of Thousand Cankers Disease (TCD). The larvae of this beetle tunnel into the twigs and branches of the tree, creating small cankers that can eventually kill the tree.

Management: To control walnut twig beetle, the use of insecticides such as bifenthrin and carbaryl can be effective. The removal of infested twigs and branches can also help in reducing the population of the beetle and the spread of TCD.

Insect pests can cause significant damage to walnut trees and reduce yield and quality of the nut. Integrated pest management (IPM) practices that combine cultural, physical, and chemical methods can be effective in controlling these pests. Careful monitoring of the population of insect pests and judicious use of insecticides.

Major Diseases and Their Management

Walnut trees are susceptible to various fungal, bacterial, and viral diseases that can significantly affect the yield and quality of the crop. Here are some of the major diseases of walnut in India and their management strategies:

Walnut blight

Walnut blight is a bacterial disease caused by *Xanthomonas arboricola* pv. *juglandis*. It primarily affects the leaves, shoots, and nuts of the walnut tree. The symptoms of the disease include water-soaked lesions on the leaves and shoots, which turn brown and necrotic. The nuts may also develop black spots or become shriveled and distorted.

Management: The best way to manage walnut blight is through the use of integrated disease management strategies. This includes the use of disease-free planting material, proper sanitation, and cultural practices such as pruning and thinning to improve air circulation and reduce humidity in the orchard. Chemical control measures such as copper-based fungicides can also be used, but it is important to follow the recommended application rates and timing.

Walnut anthracnose

Walnut anthracnose is a fungal disease caused by *Gnomonia leptostyla*. It primarily affects the leaves and shoots of the walnut tree. The symptoms of the disease include circular or irregular-shaped lesions on the leaves that eventually turn brown and necrotic. The disease can also cause defoliation and shoot dieback.

Management: The management of walnut anthracnose involves the use of a combination of cultural and chemical control measures. Cultural practices such as pruning to improve air circulation, removing infected plant debris, and avoiding overhead irrigation can help reduce the spread of the disease. Chemical control measures such as copper-based fungicides can also be used, but it is important to follow the recommended application rates and timing.

Root rot

Root rot is a fungal disease caused by various soil-borne pathogens such as *Phytophthora* spp. and Armillaria spp. It primarily affects the roots of the walnut tree, causing them to rot and die. The symptoms of the disease include yellowing and wilting of the leaves, stunted growth, and eventual tree death.

Management: The management of root rot involves the use of cultural practices such as planting disease-resistant rootstocks, avoiding waterlogged or poorly-drained soils, and maintaining proper soil pH and fertility. Fungicides can also be used to control the disease, but their effectiveness may vary depending on the severity of the infection.

Walnut husk fly

The walnut husk fly is a major insect pest that affects the quality and marketability of walnut crops. The fly lays its eggs in the developing walnut husk, which then causes the husk to blacken and become brittle. The infested nuts are also prone to rot and mold, reducing their quality and yield.

Management: The management of walnut husk fly involves the use of cultural and chemical control measures. Cultural practices such as the removal and destruction of infested husks, the use of insect traps, and the timing of irrigation and harvest can help reduce the population of the fly. Chemical control measures such as insecticides can also be used, but it is important to follow the recommended application rates and timing.

Walnut scale

Walnut scale is a common insect pest that affects the leaves and branches of the walnut tree. The scale insect feeds on the sap of the tree, causing yellowing and wilting of the leaves, stunted growth, and eventual tree death.

Management: The management of walnut scale involves the use of cultural and chemical control measures. Cultural practices such as pruning to improve air circulation, removing infected plant debris, and avoiding over-fertilization can help reduce the spread of the insect. Chemical control measures such as insecticides can also be used.

Maturity Indices

Maturity indices are important for determining the optimal time of harvest in walnut. Several parameters are used to determine the maturity of walnuts, including physical and chemical characteristics. The physical characteristics of the walnut include the kernel-to-shell ratio, the shell hardness, and the color of the kernel. The chemical characteristics include the moisture content, oil content, and fatty acid composition of the kernel (Kester *et al.*, 2011; Ferguson, 2015; Ramos *et al.*, 2015).

Physical maturity indices

Kernel-to-shell ratio: The kernel-to-shell ratio is a useful index of maturity. It is calculated by weighing the kernels and the shells separately and then dividing the weight of the kernels by the weight of the shells. A high kernel-to-shell ratio indicates that the nut is mature, while a low ratio indicates that the nut is immature.

Shell hardness: The shell hardness of the walnut can also be used as a maturity index. Immature nuts have a soft shell that is easily punctured, while mature nuts have a hard shell that is difficult to crack.

Color of the kernel: The color of the kernel is another physical characteristic that can be used as a maturity index. Mature kernels have a darker color than immature kernels.

Chemical maturity indices

Moisture content: The moisture content of the kernel is an important parameter that determines the quality of the walnut. The moisture content of the kernel decreases as the nut matures. A moisture content of 8-10% is considered optimal for storage.

Oil content: The oil content of the kernel is another important parameter that determines the quality of the walnut. The oil content of the kernel increases as the nut matures. A high oil content is desirable for nut quality.

Fatty acid composition: The fatty acid composition of the kernel changes as the nut matures. Immature nuts have a higher proportion of linoleic acid, while mature nuts have a higher proportion of oleic acid. A high proportion of oleic acid is desirable for nut quality.

Determining the optimal time of harvest in walnut is crucial for maximizing yield and nut quality. The physical and chemical maturity indices can be used to determine the optimal time of harvest. Generally, walnuts are harvested when the kernel-to-shell ratio reaches 50-55%, the shell hardness is at its maximum, and the kernel has reached its maximum size. However, the optimal time of harvest can vary depending on the variety, climatic conditions, and growing region.

Harvesting and post-harvest management

Harvesting and post-harvest management play a crucial role in maintaining the quality and shelf life of walnut fruits (Adib-Rowhani & Martinez-Gomez, 2019). Inappropriate harvesting and post-harvest handling can result in quality deterioration and a decrease in market value. Therefore, it is necessary to follow proper harvesting and post-harvest management practices to ensure maximum yield and quality of the fruits.

Harvesting

Walnut fruits are ready for harvesting when the husk begins to split and turn brown. The best way to determine the maturity of walnut fruits is by performing the hull-scrape test. In this test, the husk is scraped with a fingernail, and if it separates easily from the shell, it indicates that the fruit is mature and ready for harvest.

Walnut fruits should be harvested as soon as they are mature to avoid husk and kernel damage due to weathering or pest infestation. The fruits can be harvested by shaking the tree or by using mechanical harvesters. Care should be taken not to damage the fruits during the harvesting process.

Post-harvest management

After harvesting, the fruits should be immediately transported to the processing unit. The first step in post-harvest management is cleaning the fruits to remove any debris or dirt. The fruits should be sorted based on their size and quality, and damaged or diseased fruits should be discarded. After sorting, the fruits can be dried using various methods such as sun-drying, mechanical drying, or a combination of both. Sun-drying is the traditional method of drying walnuts, but it is a time-consuming process and depends on weather conditions. Mechanical drying is a faster and more efficient method, but it requires specialized equipment. After drying, the fruits should be stored in a cool and dry place to avoid spoilage. The ideal temperature for storing walnuts is between 0-4°C, and the relative humidity should be around 70-75%. Proper storage conditions can extend the shelf life of the fruits and maintain their quality (Singh 2020).

Grading and packing

Grading is an essential step in post-harvest management to ensure that only high-quality fruits are sold in the market. The fruits are graded based on their size, color, and quality, and are packed accordingly. The packing material should be of good quality to avoid damage during transportation. In India, the most common method of packing walnuts is in wooden crates or cartons. The fruits should be packed carefully to avoid bruising or damage during transportation. Properly packed walnuts have a longer shelf life and can fetch a higher market price.

Yield

Yield is an important aspect of any crop production, and walnut is no exception. It is a perennial crop with a long lifespan of more than 100 years. The yield of walnut trees increases gradually with age and reaches its maximum after 20-25 years of age. The yield also depends on various factors such as climate, soil type, irrigation, nutrient management, pest and disease control, pruning, and harvesting techniques. In this section, we will discuss the factors affecting yield and the methods to increase it.

Factors affecting yield

- Climate: Walnut trees require a specific climate to grow and produce high yields. A cool and humid climate during the growing season is essential for optimum growth and yield. Extreme weather conditions such as frost, hailstorms, and heatwaves can affect the yield.
- Soil type: Walnut trees grow well in well-drained soils with a pH range of 6.0 to 7.5. They prefer deep and fertile soils that are rich in organic matter. The soil should be able to retain moisture, but not waterlogged. The presence of hardpan, shallow soils, or poor drainage can affect the yield.
- Irrigation: Irrigation is crucial for the growth and development of walnut trees. Adequate and timely irrigation can increase yield and improve nut quality. However, overwatering or improper irrigation can lead to waterlogging and root rot, which can adversely affect yield.
- Nutrient management: Walnut trees require a balanced supply of nutrients for optimum growth and yield. Nitrogen, phosphorus, and potassium are the primary nutrients required by walnut trees. The deficiency or excess of any nutrient can affect the yield.
- Pest and disease control: Walnut trees are susceptible to various pests and diseases that can affect the yield. Proper pest and disease control measures should be taken to prevent yield losses.

Methods to increase yield

- Orchard management: Proper orchard management practices such as pruning, training, and thinning can increase the yield and quality of nuts. Pruning helps in maintaining the shape and size of the tree, improving light penetration, and promoting new growth. Thinning involves the removal of excess fruits, which helps in reducing competition and improving the size and quality of nuts.
- Nutrient management: Proper nutrient management practices such as fertilization, mulching, and cover cropping can increase the yield of walnut

trees. The application of balanced fertilizers can improve the nutrient status of the soil and promote healthy growth. Mulching helps in conserving soil moisture, reducing weed growth, and improving soil fertility. Cover cropping helps in improving soil structure, increasing organic matter content, and providing nitrogen fixation.

- Pest and disease control: Proper pest and disease control measures such as timely spraying of pesticides and fungicides can prevent yield losses. Integrated pest management (IPM) techniques can be used to control pests and diseases effectively while minimizing the use of chemicals.
- Harvesting techniques: Proper harvesting techniques such as timely harvesting, careful handling, and proper storage can improve the yield and quality of nuts. The nuts should be harvested when they are fully mature and have reached their optimum size and weight. Careful handling is essential to prevent damage to the nuts, which can affect their quality. Proper storage conditions such as low temperature and low humidity can help in preserving the quality of nuts.

Yield of walnut in India

The yield of walnut in India varies from region to region depending on various factors such as climate, soil type, irrigation, nutrient management, pest and disease control, pruning, and harvesting techniques. According to the National Horticulture Board (NHB), the average yield of walnut in India is around 500-600 kg per hectare. However, with proper management practices, the yield can be increased up to 2-3 tons per hectare.

Future prospect

Walnut production in India has grown significantly in recent years due to the increasing demand for nuts in both the domestic and international markets. Despite the significant progress made in the production and management of walnut orchards, there is still room for improvement in terms of crop yield, quality, and resilience to biotic and abiotic stresses. Crop improvement is a critical aspect of the production of high-quality walnuts and is essential for sustainable production in the long run. This section highlights the current state of crop improvement efforts in India and the future prospects for further enhancement of the walnut crop.

Current status of crop improvement efforts

- Walnut breeding and selection in India have been carried out by several research institutions, including the Indian Council of Agricultural Research (ICAR), the Indian Agricultural Research Institute (IARI), and state agricultural universities. The main objective of these breeding programs has been to develop new cultivars with improved yield, quality, and resistance to biotic and abiotic stresses. Some of the major cultivars released by these programs include Chandler, Howard, Serr, and Franquette.
- Several promising walnut genotypes have also been identified through germplasm evaluation and characterization. The evaluation of walnut germplasm has been carried out by the National Bureau of Plant Genetic Resources (NBPGR), which maintains a repository of more than 500 accessions of walnut germplasm. Some

of the promising genotypes identified through germplasm evaluation include Mansehra, Sindh, and Kashmiri.

- In addition to breeding and selection, there have been efforts to improve walnut production through the use of modern biotechnological tools. Genetic transformation, tissue culture, and marker-assisted selection are some of the biotechnological tools that have been used in walnut research. These technologies have the potential to accelerate the breeding process and develop new cultivars with desirable traits such as resistance to pests and diseases, tolerance to abiotic stresses, and improved yield and quality.

Future prospects for crop improvement

The future prospects for crop improvement in walnut production in India are bright, with several opportunities for further enhancement of the crop. The improvement of walnut can be achieved through various approaches such as breeding, genetic engineering, and biotechnological interventions. Some of the potential areas for improvement are highlighted below:

- Breeding is a traditional method of crop improvement that aims to develop new cultivars with desirable traits such as high yield, disease resistance, and improved quality. In India, there are several walnut breeding programs, such as those at the Indian Institute of Horticultural Research (IIHR), the Indian Agricultural Research Institute (IARI), and the Sher-e-Kashmir University of Agricultural Sciences and Technology (SKUAST). These programs are focused on the development of improved cultivars that can adapt to local climatic conditions, resist diseases, and have better nut quality. For example, the IIHR has developed the cultivar "Chandler", which is a high-yielding cultivar with good nut quality and resistance to bacterial blight.
- Genetic engineering is another approach that can be used to improve walnut. Genetic engineering involves the transfer of desirable genes into the walnut genome to confer traits such as disease resistance, stress tolerance, and improved quality. However, genetic engineering of walnut is still in its infancy, and there are only a few studies that have been conducted on this topic. One such study was conducted by researchers at the University of California, Davis, who successfully transferred a gene for resistance to the bacterial disease Xylella fastidiosa into the walnut genome.
- Biotechnological interventions such as tissue culture and somatic embryogenesis can also be used to improve walnut. Tissue culture allows for the rapid multiplication of disease-free plants, while somatic embryogenesis can be used to produce large numbers of genetically identical plants from a single parent plant. These techniques can be used for mass propagation of elite cultivars, germplasm conservation, and genetic transformation. In India, tissue culture techniques have been used for the propagation of walnut, and several laboratories such as the IIHR and SKUAST have developed protocols for tissue culture of walnut.
- Developing High-Density Planting Systems: High-density planting systems have the potential to significantly increase walnut yields and improve the

efficient use of resources such as water and nutrients. There is an opportunity to develop cultivars that are suitable for high-density planting systems and to optimize management practices such as pruning and training to achieve maximum productivity.

- Improving Post-Harvest Handling and Processing: Post-harvest losses are a significant challenge in walnut production, and there is a need to improve post-harvest handling and processing practices to minimize these losses. There is an opportunity to develop new technologies for drying, processing, and storage of walnuts that preserve quality and extend shelf life.
- Developing Organic and Sustainable Production Systems: There is a growing demand for organic and sustainable walnut production, and there is an opportunity to develop cultivars and production systems that are adapted to these practices. Organic and sustainable production systems rely on the use of natural inputs such as compost and biofertilizers and minimize the use of synthetic inputs such as pesticides and fertilizers.

References

Adib-Rowhani, B., & Martinez-Gomez, P. (2019). The walnut: A production reference. CABI.

Akram, W., Ahmad, S., Anwar, R., & Gul, S. (2021). Effects of Gibberellic Acid on Yield and Quality of Walnuts. The Journal of Animal and Plant Sciences, 31(2), 811-818.

Alasalvar, C., Bolling, B., & Shahidi, F. (2010). Tree nuts: Composition, phytochemicals, and health effects. CRC Press.

Anonymous, (2020-21). National Horticulture Board, Indian Horticulture Database. Ministry of Agriculture & Farmers Welfare, Government of India. www.agricoop.nic.in

Bolling, B. W., McKay, D. L., Blumberg, J. B., Chen, C. Y., & Kwik-Uribe, C. L. (2011). Tree nut phytochemicals: Composition, antioxidant capacity, bioactivity, impact factors. A systematic review of almonds, Brazils, cashews, hazelnuts, macadamias, pecans, pine nuts, pistachios and walnuts. Nutrition Research Reviews, 24(2), 244-275.

Chadha, K.L. (2019). Hand Book of Horticulture. Second Edition, Volume 1, Indian Council of Agricultural Research, New Delhi. pp. 454-457.

Chandra, S., Meena, R. S., Singh, S. P., & Yadav, S. K. (2012). Zinc and boron application enhances fruit set, yield, and quality of apple (Malus domestica Borkh.). *Journal of Plant Nutrition*, 35(5), 664-675.

Ferguson, L. (2015). Walnut Production Manual. UC Agriculture and Natural Resources.

Gajbhiye, R. C., & Sharma, S. K. (2015). Advances in propagation techniques of walnut (Juglans regia L.). *International Journal of Current Research in Biosciences and Plant Biology*, 2(8), 36-42.

Ghasemi, H., Rezaei, M., & Emam, Y. (2018). Effects of ethephon application on fruit quality and yield of walnut (Juglans regia L.). *Horticulture, Environment, and Biotechnology*, 59(3), 32.

Gholami, M., Rahemi, M., Sabzalian, M. R., & Dehghani, S. H. (2013). Effect of nitrogen, phosphorus and potassium fertilizers on yield, yield components and quality of walnut (Juglans regia L.) fruits. *International Journal of Biosciences*, 3(12), 19-26.

Husain, G. M., Ansari, M. A., Khan, S. A., & Ahmad, S. (2013). Omega-3 fatty acids and its role in human health. *International Journal of Pharmacy and Pharmaceutical Sciences*, 5(2), 38-44.

Jain, V., & Kumar, R. (2016). Biotechnological advances in propagation and production of walnut (Juglans regia L.) – A review. *Journal of Applied and Natural Science*, 8(1), 430-441.

Kalia, R. K., Kumar, A., Kumar, R., & Gupta, R. C. (2016). Production Technologies of Walnut. In S. Kumar, S. Kumar, & R. C. Gupta (Eds.), *Advances in Horticultural Science and Biotechnology* (pp. 39-57). Springer.

Kaur, H., Kaur, P., & Kaur, R. (2018). Evaluation of herbicides for weed control in walnut (Juglans regia L.) orchards. *Journal of Pharmacognosy and Phytochemistry*, 7(1), 3834-3836.

Kester, D. E., *et al.*(2011). Principles of Fruit Science. Jones & Bartlett Learning.

Kumar, P., Kumar, R., & Chandel, R. S. (2018). Hybrid rootstock (Juglans regia x Juglans nigra) for walnut cultivation in India. *Journal of Horticultural Science and Technology,* 1(1), 5-9.

Kumar, S., Kumar, V., & Thakur, S. (2017). Effect of pruning on growth, yield and quality of walnut (*Juglans regia* L.) cv. Chandan. *Progressive Horticulture*, 49(2), 209-214.

Kumar, S., Kumar, V., & Thakur, S. (2017). Effect of pruning on growth, yield and quality of walnut (Juglans regia L.) cv. Chandan. *Progressive Horticulture*, 49(2), 209-214.

Liu, Y., Li, W., Li, X., Zhang, J., & Li, J. (2018). Advances in Walnut Breeding. In Y. Zhang, J. Li, & R. Wu (Eds.), Advances in Walnut Research (pp. 1-18). Springer.

Martinez, M. E., Story, J. A., & Kushi, L. H. (2010). Nut consumption and health. *New England Journal of Medicine,* 362(25), 2433-2434.

Martínez-Gómez, P., Mc Granahan, G. H., & Leslie, C. A. (2016). RX1, a new Paradox hybrid rootstock for California walnut production. *Acta Horticulturae*, (1127), 105-110.

Nayyar, M. H., Malik, A. U., & Tariq, M. (2016). Effect of indole-3-acetic acid (IAA) on growth, yield and nutrient uptake of young walnut trees. *International Journal of Agriculture and Biology*, 18(5), 1051-1056.

Prasad, R. & Singh, R. (2004). Walnut varieties for north western Himalayan region of India. Acta Horticulturae, (705), 45-50

Rahman, M. M. (2017). Walnut Production and Management Practices. In The Walnut: Botany, Production and Uses (pp. 165-192). Academic Press.

Ramos, D. E., *et al.*(2015). Maturity Indices for Walnuts. University of California Cooperative Extension.

Rashid, A., Bashir, A., Wani, S. H., & Dar, R. (2020). Performance of cover crops under walnut (Juglans regia L.) orchard. *Journal of Plant Nutrition*, 43(2), 172-180.

Rathore DS 1986. Evaluation of genetic resources of walnut . In: Advances in research on Temperate fruits (eds- TR Chada,VP Bhutani and JL Kaul) UH and F Nauni Solan, Pp:-3-6.

Saini, R., Bhardwaj, A. K., Kumar, S., & Singh, R. (2020). Performance of walnut under drip irrigation and fertigation in sub-tropical Himalayas of India. *Journal of Applied and Natural Science,* 12(2), 238-244.

Sharma, R. K., Kumar, A., Sharma, R., & Abrol, R. (2019). Effect of gibberellic acid (GA3) on fruit quality, yield and physical attributes of walnut (*Juglans regia* L.) under sub-tropical region of Himachal Pradesh. *Indian Journal of Horticulture*, 76(3), 441-446.

Sharma, R. K., Kumar, S., Kumar, S., & Gupta, R. (2020). Mechanical pruning in horticultural crops: a review. *International Journal of Chemical Studies*, 8(1), 1546-1551.

Singh, A. (2020). Walnut cultivation: A comprehensive guide. Krishi Sandesh. https://krishisandesh.com/walnut-cultivation-a-comprehensive-guide/

Singh, D., & Kaur, H. (2018). Organic farming and its impact on soil health and sustainable crop production: a review. *Journal of Soil and Water Conservation*, 17(2), 156-163.

Singh, D., Gupta, S. K., Lal, B., & Sharma, R. K. (2016). Effect of honey bee (Apis mellifera L.) pollination on fruit set and yield of walnut (Juglans regia L.). *Journal of Apicultural Research,* 55(1), 1-7.

Singh, M., Kumar, R., & Sharma, S. K. (2017). Advances in walnut breeding and propagation: a review. *International Journal of Chemical Studies*, 5(4), 2351-2357.

Singh, S. K., & Singh, R. (2014). Advances in propagation and production of walnut in India. *Indian Journal of Horticulture*, 71(4), 469-482.

Singh, S. K., Singh, S., Singh, R. K., Kumar, A., & Singh, S. (2020). Precision farming in walnut orchards: a review. *Journal of Applied and Natural Science*, 12(2), 170-180.

Singh, S. P., & Bhandari, H. R. (2006). Walnut cultivars for diverse agroclimatic zones f India. *Acta Horticulturae*, (705), 51-54.

Singh, S. P., & Chauhan, D. K. (2017). Walnut production and research in India. *Acta Horticulturae,* (1170), 61-66.

USDA National Nutrient Database. (n.d.). Basic report: 12155, nuts, walnuts, english. Retrieved from https://fdc.nal.usda.gov/fdc-app.html#/food-details/171431/nutrients

9

Advances Production Technology of Pecan Nut

***Maneesh Kumar*[1], *Amit Kumar*[2], *Sajeel Ahamad*[3], *Chhail Bihari*[4] *and Vivek Saurabh*[5]**

[1]*Ph.D. Research Scholar, Department of Horticulture, G.B. Pant University of Agriculture & Technology Pant Nagar, U.S. Nagar-263145, Uttarakhand India*

[2]*Ph.D. Research Scholar, Department of Fruit Science, College of Horticulture, Sardar Vallabhbhai Patel University of Agriculture and Technology, Meerut-250110, Uttar Pradesh, India*

[3]*Ph.D. Research Scholar, Division of Food Science & Post-harvest Technology, Indian Agricultural Research Institute, New Delhi-110012 India.*

[4]*Ph.D. Research Scholar, Department of Horticulture (Vegetable Science) Nagaland Central University, SASRD Medziphema, Nagaland, India*

[5]*Ph.D. Research Scholar, Division of Food Science and Postharvest Technology, ICAR- Indian Agricultural Research Institute New Delhi-110012*

Email: maneeshk38904@gmail.com

Introduction

The pecan (*Carya illinoinensis* (Wagenh.) K. Koch), is a species of hickory that was first discovered in Cyprus and is now grown in all over the world in suitable climate. It is endemic to south-central & North America. Pecan is derived from an "Algonquian word" that means a nut that needs a stone to open. The Spanish word nuez, which is related to the English word nut, is used to refer to both pecans and walnuts in Mexico. Pecans are regarded as the queen of nuts in the USA due to their importance as both wild and farmed nuts and their ability to produce meat of a respectable standard (Woodroof

Fig. 1.1: Pecan nut

1979). The greatest multi-use trees for the northern Rivers are pecan trees, which offer a valuable commercial nut yield, a great long-term investment and fantastic beautification potential for the home gardener. Most people will agree that after devouring a delicious slice of pecan pie or licking some ice cream, pecans are among the most delectable of all nuts.

Health benefits of pecans

The health benefits of the pecan are recognized by the medical world and Table 1: nutritionists. Studies show pecans are great defiance against many modern ailments, like lowering high cholesterol and reversing type II diabetes and gallstones in women. A great antioxidant and high in unsaturated fats, pecans also help curb appetites. The nuts have a Total fat wide range of nutritional benefits, with over nineteen vitamins and minerals, as well as being a good source of vitamin E and fiber. They contain good amounts of zinc, iron, phosphorous, magnesium and calcium as well as vitamin B, C, Thiamin and Niacin. Pantothe Research proves eating about a handful of pecans (1.5 ounces) each day could have a similar effect as cholesterol lowering medications.

Culinary uses

- Similar to walnuts, raw pecans can be eaten plain, with saltor with sugar.
- Pecan nuts have a buttery texture and a mildly sweet flavour. They are excellent complements to tasty recipes like pecan caramel puddles, pie, fudge, balkava and muffins due to their high fat content.
- They are frequently added to cookies, sweets and cakes in confectionery.
- Additionally, the nuts are used in making pecan nut butter, which is commonly spread on bread and other items.

Special uses

In recent decades, consumer demand for pecan wood has been rising significantly. It is utilized for veneer, furniture, pallets, paneling and cabinets. The wood possesses favorable machining characteristics similar to those of real hickories (Adams and Thielges 1977).

Table 1: Nutritional value of per 100 g of nuts.

Principle	Nutrient value	Principle	Nutrient value
Energy	691 Kcal	Pantothenic acid	0.863 mg
Carbohydrates	13.86 g	Pyridoxin	0.210 mg
Protein	9.17 g	Riboflavin	0.13 mg
Total fat	71.9 g	Vitamin A	56 IU
Cholesterol	0 mg	Vitamin E	24.44 mg
Dietary fiber	9.6 g	Phosphorus	277 mg
Folates	22 mcg	Potassium	410 mg
Niacin	1.167 mg		

Source: USDA national nutrient database

Origin, Distribution and Systematics

In the sixteenth century, Europeans first learned about pecans. Beginning in the 16th century, Spaniards introduced the pecan throughout Europe, Asia and Africa. The only tree nut indigenous to North America is the pecan. The majority of our native pecan species, including the Cheyenne, Mohak, Choctav and Shawnee, are named for Indian tribes.

Texas produces more native pecans than any other state and it is second only to Georgia in terms of hybrid pecan production. It is currently grown in thirty more counties and is found in the majority of the state's river valleys. There is proof that the tree was present in Texas in prehistoric periods. It was exported from Galveston's ports by 1850, according to records. The tree's hardwood was utilized to produce waggon components and agricultural equipment long before its worth as an edible nut was recognized. The Texas Legislature passed a law designating the pecan as the official tree of Texas in 1919. Instead of a headstone, Governor James Hogg requested that a tree be put at his tomb.

Taxonomy

Pecan nut is classified into the following categories-

Rank	Scientific Taxonomical Details
Kingdom:	Plantae
Phylum:	Tracheophyta
Class:	Magnoliopsida
Order:	Fagales
Family:	Juglandaceae
Genus:	*Carya*
Species:	***illinoinensis***

Important Species

The Genus *Carya* has 20 species & except 4, all are native to the Americas. Pecan is only species which is important commercially while other species like *C. ovata*, *C. laciniosa* & *C. cordiformis* also produce kernel which are sweet in taste, but being thick shelled are poor in cracking quality.

C. ovata: Flavor & quality highest among native hickory nuts. Nuts are small, whitish distinctly 4 angular, thin shelled hull thick, opens to the base, shell has longitudinal ridges.

C. laciniosa: Nuts yellowish white, four angled does not open along suture lines, kernel not well filled, sweet in taste.

C. cordiformis: Shell & hull thin, kernel poor in quality due to astringency & bitterness.

C. tomenstose: Hull thick splits to the base, nuts pear shaped thick shelled, sweet & edible, difficult to shell out as halves.

C. glabra: Nut small, shell & hull thin.

Description

Pecan nut is commonly called as hickory or pecan trees. Hickories have pinnately complex leaves with 5 to 11 leaflets placed alternately along the rachis. Crushed leaves release a pleasant scent. Together with the foliage, the flowers bloom in the spring.

Around the base of the current season's growth, pendant, three-branched catkins with the male flowers are developed. Near the end of the growth, the female flowers form spiky clusters but are barely perceptible. The same tree produces both male and female blooms. The four-valved husk protects the bony-shelled nuts. The husk might be thick and split at the base or thin and split half open depending on the variety. Around October, the nuts become mature and drop off the trees. Many Carya species produce edible nuts, but the pecan nut is by far the most significant kind. Improved varieties can be eaten whole and are used in big quantities in cooking. In the south, the pecan (*C. pecan*) can reach heights of 150 feet. Even though the nuts don't mature in the northeast, it makes a fantastic shade tree there. The big, 12 to 18-inch-long leaves have a golden-yellow color in the autumn. In fruit-producing areas, the trees take four to seven years after planting to yield their first crop of nuts. They will eventually produce abundant yields of 1% to 2% nuts. Once they are ripe in the fall, the nuts will drop to the ground. It is notably appreciated in the production of some tools and equipment, axe handles and baseball bats because the wood of the Shagbark, Pignut and Mockernut Hickory is strong and sturdy.

Ecological requirements

Climate: Pecans grow in humid climates and require a minimum average annual rainfall of 760mm. Summer temperatures average as high as 27°C (81°F). It can withstand extreme temperatures from 41°C to 46°C (105" to 115°F). Average winter temperatures vary from 10 -1°C, with extremes of -18° to -29°C (0 to 20°F) (Putnam 1951, Adams and Thielges 1977).

Soils and topography: Pecan grow universally on fully drained loam soils which are not exposed to prolonged flooding. However, it does show on heavy textured soils, where it is restricted to alluvial soils of modern origin. On such land forms its best development is on the ridges and well-drained flats. It hardly grows on low and poorly drained clay flats where it is reinstated by water hickory (*Carya aquatica*) (Adams and Thielges 1977, Nelson 1965). These soils are most commonly initiate in the orders entisols inceptisols and alfisols, Pecan seedlings can survive small periods of flooding (Loucks *et al.*1973).

Commercial varieties

Early pollen-shedding varieties

Desirable: It is a superior cultivar for humid environments. Although the plant develops quickly, it takes between five and ten years to produce 15 pounds of nuts. Its light green leaves lacks the more appealing dark colour and has weak V-limb angles that require training. The modest scab resistance is manageable with fungicide applications. Desirable is not a particularly heavy bearer, but as a mature tree, it consistently produces pecan yields of the highest quality. Kernal recovery is 54 per cent.

Western: Compared to other varieties in the far west, it requires less management to grow and produce nuts. The tree grows into production in 7 to 8 years and is robust, simple to train and productive. Western should not be grown since it is particularly sensitive to scab disease. Kernal recovery is 59 percent.

Pawnee: A new variety with an incredibly early maturity period. It is a medium-sized pecan with good kernel quality. It seems to naturally resist aphids, a major pecan pest. Kernal recovery is 58 per cent.

Caddo: Strong limbs, scab resistance, early ripening and lovely foliage make this tree a great yard tree. Even though the kernels of its little nuts are of good grade. In five to six years, the trees start to bear and it typically bear well each year. Kernal recovery is 57 per cent.

Cheyenne: It is a very productive, smaller tree that takes just 5 to 7 years to start bearing high-quality nuts. Aphids are very dangerous in Cheyenne. It is cultivable in every climate. Kernal recovery is 57 per cent.

Cape Fear: Its sturdy, simple-to-train limbs, high yield and scab resistance make it a good yard tree. The trees start to develop fruit after 5 to 7 years, however the kernel quality is just fair and the foliage is very prone to fungal leaf scorch. Kernal recovery is 54 per cent.

Late pollen- shedding varieties

Sioux: An exceptional garden tree because to its robust, trainable limb structure and exceptionally excellent quality small nut. It is moderately susceptible to scab and needs fungicide applications when the humidity is high. Kernal recovery is 57 per cent.

Wichita: One of the most fruitful pecan trees in Texas is this one. The tree is quite vigorous and productive and it starts to give fruit in about 5 to 7 years, despite having severe scab problems in humid regions and frost problems in the far north. Wichita creates split V-shaped limb angles that demand precise training. If it is not properly managed, it often turns out to be an undesirable, low-bearing tree.

Forkert: This is a rare, older Mississippi type that is getting popularity owing to the superior quality of its nuts. Kernal recovery is 63 per cent.

Maramec: It has a fair quality, large pecan which has produced strong trees. Kernal recovery is 58 per cent.

Mohawk: It makes a decent yard tree. It has robust limb structure, develops before fall frostsand has lovely foliage. It bears heavily, but as the tree ages, the nuts become poorly filled, similar to other types with very large nuts. Kernal recovery is 59 per cent.

Shoshoni: This cultivar has an early maturity. The tree bears fruit in just 5 to 6 years and is quite productive. With the exception of colder climates, where later maturing cultivars are susceptible to fall freeze harm before the nuts mature, it begins alternating bearing at a young age, which limits its value as a yard tree. Kernal recovery is 54 per cent.

Mahan: It developed from as a random seedling selection of an unknown parent. The kernel percentage is 58%. It serves as Tejas, Mahan-Stuartand Maramec's female parent and Wichita and Mohawak's pollen parent.

Nugget: It is a triploid variety pecan nut and susceptible to scab.

Stuart: It is scab-resistant. Nuts are spherical in cross-section, oblong elliptic, with an obtuse apex and rounded bases. There are 51 nuts per pound, 49% of which are kernels. The kernels are golden to light brown in colour and have deep primary dorsal grooves, deep secondary dorsal grooves and a noticeable basal cleft. Protogynous with a mid-season pistillate receptivity and late pollen shed.

Rootstocks: The majority of commercial nurseries employ domestic or native open-pollinated seeds from domestic that thrive in a particular area. Elliott or Curtis rootstock seeds are the most popular choices. Elliott often germinates well and grows good caliper quite rapidly. Curtis is comparable to Elliott but more tolerant of the cold.

Propagation and Planting

Commercial pecan tree growth is accomplished through budding or grafting onto seedling rootstocks, however vegetative propagation is less effective due to high transplant mortality.

Collection, extraction and storage: Following a natural seed fall or after shaking the trees, pecan nuts can be harvested from the ground. The fruits can be run through a macerator or a maize sheller to remove rigid husks or they can be manually removed by trampling, stomping or trampling. According to several research (Adams and Thielges 1977), pecan seedlings grow larger from the larger nuts. As a result, nuts' sizes may be advantageous. According to Bonner and Maisenhelder (1974), shellbark and stagbark hickory trees can each yield between 0.75 and 1.1 hl (2 and 3 bu) and 0.75 to 0.75 hl (12 to 2 bu) of nuts.

Pecans should be chilled after being dried to low moisture concentrations. Hickories are typically thought to have embryo dormancy, but research with pecans reveals that mechanical restriction caused by the shell is the cause of that species' delayed germination (van Staden and Dimalla 1976). Nuts are typically treated by stratifying them for 30 to 150 days in a wet medium at 1 to 4 ^{0}C. Most species of imbibed nuts can be stratified in plastic bags without a medium (Bonner and Maisenhelder 1974), while pecans from southern sources can be successfully stratified by soaking them at 20^{0}C for 64 hours (Goff *et al.*1992). According to certain evidence, stratification times for stored nuts should be cut down (Bonner 1976h). If there are no cold storage facilities, stratification in a pit with a layer of roughly 0.5 m of compost, leaves or dirt will do to prevent freezing. To ensure complete imbibition, nuts should be soaked in water at room temperature for up to four days with one or two water changes daily (Eliason 1965). Gibberellin treatment of pecan has been shown to promote germination (Bonner 1976a, Dimalla and van Staden 1977), but no practical applications have been found.

Germination tests: The official North American testing regulations (AOSA 1993) provide for testing pecan and shagbark hickory for 28 days at alternate temperatures of 20°C (dark) and 30°C (light) on thick crepe paper. It is also advised to stratify for 60 days as previously said. Stratified nuts can also be tested for adequate germination on flats of sand, peat or soil at the same temperature range. Hickories can also be tested quickly using tetrazolium salts (Eliason 1965).

Growth of the rootstock: Always gather nuts that are during season and free of pests and illnesses. To activate the enzymes that promote growth, place these in a jar of water and soak them for two to three days. To assist stop the growth of fungi, a hose left to drip into the container offers oxygen. Keeping the soaked nuts between 35 to 40 degrees Fahrenheit and regularly changing the water are additional precautions against infection with the bacteria that cause seed rotting. Change the water every couple hour if you're soaking nuts at room temperature. Prior to seeding, pecan nuts need to stratify for 6 to 8 weeks. This can be done by placing the nuts in a container of peat moss or layering them with sand and peat moss. Stratified seed should be kept moist at a temperature of 35^{0}F.

When the nuts are planted, stratification doesn't speed up germination, but it does make seedlings emerge more consistently, making it easier to control weeds in the nursery. Before the soil warms up enough to hinder germination and after the threat of frost has passed, pecan seeds should be sown. Both seedbeds and containers can be used to grow seeds. Both approaches have merits and drawbacks. Even from seeds originating from the same parent tree, there is a sizable amount of variation in the rootstock, just as each seed from a specific tree is not identical to the parent. This explains a large portion of the difference in tree performance when the grafted tree grows and begins to produce in its full capacity.

Planting of seeds: Plant the seeds in rows 3 to 4 inches apart and 3 to 4 inches deep in the soil. In comparison to sandy soils, clay soils should have shallower planting depths. Round-shaped nut cultivars like Elliott and Curtis are simple to plant using a mechanical planter. Planting large or oblong nut types by hand is necessary since they have a tendency to hang up in the planter.

Graft wood collection and storage: Several grafting techniques require for stored graft wood. A productive tree should have its scions cut from the growth from the previous season. Scions should be smooth and straight, with a diameter of 1/4 to 5/8 inches. They need to be free of insects and diseases, with a length of six to eight inches long enough for three to four healthy buds. Although they can be employed, older wood scions often have less success. Because the buds are frequently young and susceptible to harm from cold, as well as having pithy wood, scions cut from the end of the shoot should be avoided. Graft wood from whip grafting can be used right away, negating the need for storage. Graft wood needs to be cut in January or February when the trees are entirely dormant when utilizing the 4-flap or bark graft techniques. Early to mid-March should be employed for the collection of bud wood for spring blooming. Scions should be placed in garbage bags with moist sawdust or peat moss and then boxed. Graft wood can be kept between 30 and 38^{0}F or up to six months. It shouldn't freeze on the wood. When kept at 35 degrees for an extended amount of time, pecan buds may enlarge.

Budding: Patch or ring budding can be employed to top-work bigger pecan trees or propagate nursery stock. Prior to the buds swelling, scion wood should be chosen. However, early buds will have more opportunities to develop a stronger union with the rootstock and will produce greater vegetative growth during the first year. Budding can be done as late as June or July. However, this is frequently a less successful technique. Budding can also be done in late July and early August utilizing mature wood from the current year's growth without storing.

Avoid exposing any cuts or strips to the sun or wind. By removing the stock two to three inches above the bud two to three weeks following bud break, spring buds can be pushed to grow. By delicately nicking the bark with a pocket knife, you can first check to see if the patch is still alive. The tissue beneath the bark is still alive if it is green.

Planting: Pecan trees can be planted as bare roots from December through the middle of March. Between nursery digging and planting, it's crucial that roots always have a light mist of moisture-not a wet mist. Although container-grown trees can be planted at any time of the year, it is preferable to plant them in the autumn or winter so they have time to adapt to their new surroundings before the summer heat arrives.

Spacing: Pecans should be planted at least 35 feet apart since they grow into huge trees with extensive root systems. Trees that are planted too closely acquire poor health, ugly forms and low nut yield as a result of crowding.

Growth: The pecan tree is a big deciduous tree that can reach heights of 20 to 40 m. rare to 44 m. It normally has a spread of 12–23 m and a trunk diameter of up to 2 m. A sapling that is 10 years old will be about 5 m tall. Alternating leaves are present, 30-45 cm long, 30-45 cm wide and pinnate with 9–17 leaflets that are each 5–12 cm long. The monoecious, wind-pollinated blooms have staminate and pistillate catkins on the same tree; the female catkins are smaller and have three to six flowers clustered together, while the male catkins are pendulous and up to 18 cm long.

Tree Architecture and Pruning

Pruning and training: Pecan tree pruning is a continual, required task that starts with the tree's original planting. Due to their inactivity, pecan trees will eventually become pecan bushes if annual pruning is not done. For pecan trees to produce a yield, sunshine must be abundant. Pecan trees should be cut to allow the upper branches to spread and absorb as much sunshine as possible in light of this. The majority of the pruning, however, takes place in the first ten to fifteen years of a pecan tree's life because mature trees need to be pruned annually to the point where it is most drastic. Mature pecan trees will always need yearly care.

When to prune your tree: Pruning of pecan trees can be done at any time without suffering significant consequences, however it is typically done in the early spring. After the leaves have grown in late April, pruning enables the tree to start recuperating right away. The pecan tree's initial five years of growth are crucial for the development of its central leader and scaffold branch structure.

Young pecan trees can be trained using a variety of trimming techniques. Heading back, branch selection, tip trimming and pinching are all effective techniques that should be used during the first few years of fresh pecan tree establishment and training. Use a central leader or modified central leader system to grow a medium-sized tree that is sturdy and wind-resistant. Tip pruning will encourage the development of the central leader and bring pecan trees into commercial bearing at a younger age. During the first 10 years, it also shrinks the size of the tree. Tip pruning trees up until their sixth or seventh year has been advised. On bigger, fruit-bearing trees, it is not required.

Training young trees: It is necessary for newly grafted pecan trees to have a single central trunk with sturdy wide-angle branches. Untrained, grafted trees typically grow several trunks with limited branching angles and trapped bark in between V-shaped limbs. When exposed to strong winds, water, ice or a dense crop, these tiny forks are unstable and prone to splitting. The easiest way to comprehend the processes of tree training is to look at each target distinctly.

Heading back the central leader: Before growth starts in April, cut off half of the central leader. A number of sprouts will grow just below the cut as a result of this trimming. The strongest of these erect, vigorous branches should be chosen and left each spring for the first few years and the remaining sprouts that emerge within 6 to 8 inches of the cutback should be pulled out. After the others are gone, this core leader will expand quickly. For food production and quick root growth, the lower trunk's small limbs are advantageous. 6 feet above the ground is where permanent branches first appear. By pinching out their developing point, keep the side shoots on the lower trunk short. Limbs with crotch angles between 45 and 90 degrees are far less likely to split out than limbs with narrow V crotch angles.

Pruning mature trees: Large pecan trees rarely require significant pruning. The typical period to remove dead wood, low-hanging limbs or branches that are in the way or harm structures or electrical lines is late winter when the trees are dormant. But pruning can be done whenever it's required.

Flowering Physiology

Growth and development: When flowering begins, a pecan tree begins to produce nuts. Every year, catkins (male flowers) often discharge a lot of pollen. If harvesting was successful the year before and alternate bearing is occurring in the orchard, female flowers may be plentiful or they may be rare. Early in the season, certain flowers may shed themselves. This will comprise (a) primitive flowers near the stem tip, (b) typical flowers that were not pollinated and (c) pollinated flowers in which nutlets failed to form because food stores were depleted during early growth or due to unfavorable moisture circumstances. The pecan grower frequently misses this shedding, which is more noticeable in certain varieties than others. Late June is often when the so-called "June drop" occurs. Inadequate fertilization (fruit set) is the root cause. Due to its double fertilization, pecan nuts are unique. The nucleus contained therein separates into two nuclei when the pollen tube takes shape. The endosperm nucleus, which will expand into the endosperm that will feed the embryo, is coincidentally produced by growth activities in the pistil. Two potential outcomes for early nut shedding are presented by this process and each one may take place at a different period. The embryo won't receive enough nutrition, which will lead to early shedding.

About 25 per cent of the total nut crop is shed during the first and second drops. Environmental stresses can increase the nut drop. Researchers have shown that less dropping occurs with cross-pollination than when self-pollination takes place. Good pollination is a must in pecan orchards. Actual nut development starts with pollination and it can be divided into two distinct phases throughout the growing season. It is difficult to put an exact date on the beginning and ending of these two phases. Because of variation in weather conditions, growing degree hours may give a better estimate.

Phase I occurs from pollination (early to mid- May) to shell hardening (or until the end of the water stage). The pecan completes the sizing of the nut during this phase. The end of the water stage may occur between the middle and end of August. *Phase II* occurs from the shell hardening (or end of water stage) until shuck splits. The kernel develops and fills out during this phase, ending when the hull splits along the sutures.

Phase I: Growth in nut size: The fruit grows to its full size during a period of approximately 90 days after pollination, initially slowly and then more quickly. The fruit has reached its maximum size by the end of August, thus the elements that affect fruit size only come into play in the first part of the growing season. It is known as the water stage because the endosperm is fully non-cellular up until the beginning of August or towards the conclusion of phase I. Phase I is notable for the slow development of the real embryo, which later develops into the edible nut kernel. Despite the genetic predisposition for pecan size, there are numerous variables that can affect nut size, including soil moisture availability, nutrient status and weather circumstances. A multitude of variables, including tree vigour, soil fertility, crop size, severe drought, hot, dry winds, insects, physical damage, etc., affect the size of the nuts that a pecan tree actually produces.

Phase 2: Nut filling: Even though the shuck gets thicker, there isn't a noticeable increase in fruit size during Phase II because the shell hardening inhibits it. Around the third week of September, the embryo (kernel tissue) reaches its full size and begins to absorb the endosperm. Assuming the shuck is green, filling of the kernel proceeds as long as conditions allow and, ideally, until the nut is mature. During the final six weeks of filling, the majority of the storage components are transferred from adjacent leaves and shoots into the nuts. As a result, the food supply in the tree is severely and occasionally completely depleted. 73–75% of the oil in a premium kernel is oil. 12–15% of the diet should include protein, 4% should be water and 1.5% should be minerals. A sizable number of interconnected elements affect how well the kernels are developed at harvest or how well the nuts are filled. The size of the crop in relation to the amount of foliage, the size of the nuts on average, the health of the leaves, the size of the crop before it and how well the nuts were filled, free from injuries from insects and diseases, weather heterosis or cross-pollination.

Pecan flowering: Understanding the tree's flowering system is useful for keeping updated on pecan flowering. The words that must be used for this will be defined below. Male and female blooms appear in various places on mature pecan trees. The compound buds, which are made up of two lateral floral or catkin buds and a middle-mixed bud, give rise to the pecan flowers. The middle-mixed bud lengthens as growth picks up in the spring to generate the vegetative shoot, which may end in the female (pistillate) inflorescence. the male (staminate) inflorescence from lateral buds, in which the vegetative shoot aborts, is produced by each of the two lateral floral buds as a group of three stalked catkins; pe, the peduncle; B: three bracteoles (bl), a bract and pistillate flowers with stigmas of various shapes (st). C: The bract (br) structure of male flowers varies between protogynous (b) and protandrous (c) cultivars. ep, epidermis; en, endothecium; mi, middle layers; ta, tapetum; sp, sporogenous tissue (pollen); sto, site of stomium (where anther splits to release pollen); D: cross section of an anther; E. Pollen grains; a distal view of a triporate pollen grain; po, pore; b a proximal view, demonstrating enfolding; c an equatorial view, demonstrating suboblate form.

Because the male and female flowers of a pecan tree mature at different periods, pecan trees have dichogamous flowering (dicho = "two parts," gamy = "sexual union"). The tree is categorized as type I and is considered protandrous (protos=first; andro-male) if male flowers release pollen before pistillate blossoms are receptive. The tree is referred to as protogynous (protos-first; gyne-female) and is categorized as type II if female flowers open before pollen is discharged from catkins. The word "heterodichogamous" (hetero-different; dichogamy two part bloom) refers to the species as a whole since distinct trees have diverse bloom patterns or dichogamies. This kind of flowering maximizes outcrossing, which promotes genetic variety. Dichogamy patterns are caused by the P gene, a single gene with simple dominance. Only one homozygous dominant (PP) cultivar, Mahan, has been identified and protogyny is the dominating trait. Since Mahan must pass on the dominant gene to all of his offspring, they are all protogynous (either PP or Pp). homozygous recessive genes are present in protandrous cultivars (pp). About half of the trees in native populations will be protogynous and half will be protandrous with typical outcrossing (Thompson and Romberg 1985).

The timing of pollen release may overlap stigma receptivity (incomplete dichogamy) or a division of male and female bloom phases for a certain tree may be perfect. A tree needs to be cross pollinated by another tree if the male and female blooms are completely separated. Nuts let go immediately after blooming without pollination and no crop is produced. The tree may be partially self-pollinated and allow for some nut set if it has incomplete dichogamy. However, self-pollination has been proven to lower nut quality (Marquard 1988), making it undesirable. It has been determined how pecan varieties differ from one another in terms of their bloom pattern, pollen shed time and pistil responsiveness. Although there are seasonal variations in the timing of flower development and the degree of bloom overlap, cultivars are consistent in the pattern of flower maturation (e.g., protogynous cultivars consistently mature female flowers first, while protandrous cultivars consistently mature male flowers first) (Grauke and Thompson 1996). In some years, a cultivar's pollen shed and pistil receptivity may be completely separate, while subsequent years of the same cultivar may have a large overlap. Growers should research cultivar flower trends (preferably locally collected) over a number of years before designing an orchard.

Differentiation of staminate inflorescence (male flowers): Prior to pollen shed, the staminate inflorescence (catkin group) begins in the spring. On the opposing ends of the shoot bud, two catkin groups are distinct and are encased in different inner scales. In the central bud's inner scales, there may occasionally develop a third pair of catkin groups that are opposite the first two and at an angle to them. The compound bud, which includes the shoot bud, lateral catkin groups and inner bud scales for each catkin group, is completely encompassed by the outer bud scale.

Protandrous and protogynous cultivars develop catkins at different stages (Wetzstein and Sparks 1984; Haulik and Holtzhausen 1988); protandrous cultivars begin anthers on catkins in the buds of the staminate inflorescence during the summer before pollen shed; protogynous cultivars begin anthers on catkins during the spring before pollen shed. According to Luza and Polito (1988), when walnut trees started growing again in the spring, staminate flower differentiation began again in protandrous clones before it began again in protogynous clones. For pecan, no comparable studies have been carried out.

Structure and development: Protogynous cultivars typically have long, thin catkins, while protandrous cultivars typically have catkins which are shorter and of greater diameter. Regardless of dichogamy class, the central catkin of a catkin group is usually the longest. Cultivars vary in the quantity of catkins produced. Some cultivars, such as Desirable and Cape Fear, are known as heavy catkin producers, while others typically produce fewer catkins. Each individual staminate flower is composed of a central bract and two lateral bracteoles. Protogynous cultivars typically have male flowers with long, thin bracts, while protandrous cultivars typically have male flowers with short, broad bracts. In pecan, the filament is so short that it is inconspicuous. When mature, each anther will have four pollen sacs or thecae.

The amount of time that each cultivar's pollen is shed varies; some, like Caddo, are known for their quick sheds, while others, like Wichita, typically shed their pollen for a longer period of time. Because of the possibility for considerable variations in different areas within the range of temperatures where pecan is cultivated, the length of pollen dehiscence for a cultivar may vary significantly in different seasons and as a result of changing weather.

Differentiation of pistillate inflorescence: In the spring, during the early phases of bud growth, female flowers are distinguished. Protandrous and protogynous cultivars differentiate pistillate flowers at essentially the same period.

Structure and development: At the end of the shoot for the current season, pistillate flowers are produced in a spike. The oldest flowers are found at the base, whilst the youngest blossoms near the apex are frequently undeveloped and sterile upon first dropping. A single inflorescence's ability to produce blooms varies depending on the season, cultivar and length of the branch.

Pistillate flowers consist of a bilobed stigma on a stigmatic disk surrounded by 3 bracteoles and a bract. The bracteoles and bract are fused at the base to form the involucres or shuck. As pistillate flowers mature, the stigmatic surface becomes more prominent and may become red in some cultivars. Receptivity has been judged by the presence of a viscous fluid on the stigmatic surface (Mullenax 1970) by adherence of applied pollen to the stigmatic surface (Madden and Brown 1973) and by calculations based on stigma drying. In general, pistillate flowers of protogynous cultivars become receptive at smaller size than those of protandrous cultivars. Dead pollen is dusted on the stigma.

Irrigation

For immature trees to survive, grow quickly and consistently produce high-quality nuts on bearing trees, irrigation is crucial. The soil should receive enough watering to become completely saturated. From the middle of March through September, water regularly. Water the plants ideally once a week, especially in the middle of summer.

In pecan trees, flowering is fairly complicated. Pecans are monoecious; the female flowers appear at the end of the growth of the current season and are made up of a spike of immature nuts, while the male flowers appear at the end of the growth of the previous season and are made up of catkins. The catkins produce a significant amount of pollen, which is then carried by the wind to the female flowers where fertilization occurs. There are two main classifications for pecan trees. Type I or protandrous, is the name given to the first type. In these trees, the pollen matures and is expelled first, while the female flowers become receptive a few days to a week later. The second category is known as Type II or protogynous and in these trees, the female flowers open initially before the pollen is discharged after the female flowers are no longer open. In the wild, Type I and Type II trees are found in almost similar quantities, assuring a continuous supply of pollen. Although the two forms of blooming are largely responsible for controlling flowering dates, other factors are equally important. Older trees frequently flower at somewhat different times than younger trees, therefore tree age can have an impact. places of flowers within the canopy have an impact on flowering as well, with inner and lower places frequently maturing more swiftly. Flowering times are influenced by bud break as well. Generally speaking, cultivars that leaf out earlier will also blossom earlier. As a result, a Type I tree that sprouts leaves swiftly will shed pollen before a Type I tree that sprouts leaves slowly. In actuality, a very late Type I variety may release pollen concurrently with a very early Type II variety. These factors can cause flowering times to vary significantly by location and year.

Nutrition, Water Management and Abiotic Stresses

Fertilization: Nutrients are necessary for pecan growth and fruit production. In addition to supplying nutrients to soils, fertilizers can help to remedy nutritional deficits. The foliar or leaf analysis is the single factor most capable of predicting the nutritional requirements of the pecan tree. In mature orchards, soil tests are less effective at identifying nutritional shortages but are very helpful in tracking soil pH and lime needs. Between July 7 and August 7, when nutrient contents in the leaves are at their most stable, leaves should be sampled for analysis. A pecan tree's root system is at least twice as wide as its tree canopy. To encourage the greatest amount of development and output, water the entire root soil region. Drip and micro sprinkler irrigation systems can be used to deliver water efficiently, but use sufficient drip emitters or sprinklers to provide water to at least 50 percent of the soil root area under each tree.

Table 2: Optimum nutrient level for pecan nut cultivation.

Nutrient	Optimum dose	Nutrient	Optimum dose
Nitrogen (%)	2.5-3.3	Zinc (ppm)	50-100
Phosphorus (%)	0.18-0.2	Iron (ppm)	50-300
Potassium (%)	1.3-2.5	Manganese (ppm)	100-800
Magnesium (%)	0.35-0.6	Boron (ppm)	15-50
Calcium (%)	1.3-1.75	Copper (ppm)	6-30
Sulfur (%)	0.25-0.5	Nickel (ppm)	5-15

Weed control

Young pecan trees frequently suffer from poor development or even die as a result of weed competition. Many other grasses and weeds can cause very poor tree development, but Bermudagrass and Johnsongrass are particularly fierce rivals. If a circle 6 feet in diameter is kept weed-free with glyphosate herbicide, hand cultivation or mulch, pecan trees will grow at their fastest rate in the first few years after being planted. Herbicides should only be used as directed on the label and with the utmost care.

Physiological Disorders and their Management

Alternate bearing: Alternate or biennial bearing refers to a fruit tree's tendency to produce enormous crops followed by small or no crops every two years. Almost all tree fruit crops have alternate bearing, but pecans are particularly susceptible to it. Three characteristics of pecan trees contribute to this harshness:

1. The time of fruit maturity
2. The nature of fruit growth; and
3. The chemical composition of the kernel

Contrary to many other fruit crops, pecans mature late in the growing season, allowing little time before leaf fall for carbohydrate storage to sustain the upcoming season's developing blooms and fruit. Additionally, at the conclusion of the season, the majority of the dry nut matter collects. And finally, lipids-which make up around 70% of kernels-are produced using a lot of energy. These characteristics, when coupled with a heavy fruit set, may cause trees to lose their carbohydrate stores at the end of the season, especially in the roots and may also cause poor fruit set the following year.

Major Diseases and their management

Scab (*Cladosporium effusum*): The fungus attacks developing nuts, leaves and emerging shoots. Nuts that have been severely diseased fall or never fully grow. During seasons of regular rainfall and moderate temperatures, infected foliage prematurely drops. Infected shoots, aged shucks and leaf petioles are all places where the scab fungus overwinters. The fungus becomes active in the spring when temperature and moisture conditions are favourable and within a few days, spores form at the surface. Wind and rain carry them to developing leaves and nutlets. The spores germinate and enter the host tissue if the weather is suitable. New lesions develop on the host in 7–10 days.

Lesions that develop on the bottom leaf surface have the distinctive colour and shape of olive brown and are a little bit elongated. Adjacent lesions frequently merge.

creating substantial, unevenly formed patches in chocolate brown. Scab sores can be present between the veins as well as on or along the leaflet veins. Small black specks that turn into sunken sores as nuts mature are the scab lesions that affect them.

To enable for air circulation between the trees, trees should be spread apart. This promotes the foliage's quick drying. Trees that are properly spaced will have more of their leaves exposed to the sun. A tree's lower limbs can be cut down to improve airflow and shorten the amount of time leaves stay wet after heavy dews or rain. Fungicides have a significant role in the management of pecan disease. Although some of the more recent fungicides have post infection action, preventive applications are where they perform best. To avoid infections, it is necessary to maintain complete fungicide coverage on the surfaces of leaves, nuts and shoots.

Brown leaf spot: Only mature leaves are impacted by *Cercospora fusca's* brown leaf spot, which doesn't manifest itself until the latter half of May or mid-June. On the lower leaf surfaces, primary lesions begin as tiny spots that gradually get larger and turn reddish brown with a grey tinge. Lesions can have a round or irregular shape, particularly when several lesions form next to one another. If the fungus is not treated, pecan trees may entirely lose their leaves during seasons that are conducive to the development of brown leaf spot. This fungus only affects the foliage.

Vein spot (fungus *Gnomonia nerviseda*): The lesions caused by the scab fungus resemble the disease's symptoms in the foliage. Vein spot, as opposed to scab, solely impacts the leaves. Vein spot disease causes lesions that typically range in colour from dark brown to black and appear on the veins of leaflets. Leaflets that have been severely affected shed immediately after symptoms are noticed. Under the tree, under the fallen leaves, the fungus overwinters. Sanitation practices and fungicides will be used to control this fungus. The vein spot fungus can be controlled by fungicides used for scab control.

Leaf blotch (*Mycosphaerella dendroides*): This disease mostly affects weak trees. Infected leaves on the orchard floor serve as its winter home. Early in the summer, symptoms first develop as tiny, velvety dots of olive-green colour on the underside of mature leaves. By July, the patches have developed black, pimple-like specks. When the infection is severe, infected leaflets die, resulting in leaf fall in the late summer or early fall. Premature shedding contributes to the plant's decreased vigour. Following the pecan scab spray program, fungicides will be used to manage this disease.

Downy spot (*Mycosphaerella caryigena*): Soon after symptoms are noticed, the foliage on infected trees sheds. In the spring, spores formed on the foliage from the previous year infect newly emerging leaves. On the undersides of leaflets, symptoms typically arise in the summer. Lesions have a diameter of one-eighth to one-fourth of an inch and are greenish-yellow. Due to the diseased area's leaf cells dying later in the season, the lesions appear brown. Rain and wind carry spores to surrounding leaves and other trees. All pecan types, however, are only marginally to moderately sensitive. The three cultivars with the highest susceptibility to the downy spot fungus are Pawnee, Moneymaker and Stuart. Before the leaf buds start to swell, rake or disc under fallen leaves in the early spring or fall. Spore discharge is halted by clearing the leaves from the soil's top layer. The most crucial period for fungicide applications in the management of downy spot is between bud break and case bearer.

Powdery mildew (*Microsphaera alni*): Virus-infected nuts have a white powdery coating on them. Most frequently, the fungus appears between mid-July and mid-August. During months of heavy rain, the fungus rarely becomes an issue since the rain washes the spores off the nuts or leaves. When the powdery mildew fungus is cleaned by rain, pecans that have previously been afflicted by it turn brown. The fungus is occasionally connected to young trees or sucker growth at the base of established trees. Infected leaves have slightly deformed shapes and are dusted with a light grey dust. Nuts or leaves with powdery mildew infection are not seriously harmed by the fungus. Only the shuck's outer cells are where it feeds. This does not limit the movement of nutrients into the developing kernel. Fungicides applied in the course of the even scab spray program will control powdery mildew.

Fungal leaf scorch: Along the leaf border or tip, infected leaves become brown. The necrotic patches can cover the entire leaf margin or just a tiny portion of it near the leaf margin. The lesion turns grey as it ages. Leaf shedding due to infection can happen between mid-July and mid-August. Early to mid-summer is the best time to apply fungicides for scab treatment. The cultivars Western, Cheyenne, Chickasaw, Shoshoni, Stuart and Cape Fear are among those that are more vulnerable to this particular fungus group.

Bunch disease (mycoplasma). Even though the exact etiology of bunch disease is unknown, research indicates to a mycoplasma (MLO). Trees with bunch disease display the usual bunching symptom, which is brought on by excessive lateral bud growth. Typically, only the most distant bud will emerge from dormancy, grow into a twig and then mature into a limb. On infected bunch trees, a significant portion of the buds will emerge from sleep at once. Small twigs develop densely as a result of this. On bunch infected trees, nut production is severely constrained and the size of the nuts produced by these trees is constrained. Except for the removal of trees, there is no effective treatment for bunch disease.

Articularia leaf mold (*Articularia quercina*): The disease most frequently appears on the leaves of weak-vigor trees, after rainy times, particularly in regions with high relative humidity. The fungus, which appears as a development of white tufts containing spore masses, develops on the underside of the leaves. Trees in orchards where disease control pesticides have been applied often do not develop *Articularia* leaf mould. *Articularia* leaf mould disease may typically be controlled with a single application of fungicide at the first sign of disease. The substances used to treat scabs will also offer some defense against this fungus.

Pink mold (*Trichothecium roseum*): Pink mold infection occurs on scab infected nuts. White mould grows on infected nuts in old scab sores. The white area turns pink in locations that aren't exposed to air and sunshine. The kernel may be infected by the fungus. The kernel turns greasy and drips onto the neighboring nuts. The kernel is unusable because it is rotten. Pink mould is not a concern in places where scab disease control is consistently practiced. It could be required to remove the oily nuts or to segregate the disease-free nuts from the problem load if pink mould is a serious issue.

Cotton root rot (*Phymatotrichum omnivorum*): A pathogenic fungus that lives in the soil and damages a variety of host plants, including pecans, is the cause of cotton root rot. Pecan tree roots can be attacked at any time, with the exception of the winter months.

Major insect pests and their management

Stem borer: On pecan nut trees, this stem borer is sporadically seen. Reddish-brown granular excretions at the base of the trunk are the first indication of an infestation. The pink-colored larvae that have tunneled into the tree's trunk and branches are the source of this flow. The tunnels can be as thick as a pencil and vary in size depending on the age of the larvae. One tree trunk contains many tunnels. From the middle of December to the end of February, young larvae hatch. In the spring and early summer, the larvae grow to a maximum size of 40 mm. The larvae at this stage stop moving and develop into pupae in the tunnels. The pupal phase lasts about 6 weeks and it seems that only 1 generation happens per year. The larvae persist in the trunk for approx11 months.

Control: It is possible to achieve effective chemical control of the larvae in the tunnels. The producer can later ascertain whether part of the tunnels was skipped during spraying by extracting larval excreta from the area around the stem right after spraying. A soft piece of wire can be used to destroy the larvae in the tunnels in young trees. Although crude, this technique is quite successful and must be used in the winter when the tunnels and excrement are more obvious around the tree stem.

Bark borer: Pecannut trees' live bark is consumed by bark borer larvae, especially in newly planted trees. Later, they drilled into the sturdy wood. Penetration can happen in branches of any thickness and typically occurs where branches are produced. When the larvae are at their largest, the holes in the branches where they live are around 70 mm long and 5 mm in diameter. Larval excreta that have been spun together with threads over feeding marks on the bark allows the larvae to travel freely underneath the threads. This home of spun threads and excrement, which can be seen around a limb while the larva feeds, grows larger. Although infested trees do not die, the branch is ring barked and it could die back.

Control: If only the lesions on the branches are treated, a registered chemical can still provide effective control. Before spraying, there is no need to clean the waste from the branches. It is not advised to spray the tree entirely.

Parasitic plants: In the majority of South Africa's pecannut-producing regions, parasitic plants called *Tapinanthus spp.* (bird-lime) grow. These plants parasitize the host plant and lack a root system. They weaken the tree and decrease its bearing capacity. The plants' red and yellow flowers are visible in the pecan trees' crowns, particularly in the winter and in September.

Control: These parasitic plants cannot be controlled chemically. The parasitic plants must be pruned as the only solution. The bird lime branch needs to be cut off and removed from the orchard.

Ripening and Storage of Fruits

Maturity indices and harvesting: In order to capture early market and to reduce bird damage, early harvest of pecan has been recommended but shucks do not dehisce and easily separate from the shell in immature fruits. Appearance of colored dots and streaks markings on the shells indicates the maturity of nut. The shell colour changes from white to brown along the sutures with the onset of shuck dehiscence colour markings on shell and clean separation of packing tissues from kernel indicates that the fruit is ready for picking (Eddy and Storey 1988). The nuts should be harvested at proper time as the quality of nuts is influenced by the time of harvest. Mature nuts do not fall at one time, so the nuts either thrashed down from the trees with long sticks or poles or allowed to fall naturally on the ground where from these are gathered manually. Mechanical harvesting is advocated for large plantations which involve harvesting, picking and cleaning of nuts.

Optimum storage conditions: To maintain quality and stop mould growth, mechanically harvested in-shell pecans should be dried to 4.5% moisture before to storing (Heaton *et al.*1977). After being shelled, pecan nut meats have 7–9% moisture; this moisture level should be reduced to 3–4% to retain quality. Drying should be avoided at temperatures above 38°C (100°F), since these raise the risk of nut flesh discoloration. Pecans should be stored at a temperature of 2°C (36°F) with 70% relative humidity. Long-term storage should be kept at or just below 18°C (0°F) with 70% RH. Pecans that are still in their shells can be kept for up to 6 months at 22°C (72°F), 12 months at 0°C (32°F)or 24 months at -18°C (0°F), however shelled nuts can only be kept for 3 months at 22°C (72°F). 18 months at 18 °C or 9 months at 0 °C.

Modified atmosphere considerations: Storage of pecans in 2 to 3 percent Oxygen and less commonly in CO, as the balance gas, may extend their shelf life. 52 days of storage at 2% Oxygen can result in the development of a fruity flavors (Santerre *et al.*1990). According to Dull and Kays (1988), packaging materials should have Oxygen transmission rates greater than 0.08 ml/100 cm every 24 hours. Another advantage of vacuum-packaging is its ability to prevent breakage.

Future prospect

The pecan nut holds great importance and future prospects in crop improvement. With its nutritional composition and health benefits, pecans have gained popularity among consumers. The demand for pecans continues to rise, creating economic opportunities in the food and agriculture industry. To meet these demands, crop improvement strategies are being explored, including breeding programs to develop disease-resistant and high-yielding varieties. Genetic engineering techniques can also be utilized to introduce desirable traits. Additionally, research on agronomic practices such as irrigation and soil management can optimize pecan cultivation. The focus on improving pecan production and quality ensures its continued significance as a valuable crop with promising prospects for the future.

References

Adams JC and Thielges BA 1977. Research underway on pecan timber improvement. Louisiana Agriculture 20(2): 14-15.

AOSA 1993. Rules for testing seeds. *J Seed Technology* 16(3): 1B113.

Bonner FT 1976b. Storage and stratification recommendations for pecan and shagbark hickory. Tree Planters Notes 27(4): 3B5.

Bonner FT, Maisenhelder LC 1974 Carya Nutt., hickory. In: Schopmeyer CS, tech, coord. Seeds of woody plants of the United States. Agric. Handbk. 450 Washington, DC: USDA, Forest Service: 269B272

Bonner FT 1976a. Effects of gibberellin on germination of forest tree seeds with shallow dormancy. In: Hatano I, Asakawa S, Katsuta M, Sasaki S, Yokoyama T, eds. Proc 2nd Inter Symp on Physiology of Seed Germination; October, 18B30; Fuji, Japan Tokyo: Government Forest Experiment Station: 21B32.

Dimalla GG, van Staden J 1977. The effect of temperature on the germination and endogenous cytokinin and gibberellin levels of pecan nuts. Zeitschrift fürflanzenphysiologie 82: 274B280.

Eliason EJ. 1965. Treatment of forest tree seed to overcome dormancy prior to direct seeding. In: Proc Direct Seeding in the North east CA Symp Exp Sta Bull Amherst: Univ of Massachusetts: 87B90.

Goff WD, Brasher LR, McGuire JA. 1992. Germination of un- stratified pecans is affected by exposure to high temperature and by soaking. Scientia Horticulturae 50(1): 159B163.

Graukc LJ and Tommy ET 1996. Variability in pecan flowering. *Fruit Varieties J* 50:140-150.

Haulik TK and Holtzhausen LC 1988. Anatomy of staminate flower ontogeny of the pecan as determined by scanning electron microscopy. *S Afr J Plant Soil* 5(4): 205-208.

Heaton EK, Shewfelt AL, Badenhop AE and Beuchat LR 1977. Pecans: Handling, storage, processing and utilization. Univ. Georgia, Georgia Expt Sta Res Bull 197.

Little EL 1979. Checklist of United States trees (native and naturalized). Agric Handbook. 541. Washington, DC: USDA Forest Service. Pp. 375.

Loucks WL and Ray AK 1973. Submersion tolerance of selected seedling trees. *J Forestry* 71 (8) 496-497.

Luza and Polito 1988. Micro sporogenesis and anther differentiation in Juglans regia L. A developmental basis for heterodichogamy in walnut. Bot Gaz 149: 30-36.

Madden GD and Brown EJ 1973. Blossom dates of selected pecans. Pecan Quarterly 7(1): 17-19.

Marquard RD 1988. Out crossing rates in pecan and the potential for increased yields. *J Amer Soc Hort Sci* 113: 84-88.

Mullenax RH 1970 Bud ontogeny, flowering habits, and disease resistance studies of pecan, Carya illinoensis (Wang) K Koch Ph D dissertation, Louisiana State Univ. Baton Rouge, LA.

Nelson TC 1965, Silvical characteristics of the commercial hickories. USDA Forest Service, Hickory Task Force Report 10, South eastern Forest Experiment Station. Asheville, NC, Pp: 16.

Santerre CR, Scouten AJ and Chinnan MS 1990. Room temperature of shelled pecans: Control of oxygen. Pro South eastern Pecan Growers' Assoc 83: 113-121.

Wetzstein HY and Sparks D 1984. The morphology of staminate flower differentiation in pecan. *J Amer Soc Hort Sci* 109: 245-252.

Woodruff JG 1979. Tree nuts. AVI Publishing Company, INC., Westport, Connecticut.

van Staden J, Dimalla GG 1976. Regulation of germination of pecan. Carya illinoensis Zeitschrift für Pflanzen- physiologie 78 66B75

Thompson. TE and Romberg LD 1985. Inheritance of hetero dichogamy in pecan. *J Heredity* 76: 456-458

Sargent CS 1965. Manual of the trees of North America (exclusive of Mexico). 2nd ed., corrected and reprinted. New York. Dover Publications, Pp: 934.

10

Advances Production Technology of Pistachio Nut

Sajeel Ahamad[1]* Ganesh Kumar Choupdar[2], Chhail Bihari[3] Amit Kumar[4], Maneesh Kumar[5] and Vinod BR[6]

[1,2,6] Ph.D. Research Scholar, Division of Food Science and Postharvest Technology, ICAR- Indian Agricultural Research Institute New Delhi 110012
[3]Ph.D. Research Scholar, Department of Horticulture(Vegetable Science), Nagaland Central University, SASRD Medziphema, Nagaland, India
[4]Ph.D. Research Scholar, Department of Fruit Science, Sardar Vallabhbhai Patel University of Agriculture and Technology, Meerut, Uttar Pradesh, India
[5]Ph.D. Research Scholar, Department of Horticulture, G.B. Pant University of Agriculture & Technology Pant Nagar, U.S. Nagar, Uttarakhand, India
Email: sajeelahamad04@gmail.com

Introduction

The pistachio nut (*Pistacia vera* L.), a small tree native to Afghanistan, Central Asia, and Iran. The *Pistacia* genus is a member of the Anacardiaceae family, which also includes the cashew, mango, sumac, and poison ivy plants. The tree yields seeds, which are frequently eaten as food. The genus *Pistacia* has about 20 species including *Pistacia vera*, *Pistacia atlantica*, *Pistacia terebinthus*, *Pistacia khinjuk*, and *Pistacia lentiscus* (Parizi *et al.*, 2016). Western Asia is the pistachio nut home region, and traders brought them to the Middle East, the Mediterranean nations, and Europe. *Pistacia vera* L. is the only species in this genus commercially grown for edible nuts. It was probably first grown in areas close to where it grew wild, as remains of pistachio nuts dating to the 6 millennium BC have been discovered in both Afghanistan and south-eastern Iran in recent times. Pistachio cultivation progressively spread to the West from the old Persian Empire, where it was widely grown. the word "Pistachio" etymology most likely stems from the Avestan word "pistak" from ancient Persian (Mandalari *et al.*2022).

Fig. 1.1: Pistachio Nut

Health benefits

In pistachio nuts more than 50% of the fat is unsaturated. There are also 20% proteins, 20% complex carbs, and pro-vitamin A. Dried fruit pistachios nut has the highest potassium content that is very good for health. It is also great supplier of energy, magnesium, phosphorus, and iron. The heart's functioning can benefit from the consumption of this fruit. Moreover, it participates in cell metabolism, and the potassium it contains controls the water content of the cells. The production of hemoglobin and myoglobin depends on iron, consumption of pistachio nuts fulfill the iron requirements. It contributes to cellular respiration, intellectual growth, and immune system amplification. A crucial component of bones and teeth is phosphorus. Moreover, magnesium aids in the formation of bones and the production of proteins. This nut is good source of phosphorus and magnesium also.

Area and production

The area and production are limited to only temperate countries. In 2020, global production of pistachios was about 1.1 million metric tons with the United States and Turkey as leading producers, together accounting for 68% of the total followed by producers Turkey, Iran, China, and Syria (FAOSTAT 2020).

Botany and taxonomy

The tree can reach a height of 10 m which is around (33 ft). It has pinnate, deciduous leaves that are 10–20 cm (3.9–7.9 inches) long. The plants have distinct male and female trees and are dioecious. The blooms are unisexual, apetalous, and produced in panicles. The edible part is an elongated seed inside the drupe. Although the seed is frequently mistaken for a nut, it is a culinary nut. Pistachio nut has a wide range of species that are mainly used as rootstocks or for oil, agro-forestry, timber production and carpentry and include *P. atlantica, P. cabulica, P. chinensis, P. falcata, P. integerrima, P. kinjuk, P. kurdica. P. lentiscus, P. mutica, P. palaestina and P. terebinthus* (Mandalari *et al.*2022).

Soil and climate

The deciduous pistachio tree grows well in dry areas and unfavorable soil. Pistacia plants thrive in desert and semi-desert environments found in temperate and sub-tropical climates. While being able to grow in a broad variety of soil types, these plants favor deep, light, or dry sand loams with a lot of lime. (Mandalari *et al.*, 2022)

Nutrient and bioactive compounds

Like all nuts, pistachios have a high-fat content that is primarily made up of mono- and poly-unsaturated fatty acids with smaller levels of saturated fatty acids. Among these fatty acids, oleic and linoleic acids account for more than half of the total fat content of pistachios. In addition, pistachios provide a healthy amount of protein (21% of their weight). Pistachios are a significant source of protein, fiber, copper, manganese, vitamin B6, thiamin, potassium, phosphorous, chromium, vitamins E and K (phylloquinone), riboflavin, folate, magnesium, iron, zinc, and selenium. They are also a good source of fiber, with 10% by weight of insoluble fiber and 0.3% of soluble fiber. Moreover, pistachios contain a variety of beneficial phenolic compounds, including lutein and zeaxanthin (xanthophyll carotenoids), (Mandalari *et al.*2022).

Pistachio (raw, unsalted) composition.

S.N.	Macronutrient and Energy Content	g/100 g
1	Protein	20.2
2	Total lipid (fat)	45.3
3	Saturated fatty acids	5.9
4	Monounsaturated fatty acids	23.3
5	Polyunsaturated fatty acids	14.4
6	Carbohydrate, by difference	27.2
7	Fiber, total dietary	10.6
8	Sugars, total	7.66
9	Starch	1.67
10	Energy	2340 kJ

Source: U.S. Department of Agriculture Food Data Central 2019. (Mandalari *et al.*2021)

Commercial varieties

The three most common varieties of pistachio are the smaller tunis, the yellow pistachio of levante, which is less popular commercially, and the green Pistachio of Sicily. Pistachio variations are categorized based on where they were grown or by culture. Each nation has it own options, with the main variations being the colour, flavour, size, harvesting season, and quality. The male cultivar standard is Peter, while the female cultivar Kerman is the most widespread in California and well-liked everywhere in the world. A selection of plants from the Kerman region (Iran) were brought in the United States for commercial usage in 1957. Iran is one of the most significant sources of pistachio germplasm in the world, with more than 70 cultivars and genotypes of female nuts and several male genotypes. Koleghoochi, Akbari, Mumtaz, Badami Zarand, Sefid pistachio Nogh, Ahmad Aghaei, Ouhadi, Khanjari Damghan, Shah Pasand Damghan, and Qazvini pistachio are the most well-known varieties of pistachios in Iran. The main pistachio cultivars in Turkey include Antep, Kirmizi, Uzun, Halebi, and Siirt (Mandalari *et al.*, 2021). Low yield and irregular output caused by alternate bearing are two of the biggest problems in pistachio farming.

Kerman Large, high-quality pistachio nuts. It was grown commercially in Iran, introduced to the United States, and is also grown in Castilla-La Mancha, Spain, where the fruit ripens during the first two weeks of September.

Peter It is used as a male cultivar with Kerman because it produces a lot of pollen and because their flowering times partially overlap. in California, chosen.

Uzun Long, transparent, and average size green. It is grown in Turkey.

Kirmizi the typical size and ruddy hue. It is the most cultivated variety in Turkey along with the cultivar Uzum.

Miwahi Abiad The white color, average size, and superb quality. produced in Turkey.

Achouri The average size, the color of red, superior quality, and high level of productivity are mostly grown in Syria.

Batouri Nuts having good quality are thick and white in color. Syrian cultivar of importance.

Sefideh-Montaz and Imperiale de Dameghan The fruit of the is spherical, thick, and yellowish. highly regarded in Iran.

Kouchka is Quite thick, cream-colored, and of high-grade pistachio.

Mateur Long, average-sized, yellowish-green fruit with excellent flavour. It was chosen in Tunis and performs well in Spain. Around the end of August, it ripens in Castilla-La Mancha.

Larnaka Pistachio of average size that is shorter than "Mateur." Cyprus-born original. It produces good results when grown in Spain and Greece.

Aegina Fruit of average size, length, and resemblance to " Mateur " It originates in Greece and does well in Spain as well.

Commercial rootstocks

Rootstock plays a significant role in plant growth and is resilient to biotic and abiotic stress. The following list includes some well-known pistachio nut roots (Vahdati *et al.*, 2021)

Badami-Riez Zarand (*P. vera*) a seedling Strong, *Phytophthora* spp. tolerant, and salt tolerant.

Sarakhs seedling (*P. vera*) Salinity tolerant and *Phytophthora* spp. vulnerable

Qazvini seedling (*P. vera*) Salinity-tolerant

Beneh (*P. atlantica* Desf. ssp. Mutica): Yield is negatively impacted by scion-rootstock incompatibility, is less vigorous and difficult to bud than "Badami-Riez Zarand," and is resistant to the root-knot nematode.

Terebinthus (*P. terebinthus*): High rusticity, less vigorous and uniform than other common rootstocks, resistant to Armillaria root rot, vulnerable to Verticillium dahliae, effective zinc and copper absorption.

Atlantica (*P. atlantica*): Greater root hypoxia tolerance and lower vigour than *P. integerrima*, is vulnerable to *Verticillium dahliae*.

P. integerrima (Integerrima): is a vigorous rootstock that is resistant to *Verticillium* wilt but is least cold tolerant of the regularly used rootstocks.

Propagation

Typically, pistachio trees are multiplied by budding onto an appropriate pistachio rootstock. In nurseries, these rootstocks (seedlings) can be grown. Typically, budding takes place in the fall, and depending on the size of the rootstock, the trees are planted the same year or the year after seedling.

Alternate bearing

One of the main problems with pistachio production is the uneven yield brought on by alternate bearings and the low yield. As the flower buds separate during the years of abundant crop production yet then disappear, the alternate bearing phenomena is odd. The bud drop begins precisely at the end of June, shortly after the zygote starts to multiply, and it gets worse in July and August, when the embryo is growing the fastest. Certain growth regulators are directly engaged in bud abscission, according to the "hormonal" hypothesis. Despite all of these scientific proofs, pistachio alternating fruiting and bearing patterns may be more complex than previously thought (Rosenstock *et al.*, 2010; Martinelli *et al.*, 2018) When critically analyzing the effects of boron and NPK fertilizer, in addition to the standard grower practices of pruning and training to reduce alternate bearing, it is crucial to understand the mechanisms causing alternate bearing in nut tree crops.

Training and pruning

During their second spring, young pistachio trees are typically trained in a way that will allow them to eventually bear heavy crops that are simple to harvest. As soon as the scion (the top portion of the tree that was budded to the rootstock) reaches a height of 40 inches, we can tip the tree's top to control its shape (1 meter). At this point, this tipping will encourage the emergence of 3–4 major branches. These will be chosen as the tree's main branches. They must not cross each other because they will make up the tree's core skeleton. When the main branches are 15 inches long, we then tip each one (38 cm). A mature, healthy pistachio tree should generally only undergo mild pruning (remove no more than 10% to 15% of the tree's crown), as studies have shown that heavy pruning (more than 30% of the crown) increases the proportion of closed fruits. Nonetheless, severely pruning should be done to very weak trees. Pruning should always take place while the tree is still dormant, which is often between late winter and early spring. In order to prevent them from potentially bending downward over the summer, new shoots with a length of 30 inches (75 cm) or more can be removed. Shoots should be positioned at a 45-degree angle or less. The growth of the crown skeleton is accelerated by this procedure, which also encourages side vegetation (Küden *et al.*, 1997).

Planting and spacing

Pistachio trees should typically be planted 1 inch lower than they were when they were nurtured in nurseries or containers. The spacing of the plants is influenced by irrigation. Plants in irrigated gardens need to be spaced 6 x 6 metres apart in a grid layout. Trees can be spaced 8 × 10 metres apart in locations without irrigation. One male tree should be planted for every eight female plants in order to produce pistachio nuts, and the ratio should be 1:8 to 1:10. (one male and ten female trees).

Irrigation management

Droughts are not a problem for pistachio trees, but they should always have enough moisture. The best method for retaining water is mulching. For effective water management, drip irrigation can be used. Steer clear of any water logging situations. No irrigation is required during the rainy season (Kirnak *et al.*, 2001).

Manure and fertilizer

Pistachio plants need the same amount of nitrogen as other nut trees need because nitrogen is an essential nutrient for all nut crops. But, the following year after planting should be when fertilizers are applied. Throughout the growing season, 450 gm of ammonium sulphate should be given to each pistachio tree in two divided doses. Later years should give actual nitrogen (N) of 45 to 65 kg per acre. During the growing season, nitrogen should be applied in two separate dosages (Malakouti, M. J. 2005)

Intercultural operation

Pistachio trees need to be educated to develop into modified open-vases-shaped trees that grow upward and outward. For optimum fruit set and flower development, which may be required by the fourth or fifth winter season, the tree's center should remain open to the sun. The thinning process should include the removal of secondary branches. After the tree's framework is in place, only a few small pruning cuts ought to be necessary. For healthy tree growth and the production of high-quality nuts, weed management is another duty in the pistachio industry. Clean up the area between trees since weeds there compete with them for nutrition. Pre-emergent and/or post-emergent herbicide treatments should be applied primarily on the berms. Mulching the trees prevents weed growth and keeps the soil's moisture content high.

Major Insect pest and their management

Navel orangeworm (*Amyelois transitella*) It enters pistachio orchards from almond orchards after the almond crop has been harvested. This problem arises after nut fill. To get to the nut, an orangeworm burrows into fragile or broken hulls. Crop losses arise from the larvae eating the nut and leaving behind frass and webbing. As a result of this insect's damage, mould can grow and produce aflatoxin, rendering the pistachios unusable. There are two navel orangeworm parasitic wasps, however neither one significantly reduces the infestation. To control this pest, pyrethroids and organophosphates are frequently used. The carbamate pesticide carbaryl can also be used to treat navel orangeworms (Mourikis *et al.,* 1997).

Oblique-banded leafroller (*Choristoneura rosaceana*) Defoliation and stem damage are the results, which happen between late April and about August. According to field studies, oblique-banded leafroller infestations of pistachios result in larger navel orangeworm populations. In order for Bacillus thuringiensis (*Bt*) to be effective, it may need to be administered more than once to control oblique-banded leafroller. To control this bug, pyrethroids or organophosphates are commonly treated in May and between July and August (Mourikis *et al.,* 1997)

Citrus flat mite (*Brevipalpus lewisi*), which damages citrus trees in the summer, is most active from June to August. The navel orangeworm needs a location to overwinter, and the citrus flat mite damages the nuts and stems as well as having a chance to keep the nuts on the tree. Although there are predators of the citrus flat mite, they might not be able to substantially manage the population to lessen pest-related damage. If necessary, sulphur might be administered once during the season.

In the spring and summer, soft scales especially European fruit lecanium (*Parthenolecanium corni*) and frosted scale (*P. pruinosum*) because damage until the temperatures rise too high (Mourikis *et al.,* 1997)

"Major Diseases and their management

Botryosphaeria panicle and shoot blight (*Botryosphaeria dothidea*) It appears towards the end of May and can cause the tree damage up until October. Before harvest, damage is frequently the worst. Together with cankers and black lesions, this illness also causes nuts to stay on the tree, which the navel orangeworm uses as a place to overwinter. F benomyl for Botrytis control aided in the control of Botryosphaeria but is currently being discontinued. Fungicides are administered by air blast ground sprayers, although under wet field circumstances, aerial sprays can be required (Michailides *et al.*, 1994)

Botrytis shoot blight (*Botrytis cinera*) It occurs between the middle of April and the beginning of May, and if it's rainy and cool, it might last longer. Young branches that are affected by the illness wilt and die. Male flowers that are affected by the disease may produce less pollen, which will reduce output. Managing Botrytis shoot blight will involve pruning diseased shoots (Michailides *et al.*, 2004)

Alternaria late blight (*Alternaria alternata*) Mid-summer through the fall are the times when it causes very serious damage manifests. Lesions brought on by Alternaria late blight can cause defoliation, future yield loss, fruit discoloration, and losses from mould in the nuts. Moreover, this illness can make the hull more vulnerable to navel orangeworm damage. This disease can be controlled by changing irrigation practises, such as ceasing irrigation for the first two weeks of August in cases of significant outbreaks (Chao *et al.*, 2001).

Harvesting: One of the most significant elements impacting the quality of pistachios is the right timing of harvest. Pistachios are mechanically shaken from the trees in the U.S. beginning in late August or early September. If pistachios are left on the tree for too long or if the hulls come into touch with the shells for a long time after harvest, the shells may become stained. As a result, during the harvest season, the sector concentrates its harvesting operations for six to eight weeks. It takes longer for pistachio plants to yield nuts. Five years after being planted, these budded trees are ready to start producing fruit, and it will take them until the 12th year to attain their optimum yield. The fifth year after budding is when pistachio trees start to produce. But, the seventh or eighth season is when a sizable crop is really gathered. Until around the 12th year does the first full production year begin. When the hull separates readily from the shell, maturity can be easily identified. This time frame typically lasts between six and ten days. During harvesting, exercise caution by avoiding undeveloped kernels (Mandalari *et al.*2022).

Postharvest management: The pistachio nut is one of the main food commodities impacted by food safety and trade difficulties related to aflatoxin contamination. *Aflatoxins* are regarded as strong teratogens and carcinogens for both people and farm animals. A number of *Aspergillus* species, primarily *A. flavus* and *A. parasiticus*, produce aflatoxins. This fungi's spores are frequently found in the air and soil in temperate and tropical agricultural areas. The overall aflatoxin action threshold level for major trading partners of American agricultural products is established at 4 ng/g (ppb). The 20ppb limit suggested by the US Food and Drug Administration for domestic foods is much lower than this action level. All of California's commercial domestic production of these tree nuts occurs in the United States. Moreover, 50 to 75 percent of domestically produced tree nuts are exported, primarily to EU nations that follow the 4ppb action threshold level. The moisture content in pistachio nuts should not exceed more than 10 % during storage, (Campbell *et al.*, 2003).

Yield

The yield of pistachio nut varies from cultivar, climate, and crop management techniques all have a significant impact on the yield. After 10 to 12 years of budding, a yield of 8 to 10 kg can typically be obtained per fully developed tree.

Future prospects

Pistachio nut crops have limited availability due to not lack of good planting materials. To develop good planting material or true-to-type plants that give higher yield in temperate region. Moreover, to develop good pollinizer cultivars that give higher yields compare to existing old varieties and also need to develop good postharvest management practices during handling, transportation as well as during storage it reduces the damage to commodity and provide better quality to consumers.

References

Arpaci, S., Tekin, H., & Aksu, Ö. (1994, September). Improvement of pruning techniques for bearing pistachio nut trees. In *I International Symposium on Pistachio 419* (pp. 253-258).

Campbell, B. C., Molyneux, R. J., & Schatzki, T. F. (2003). Current research on reducing pre- and post-harvest aflatoxin contamination of US almond, pistachio, and walnut. *Journal of Toxicology: Toxin Reviews*, *22*(2-3), 225-266.

Chao, C. C. T., Parfitt, D. E., & Michailides, T. J. (2001). Alternaria late blight (Alternaria alternata) resistance in pistachio (Pistacia vera) and selection of resistant genotypes. *Journal of the American Society for Horticultural Science*, *126*(4), 481-485.

Ferguson, Louise, Adel Kader, and Jim Thompson. "Harvesting, Transporting, Processing and Grading." Fruits and Nuts. UC Davis. 164-169. n.d. Web. 10 Mar. 2015.

Kirnak, H., Ak, B. E., & Acar, I. (2001). Irrigation and irrigation management strategies of pistachio orchards. *Cahiers Options Méditerranéennes*, (56), 271-278.

Klein, Bob. Research director, California Pistachio Commission, Personal communication. 19 November 2001.

Küden, A., Ikinci, A., Küden, A. B., & Tekin, H. (1997, August). Different pruning applications on pistachio and almond. In *II International Symposium on Pistachios and Almonds 470* (pp. 477-480).

Malakouti, M. J. (2005, May). Increasing the yield and quality of pistachio nuts by applying balanced amounts of fertilizers. In *IV International Symposium on Pistachios and Almonds 726* (pp. 293-300).

Mandalari, G., Barreca, D., Gervasi, T., Roussell, M. A., Klein, B., Feeney, M. J., & Carughi, A. (2021). Pistachio nuts (Pistacia vera L.): Production, nutrients, bioactives and novel health effects. *Plants*, *11*(1), 18.

Martinelli F., Marchese A., Balan B., Giovino A., Caruso T., Fretto S. Marra F.P. 2018. RNA-Seq analysis to investigate alternate bearing mechanism in Pistacia vera L. *Acta Horticulturae* 1229: 71-78.

Michailides, T. J., & Morgan, D. P. (2004). Panicle and shoot blight of pistachio: A major threat to the California pistachio industry. *APSnet Feature Story. Published online.*

Michailides, T. J., Morgan, D. P., & Doster, M. A. (1994, September). Diseases of pistachio in California and their significance. In *I International Symposium on Pistachio 419* (pp. 337-344).

Mourikis, P. A., Tsourgianni, A., & Chitzanidis, A. (1997, August). Pistachio nut insect pests and means of control in Greece. In *II International Symposium on Pistachios and Almonds 470* (pp. 604-611).

Popa, D.-S.; Bigman, G.; Rusu, M.E. The role of vitamin K in humans: Implication in aging and age-associated diseases. Antioxidants 2021, 10, 566. [CrossRef]

Rosenstock T. S., Rosa U. A., Plant R. E., Brown P. H. 2010. A revaluation of alternate bearing in pistachio. Scientia Horticulturae, 124(2), 149-152

Saadlou Parizi, M., Jahanbakhsh Godehkahriz, S., Dashti, H., Saberi Riseh, R., & Hashemi Nasab, H. (2022). Genetic Relationships of Pistacia Species and Cultivars by SCoT Markers. *Agricultural Biotechnology Journal*, *14*(4), 1-20.

Statista. Pistachios: Leading Global Producers 2020/2021. Available online: https://www.statista.com/statistics/933042/globalpistachio-production-by-country/ (accessed on 16 September 2021).

U.S. Department of Agriculture, Agricultural Research Service. FoodData Central. Available online: fdc.nal.usda.gov (accessed on 16 September 2021).

University of California Pest Management Guidelines: Pistachios, Feb. 2000.

Vahdati, K., Sarikhani, S., Arab, M. M., Leslie, C. A., Dandekar, A. M., Aletà, N., ... & Mehlenbacher, S. A. (2021). Advances in rootstock breeding of nut trees: objectives and strategies. *Plants*, *10*(11), 2234.

11

Advances Production Technology of Chestnut

Ajay Kumar[*1], Amit Kumar[2], Shalini Singh[2], Khursheed Alam[3] and Vishal Gangwar[1]***

[1]Ph.D. Research Scholar, Division of Fruit Crops, ICAR-IIHR, Hessarghatta, Bengaluru-560089, Karnataka, India
[2]Research Scholar, Department of Fruit Science, College of Horticulture, Sardar Vallabhbhai Patel University of Agriculture & Technology, Meerut-250 110 (U. P.), India
[3]Research Scholar, Department of Vegetable Science, College of Horticulture, Sardar Vallabhbhai Patel University of Agriculture & Technology, Meerut-250 110 (U. P.), India
**E-mail: ajayaadihorti@gmail.com*

Introduction

Chestnut (*Castanea sativa* Mill.) belongs to the family Fagaceae and is a deciduous tree. It is also known as sweet chestnut and has a chromosome number of 2n=24. Chestnuts have been used for centuries for various purposes such as food, timber, and medicine. They are an important source of carbohydrates and contain several nutrients such as vitamins, minerals, and fiber (Laribi *et al.*, 2020). In India, chestnuts are mainly grown in the hilly regions of Jammu and Kashmir, Himachal Pradesh, and Uttarakhand.

Origin and distribution

Chestnut is believed to have originated in the Mediterranean region and was later introduced to other parts of the world. It is widely grown in Europe, Asia, and North America. In India, chestnuts were introduced by the British in the 19th century and were mainly grown as an ornamental tree. It was only after Independence that chestnuts were recognized as a potential crop for the hilly regions of the country (Bhat *et al.*, 2016).

Fig. 1.1: Chestnut

Area and production

The total area under chestnut cultivation in India is estimated to be around 1000 hectares, with an annual production of around 1000 metric tons (Bhat *et al.*, 2016). The major states growing chestnuts are Jammu and Kashmir, Himachal Pradesh, and Uttarakhand.

Botany

Chestnut is a large deciduous tree that can grow up to 35 meters in height. The leaves are oblong and have serrated margins. The flowers are monoecious, with male and female flowers borne on the same tree. The male flowers are borne in long catkins, while the female flowers are borne in groups of two to three at the base of the catkins (Russo *et al.*, 2019).

Flowering, pollination and fruit set

Chestnut trees start flowering in May-June, depending on the altitude and temperature. The male flowers release pollen, which is carried by the wind to the female flowers. The fruit set is dependent on the number of female flowers that get pollinated. The fruit is a spiny, bur-like capsule that contains one to three nuts.

Soil and climatic requirements

Chestnuts prefer well-drained, deep, and fertile soils. They grow best in soils with a pH range of 6.0-7.5. The ideal temperature range for chestnut growth is 18-22°C. They require a cold winter period for good growth and fruiting (Bhat *et al.*, 2016).

Propagation and rootstocks

Chestnuts are propagated by seed or grafting. The seedlings take around 6-7 years to start bearing fruits. Grafting is preferred for commercial orchards as it helps in early fruiting and production of uniform trees. The most common rootstocks used for chestnut grafting are *Castanea sativa*, *Castanea crenata*, and *Castanea mollissima* (Sousa *et al.*, 2020).

Use of growth regulator

Growth regulators such as gibberellins and cytokinins are used to induce early flowering and increase fruit set in chestnut trees (Rugini *et al.*, 2019)

Planting and orchard establishment

Chestnuts are usually planted in pits dug at a spacing of 10-12 meters. The pits should be filled with topsoil and well-rotted manure. The best time for planting is during the monsoon season. The orchard should be established on a sloping terrain to ensure good drainage. The orchard floor should be kept weed-free and a cover crop can be grown to improve soil fertility and prevent soil erosion (Bhat *et al.*, 2016).

Commercial varieties

In India, some of the popular varieties of chestnuts grown include the Bouche de Betizac, Marron Comballe, Marron De Lyon, and Marron Impérial. These varieties are known for their good yield and quality (Bhat *et al.*, 2016).

Training and pruning

Chestnut trees need proper training and pruning to ensure good tree shape and maximum fruit production. Pruning should be done during the dormant period to remove dead and diseased wood, and to open up the canopy for better sunlight penetration. Training should be done to develop a strong central leader and well-spaced lateral branches (Bhat *et al.*, 2016).

Nutrition management

Chestnut trees require adequate nutrition for good growth and fruiting. The fertilizer application should be based on soil and leaf analysis. Nitrogen, phosphorus, and potassium are the primary macronutrients required by chestnut trees. Micronutrients such as zinc, manganese, and iron are also essential for good growth and fruiting (Bhat *et al.*, 2016).

Water management

Chestnut trees require regular irrigation during the dry season to ensure good growth and fruiting. The amount of water required depends on soil type, climate, and tree age. Overwatering should be avoided as it can lead to root rot and other diseases (Bhat *et al.*, 2016).

Intercultural operations

Intercultural operations such as weed control, mulching, and cover cropping are important for chestnut orchard management. Weeds compete with the chestnut trees for nutrients and water and can reduce tree growth and yield. Mulching helps to conserve soil moisture, suppress weeds, and improve soil fertility. Cover cropping helps to improve soil fertility and prevent soil erosion (Bhat *et al.*, 2016).

Weed management

Weed management in chestnut orchards can be done through mechanical, chemical, or cultural methods. Mechanical methods include hand weeding, hoeing, and plowing. Chemical methods include the use of herbicides, but care should be taken to avoid damage to the chestnut trees. Cultural methods include mulching, cover cropping, and intercropping (Bhat *et al.*, 2016).

Physiological disorders and their management

Chestnut trees are prone to several physiological disorders such as internal breakdown, brown spots, and ink disease. The internal breakdown is caused by inadequate pollination or high temperatures during fruit development. Brown spot is caused by a fungal infection, while ink disease is caused by a bacterial infection. Proper orchard management and timely application of fungicides and bactericides can help prevent these disorders (Bhat *et al.*, 2016).

Major insect-pests and their management

Chestnut trees are susceptible to several insect pests such as chestnut weevil, leaf miner, and gall wasp. Chestnut weevil and leaf miner damage the fruits, while gall wasp causes the formation of galls on the leaves and twigs. Insect pest management can be done through cultural, mechanical, and chemical methods. Cultural methods include pruning, removal of infested plant material, and use of pheromone traps.

Mechanical methods include handpicking of insects, while chemical methods include the use of insecticides (Bhat *et al.*, 2016).

Major diseases and their management

Chestnut trees are prone to several fungal and bacterial diseases such as chestnut blight, ink disease, and root rot. Chestnut blight is caused by the fungus *Cryphonectria parasitica,* while ink disease is caused by the bacterium *Pseudomonas syringae pv. syringae*. Root rot is caused by the fungus *Phytophthora innamomic*. Disease management in chestnut orchards can be done through cultural, mechanical, and chemical methods. Cultural methods include sanitation practices such as removal of infected plant material, crop rotation, and planting resistant cultivars. Mechanical methods include pruning, while chemical methods include the use of fungicides and bactericides (Bhat *et al.*, 2016).

Maturity indices

Chestnuts are harvested when they are fully mature, and the nut inside the spiny husk has reached its maximum size. The maturity of chestnuts can be determined by their weight, size, and color. Chestnuts are ready for harvest when the spiny husk starts to open up and the nut inside can be easily removed (Bhat *et al.*, 2016).

Harvesting and post-harvest management

Chestnuts are harvested by handpicking when they are fully mature. The spiny husks are removed, and the nuts are graded based on their size and quality. Chestnuts should be handled with care during harvesting and post-harvest management to prevent bruising and damage to the nuts. Chestnuts can be stored in a cool, dry place for several weeks, but they are best consumed fresh (Bhat *et al.*, 2016).

Yield

The yield of chestnuts varies depending on the cultivar, orchard management practices, and environmental factors. In India, the average yield of chestnuts is around 8-10 kg per tree per year (Bhat *et al.*, 2016).

Future prospects

There is a lot of scope for crop improvement in chestnut production in India. Efforts are being made to develop new cultivars with improved yield, quality, and disease resistance. The use of biotechnology tools such as genetic engineering and molecular markers can help accelerate the crop improvement process. Future prospects for chestnut production in India look promising, as there is a growing demand for high-quality, nutritious nuts in the domestic and international markets (Bhat *et al.*, 2016).

References

Bhat, J. A., Kumar, M., Bussmann, R. W., & Ahmad, M. (2016). Ecological status and traditional knowledge of medicinal plants in Kedarnath Wildlife Sanctuary of Garhwal Himalaya, India. *Journal of Ethnobiology and Ethnomedicine*, 12(1), 47.

Bhat, Z.A., Rather, A.H., and Mir, M.A. (2016). Advances in production technology of chestnut in India. *Indian Journal of Horticulture*, 73(2), 198-205.

FAOSTAT. (2021). Crops/Regions/World list/Chestnut, with shell. Retrieved from http://www.fao.org/faostat/en/#data/QC/

Laribi, R., Kouki, K., Bettaieb, I., & Mougou, A. (2020). Castanea sativa Mill.: Phytochemical screening and biological activities. *Plants*, 9(6), 707.

Rugini, E., Cristofori, V., Silvestri, C., & Tacchini, M. (2019). Effect of different plant growth regulators on in vitro multiplication and rooting of Castanea sativa Mill. and Castanea crenata Sieb. et Zucc. *Scientia Horticulturae*, 256, 108597.

Russo, G., Vivaldi, G. A., Macchioni, N., & Vendramin, G. G. (2019). Castanea sativa Mill. chestnut genetic resources: from a living germplasm bank to the reconstruction of the history of cultivated varieties. *Frontiers in Plant Science*, 10, 1143.

Sousa, A. C. B., Baptista, P., Chaves, I., Pinto, G., & Lopes, J. (2020). Evaluation of vegetative compatibility in chestnut trees (*Castanea sativa* Mill.) in Portugal. *Forests*, 11(4), 445.

12

Advances Production Technology of Olive

Amit Kumar*[1], Om Prakash[2], Rahul Dev[3], Chhail Bihari[4] and Deepak Kumar[5]

[1]Ph.D. Research Scholar, Department of Fruit Science, College of Horticulture, Sardar Vallabhbhai Patel University of Agriculture and Technology, Meerut-250110, Uttar Pradesh, India
[2]Assistant Professor, Department of Fruit Science, College of Horticulture, Banda University of Agriculture & Technology, Banda-210001, (U. P.), India
[3]ARS-Scientist, Crop Improvement Division at ICAR-Vivekananda Parvatiya Krishi Anushandhan Sansthan Almora-263601 Uttrakhand
[4]Ph.D. Research Scholar, Department of Horticulture (Vegetable Science) Nagaland Central University, SASRD Medziphema-797106, Nagaland
[5]Ph.D. Research Scholar, Department of Vegetable Science, College of Horticulture, Sardar Vallabhbhai Patel University of Agriculture and Technology, Meerut-250110, Uttar Pradesh, India
*Email: amitworld701@gmail.com

Introduction

The olive fruit, scientifically known as *Olea europaea* L., is the edible fruit of the olive tree. It belongs to the Oleaceae family, which encompasses various other plants such as jasmine and lilac (Bartolini & Prevost, 2002). The scientific name *Olea europaea* reflects its European origin, with the genus name, Olea, derived from the Latin word "oleum" meaning oil, emphasizing the fruit's high oil content. The species name, europaea, signifies its prevalence in Europe (Vossen, 2007). In India, the olive fruit is commonly referred to as "Jaitoon" or "Jaitooni" (Bhardwaj & Gupta, 2019). These names are derived from the Hindi language and are used to denote both the tree and its fruit. The

Fig. 1.1: Olive Fruit

olive tree has a diploid chromosome number of 2n = 46 (Besnard *et al.*, 2008). This indicates that each cell of the olive tree contains 46 chromosomes, arranged in 23 pairs. Understanding the chromosome number is essential for genetic studies and breeding programs as it helps in comprehending the genetic diversity and variability of the olive fruit.

Importance

The production of olive fruit in India has gained significant importance in recent years due to various factors. Firstly, olive fruit is a valuable source of edible oil. Olive oil is renowned for its health benefits, being rich in monounsaturated fats, antioxidants, and vitamins (Vossen, 2007). The cultivation of olive fruit provides economic opportunities for farmers and contributes to the country's agricultural diversification.

Uses

Olive fruit has diverse uses that contribute to its value and demand. The primary use of olive fruit is for the production of olive oil. Olive oil is widely used for cooking, salad dressings, and various culinary purposes. It is also utilized in the food processing industry for the production of snacks, bakery products, and confectionery items. Olive oil's unique flavor profile and health benefits make it a sought-after ingredient (Vossen, 2007). Apart from oil production, olive fruit is consumed as table olives. These olives undergo processing methods to reduce their bitterness and enhance their flavor. Table olives are popular as appetizers, toppings in salads, and ingredients in various dishes. Their unique taste and texture make them a versatile and flavorful addition to Mediterranean-inspired cuisine.

Composition

Olive fruit possesses a unique composition that contributes to its nutritional value and health benefits. The fruit contains varying levels of monounsaturated fats, particularly oleic acid, which is the predominant fatty acid in olive oil (Vossen, 2007). These fats are known for their positive impact on heart health and can help in reducing LDL cholesterol levels. Additionally, olive fruit is rich in antioxidants, including phenolic compounds. These compounds have been associated with various health benefits, such as anti-inflammatory and antioxidant properties (Barranco, Fernandez-Escobar, & Rallo, 2009). The phenolic compounds in olive fruit contribute to its characteristic flavor and provide additional nutritional value.

Origin and Distribution

The olive fruit has a rich history that dates back thousands of years. It is believed to have originated in the Mediterranean region, specifically in the area known as the Levant, which includes modern-day countries such as Lebanon, Syria, Israel, and Palestine (Besnard *et al.*, 2008). The cultivation of olives can be traced back to ancient civilizations such as the Egyptians, Greeks, and Romans, who recognized the fruit's value for its oil and nutritional properties (Bartolini & Prevost, 2002). The olive tree's ability to adapt to various climatic conditions and its resilience to drought contributed to its spread and cultivation in different parts of the world. Olive fruit cultivation gradually extended to regions in Europe, North Africa, and parts of Asia, where suitable environmental conditions and farming practices were favorable (Bartolini & Prevost, 2002).

In India, olive fruit cultivation is predominantly concentrated in the northwest region, specifically in the states of Rajasthan and Gujarat. These regions provide suitable agro-climatic conditions for olive cultivation, characterized by a semi-arid climate with hot summers and mild winters (Bhardwaj & Gupta, 2019). The regions' sandy loam and loamy soils with good drainage are well-suited for olive tree growth. The introduction of olive fruit cultivation in India can be attributed to collaborative efforts between Indian and foreign research institutions and the establishment of olive orchards in the early 2000s (Bhardwaj & Gupta, 2019). The cultivation of olive fruit in India has shown promising results, with increased acreage and the adoption of advanced production technologies in recent years.

Area and production

The cultivation of olive fruit in India has witnessed steady growth over the years. Olive orchards are primarily concentrated in the northwest region, particularly in the states of Rajasthan and Gujarat. These regions offer suitable climatic and soil conditions for olive tree cultivation (Bhardwaj & Gupta, 2019). The sandy loam and loamy soils, coupled with the semi-arid climate characterized by hot summers and mild winters, provide favorable conditions for olive tree growth. The area under olive fruit cultivation in India has expanded significantly in recent years. With the adoption of advanced production technologies and increased awareness of the crop's potential, more farmers and entrepreneurs have ventured into olive farming. The establishment of new orchards and the expansion of existing ones have contributed to the growth of the olive fruit cultivation area in the country.

The production of olive fruit in India has shown a positive trend, reflecting the increased focus on olive cultivation. While the production volume is still relatively small compared to traditional crops, it has been steadily increasing over the years. The availability of improved varieties, the adoption of good agricultural practices, and the introduction of advanced production technologies have contributed to higher yields and enhanced productivity in olive orchards (Bhardwaj & Gupta, 2019). The production of olive fruit in India serves both domestic and export markets. The demand for high-quality olive products, including olive oil and table olives, has been rising, both within the country and internationally. The increasing recognition of the health benefits and culinary value of olive products has created market opportunities for Indian olive growers.

Botany

Olive fruit belongs to the botanical genus Olea and the species *Olea europaea*. It is a member of the Oleaceae family, commonly known as the olive family (Besnard *et al.*, 2007). The olive tree is an evergreen tree that can reach a height of 10 to 15 meters and has a dense canopy of grayish-green leaves. The leaves are opposite, simple, and elliptical in shape, with a leathery texture. The olive fruit itself is a drupe, characterized by a fleshy outer layer (the mesocarp) and a hard inner pit (the endocarp) enclosing the seed (Besnard *et al.*, 2007). The fruit is small and oval-shaped, with a smooth skin that turns from green to black as it ripens. The inner flesh of the olive fruit is rich in oil and has a distinct bitter taste, which is reduced through various processing methods.

Soil and climatic requirements

Olive trees prefer well-drained soils with good moisture-holding capacity. Sandy loam and loamy soils are considered ideal for olive cultivation due to their ability to retain moisture while allowing excess water to drain effectively (Bhardwaj & Gupta, 2019). These soil types provide a favorable environment for the development of the olive tree's root system. The pH level of the soil is also important for olive fruit cultivation. Olive trees thrive in slightly acidic to neutral soils, with a pH range of 6.0 to 7.5 (Bhardwaj & Gupta, 2019). Soil testing is recommended to determine the pH level and nutrient content of the soil. Based on the test results, necessary soil amendments can be made to ensure optimal soil conditions for olive tree growth.

The crop is well-suited to regions with a Mediterranean-type climate, characterized by hot, dry summers and mild, wet winters. The ideal temperature range for olive tree growth is between 20°C and 30°C (Bhardwaj & Gupta, 2019). Extreme temperatures below freezing or above 40°C can adversely affect the growth and productivity of olive trees. In terms of rainfall, olive trees are drought-tolerant and can survive with limited water availability. However, sufficient moisture during critical growth stages, such as flowering and fruit development, is essential for optimal yield. Rainfall of around 500-800 mm per year is considered suitable for olive fruit cultivation (Bhardwaj & Gupta, 2019). Supplementary irrigation may be required during periods of drought to ensure proper tree growth and fruit development.

Propagation and rootstocks

In India, olive trees are commonly propagated through various methods, including seed propagation, hardwood cutting propagation, and grafting (Bhardwaj & Gupta, 2019). Seed propagation involves collecting and germinating olive seeds. However, this method is not commonly used for commercial production due to the variability in seedling traits and the lengthy time required for trees to bear fruit. It is more commonly used for establishing rootstocks for grafting purposes. Hardwood cutting propagation involves taking stem cuttings from desired olive tree varieties and rooting them under controlled conditions. This method allows for the propagation of genetically identical trees and is commonly used for propagating specific cultivars with desirable characteristics. Grafting is another widely used method for olive tree propagation. It involves joining a desired scion (the upper portion of the plant with the desired traits) to a compatible rootstock (the lower portion of the plant with desirable root characteristics) (Bhardwaj & Gupta, 2019). This method ensures that the resulting tree inherits the desired traits of the scion while benefiting from the rootstock's vigor and adaptability.

The selection of appropriate rootstocks is crucial for the success of olive fruit cultivation. Rootstocks play a significant role in determining the tree's vigor, growth, and adaptation to different soil and climatic conditions. In India, various olive rootstocks are used, including local varieties as well as imported rootstocks from countries like Spain and Italy (Bhardwaj & Gupta, 2019). The choice of rootstock depends on factors such as soil type, climate, disease resistance, and desired tree size. Rootstocks influence the tree's tolerance to different soil conditions, such as salinity and alkalinity, and its resistance to pests and diseases (Bhardwaj & Gupta, 2019). Some common rootstocks used in India include *'Olea europea* var. *sylvestris*,' *'Olea europea* var. *cuspidata*,' *Olea europea* var *hispida* and *'Olea europea* var. *europaea*.' Rootstock selection should be based on thorough research and consultation with experts to ensure compatibility with the desired scion, adaptability to local conditions, and resistance to prevalent pests and diseases. It is essential to consider the specific requirements of the region and the objectives of the olive fruit production.

Use of growth regulator

Growth regulators play a crucial role in optimizing the growth and productivity of olive trees in India. They help in regulating various physiological processes, such as flowering, fruit set, and fruit development, leading to improved yield and fruit quality (Bhardwaj & Gupta, 2019).

One commonly used growth regulator in olive fruit cultivation is gibberellic acid (GA). GA promotes bud break, stimulates flowering, and enhances fruit set in olive trees (Bhardwaj & Gupta, 2019). The recommended dosage for GA application in olive orchards is around 20 to 40 ppm (parts per million) during the bud break stage (Bhardwaj & Gupta, 2019). Proper timing and dosage are critical to ensure effective results without any adverse effects. Another growth regulator used in olive fruit production is cytokinin. Cytokinins promote cell division and differentiation, leading to increased shoot and root growth. They are often applied to stimulate lateral shoot development and enhance branching in olive trees (Bhardwaj & Gupta, 2019). The recommended dosage for cytokinin application is typically around 10 to 20 ppm during the active growth phase. It is important to note that the use of growth regulators should be based on scientific recommendations, dosage guidelines, and proper timing to avoid any negative impacts on tree health and fruit quality. Regular monitoring and evaluation of the tree's response to growth regulators are crucial to make any necessary adjustments in the application.

Planting and orchard establishment

Site selection

The success of olive fruit production in India greatly depends on appropriate site selection for planting orchards. Several factors need to be considered to ensure optimal growth and productivity of olive trees.

Firstly, the site should have good sun exposure as olive trees require full sunlight for healthy growth and fruit development. A minimum of 6 to 8 hours of direct sunlight per day is recommended for optimal performance (Bhardwaj & Gupta, 2019). Areas with partial shade or excessive shading should be avoided. Secondly, proper spacing between olive trees is crucial to ensure adequate air circulation and light penetration, reducing the risk of diseases and ensuring optimal growth. The recommended spacing for olive tree planting is typically around 6 to 8 meters between rows and 4 to 6 meters between trees within a row (Bhardwaj & Gupta, 2019). However, the spacing may vary depending on the variety, soil conditions, and management practices.

Orchard establishment and planting techniques

Proper orchard establishment and planting techniques are essential for the long-term success of olive fruit production. The following steps should be followed:

Land preparation: The land should be cleared of any existing vegetation and prepared by plowing and leveling. Soil testing should be conducted to determine nutrient status and pH level, and necessary amendments should be made accordingly.

Planting layout: The planting layout should be well planned, considering factors such as row spacing and tree spacing. Adequate spacing allows for proper air circulation, light penetration, and ease of orchard management.

Plant selection: Healthy and disease-free nursery plants should be selected for planting. The selection should consider the specific requirements of the region, including the desired variety and rootstock combination.

Planting technique: The planting hole should be dug wide and deep enough to accommodate the root system of the plant. Care should be taken not to damage the roots during planting. After planting, the backfilling should be done with a mixture of soil and organic matter to provide adequate nutrients and ensure good root-soil contact. It is important to water the newly planted trees thoroughly to settle the soil and provide moisture for root establishment.

Regular irrigation, weed control, and mulching are essential for the establishment and growth of olive orchards. Irrigation should be carefully managed to ensure that the trees receive sufficient water during critical stages, such as flowering and fruit development, while avoiding waterlogging or excessive moisture in the root zone (Bhardwaj & Gupta, 2019). Mulching helps conserve soil moisture, suppress weed growth, and maintain stable soil temperatures.

Proper care and maintenance of the orchard during the establishment phase are crucial for the long-term success of olive fruit production in India. Regular monitoring, pruning, and nutrient management practices should be followed to promote healthy growth, optimize yield, and ensure the longevity of the orchard.

Commercial Varieties

The selection of appropriate olive varieties is a crucial aspect of olive fruit production in India. Different commercial varieties offer unique characteristics and are specifically suited for either oil production or pickle production. By choosing the right commercial varieties, farmers can maximize their yield potential and meet market demands (Bhardwaj & Gupta, 2019).

Commercial varieties for oil

Frantoio

Frantoio is a prominent olive variety widely cultivated for oil production. Originally from Italy, it is known for its high-quality oil with a distinct fruity flavor and a hint of bitterness and spiciness. Frantoio olives are medium to large-sized, and they have shown promising results in regions with a Mediterranean climate. This variety is suitable for states such as Maharashtra and Karnataka, where it has been successfully grown (Bhardwaj & Gupta, 2019).

Corolea

Corolea is an olive variety originating from Australia that has gained popularity in India for oil production. It is known for its high oil content and excellent fruit quality. Corolea olives have a medium size, and the resulting oil is characterized by its fruity aroma and mild flavor. This variety has shown adaptability to various regions in India and has been successfully cultivated in states like Tamil Nadu (Bhardwaj & Gupta, 2019).

Coratina

Coratina is an Italian olive variety widely recognized for its high oil yield and excellent resistance to diseases. The olives are medium-sized, and the resulting oil is highly prized for its strong, fruity flavor and rich antioxidant content. Coratina has demonstrated good performance in regions with a Mediterranean climate, making it suitable for cultivation in states like Maharashtra and Karnataka (Bhardwaj & Gupta, 2019).

Pendolino

Pendolino is an olive variety originating from Italy that is primarily grown for oil production. It is valued for its high oil content and low bitterness, resulting in a delicate and fruity oil. Pendolino olives are small to medium-sized, and the variety has shown adaptability to different climatic conditions. In India, Pendolino has been successfully cultivated in states such as Tamil Nadu and Maharashtra (Bhardwaj & Gupta, 2019).

Commercial varieties for pickle

Ascolano

Ascolano is an olive variety native to Italy and is predominantly used for table olive production, particularly for pickling. The olives are large in size with a mild and buttery flavor, making them suitable for pickling in brine or vinegar. Ascolano olives have been successfully cultivated in various regions of India, including Tamil Nadu and Karnataka (Bhardwaj & Gupta, 2019).

Mission

Mission is an olive variety known for its versatile use, both for oil production and pickling. Originating from Spain, it is primarily cultivated for table olive production, including pickling in various forms. Mission olives are small to medium-sized with a rich, fruity flavor. This variety has shown adaptability to different regions of India, making it suitable for pickle production (Bhardwaj & Gupta, 2019).

Grosseune

Grosseune is a French olive variety commonly used for table olive production, particularly for pickling. The olives are large-sized with a firm texture and a rich, nutty flavor. Grosseune olives have shown good adaptability to regions in India, including Tamil Nadu and Maharashtra, making them suitable for pickle production (Bhardwaj & Gupta, 2019).

Cornicabra

Cornicabra is an olive variety native to Spain, and it is well-known for its suitability for pickling. The olives are medium-sized with a distinct horn-like shape, hence the name "Cornicabra." They have a slightly bitter and tangy flavor, making them ideal for brine curing or marinating. Cornicabra olives have demonstrated successful cultivation in regions such as Tamil Nadu and Maharashtra, making them a suitable choice for pickle production (Bhardwaj & Gupta, 2019).

Training and pruning

Training and pruning are essential cultural practices in olive cultivation that contribute to the development of a productive and healthy olive orchard. Proper training and pruning techniques help shape the tree structure, promote optimal light penetration, facilitate air circulation, and improve fruit quality and yield.

Training young olive trees is crucial for establishing a strong framework and promoting desirable tree architecture. Several training techniques have been employed to optimize tree growth and facilitate efficient orchard management.

Open-center training system

The open-center training system, also known as the vase or goblet system, is commonly used for olive trees in India. This system involves the removal of the central leader during the early stages of tree development to encourage the growth of multiple lateral branches. It promotes sunlight penetration and facilitates easier harvesting and maintenance tasks (Gucci *et al.*, 2014).

Central leader training system

The central leader training system involves maintaining a dominant central leader, allowing the tree to grow in a pyramidal shape. This training technique is suitable for areas with high winds and helps ensure a balanced distribution of branches and fruiting wood (Lavee *et al.*, 2007).

Pruning techniques

Pruning plays a crucial role in maintaining the health and productivity of olive trees. It involves the removal of dead, diseased, or damaged wood, as well as unproductive branches to stimulate new growth and maintain an open canopy structure.

Winter pruning

Winter pruning, performed during the dormant season, is a common practice in olive orchards. It involves the removal of excess shoots, thinning of branches, and shaping the canopy. This pruning method helps maintain tree vigor, improve fruit quality, and facilitates the entry of sunlight into the canopy (Sánchez-Rodríguez *et al.*, 2021).

Thinning

Thinning in olive cultivation refers to the practice of selectively removing excess fruit from the trees. It helps regulate fruit load, promote better fruit development, and improve overall fruit quality and yield. Thinning is done when the olives are small and immature, allowing the remaining fruits to have more space and resources to grow. This practice optimizes resource allocation and helps mitigate alternate bearing.

Thinning is an important technique for enhancing olive production and ensuring high-quality fruit.

Renewal pruning

Renewal pruning is essential for rejuvenating old or unproductive olive trees. It involves the removal of older branches close to the ground level to stimulate the growth of new, productive shoots. This technique helps extend the productive life of the tree and ensures continuous fruit production (Barranco *et al.*, 2013).

Nutrition management

Nutrient requirements may vary depending on the age of the plants and the specific cultivar being grown. For example, young olive trees generally require higher amounts of nitrogen (N), phosphorus (P), and potassium (K) to support their growth and establishment (Barranco *et al.*, 2013). As the trees mature, the nutrient requirements may shift towards more balanced ratios to promote fruiting and oil production. The application of farmyard manure (FYM) in olive cultivation can provide essential nutrients, improve soil fertility, and enhance overall plant growth. The quantity of FYM to be applied depends on various factors, including soil characteristics, nutrient requirements, and age of the olive trees. For young olive trees, it is recommended to apply approximately 10-15 kilograms of FYM per tree per year (Nayak *et al.*, 2019). The FYM should be spread evenly around the base of the tree, ensuring that it does not come into direct contact with the trunk. It is advisable to apply FYM during the pre-monsoon period to allow the organic matter to decompose and release nutrients gradually over time. For mature olive trees, the quantity of FYM can be increased to around 20-25 kilograms per tree per year (Nayak *et al.*, 2019). The application can be done in multiple doses throughout the year, preferably during the pre-monsoon and post-monsoon periods.

To maintain optimal nutrition levels, fertilizers should be applied in appropriate quantities and ratios. The recommended quantities of N, P, and K may vary based on the stage of tree development. For young olive trees (1-3 years), a common recommendation is to apply approximately 200-300 grams of nitrogen, 100-150 grams of phosphorus, and 200-250 grams of potassium per tree per year (Barranco *et al.*, 2013). As the trees reach maturity (4 years and older), the nitrogen requirement may decrease to around 100-200 grams, while phosphorus and potassium quantities can be adjusted based on soil analysis results and specific cultivar requirements. In addition to macronutrients, the availability and uptake of micronutrients such as iron, zinc, manganese, and boron should also be considered. These micronutrients play crucial roles in various physiological processes of olive trees.

Water management

Efficient water management is crucial for olive cultivation, particularly in regions with limited water availability. Drip irrigation is a commonly used technique in olive orchards as it allows precise water application near the root zone, minimizing water loss through evaporation and runoff. It provides better control over water application and reduces the risk of overwatering or underwatering. Regular monitoring of soil moisture levels is important to adjust irrigation schedules based on the growth stage of the trees, weather conditions, and specific water requirements (Gucci *et al.*, 2014). Proper water management is crucial to mitigate water stress in olive trees. While olive trees are known for their drought tolerance, water stress at critical growth stages, such as flowering and fruit development, can significantly impact yield and fruit quality. Techniques such as mulching and soil moisture conservation help retain soil moisture, reduce evaporation, and prevent weed competition. Implementing efficient irrigation scheduling based on the specific growth stage and weather conditions helps maintain optimal soil moisture levels and minimize water stress (Gucci *et al.*, 2014).

Weed management

Weed management is a critical aspect of olive fruit production in India to ensure optimal tree growth and maximize yield. While glyphosate is a commonly used herbicide for weed control in olive orchards, there are other chemical options available as well. One alternative herbicide for weed control in olive orchards is diuron. Diuron is a pre-emergent herbicide that provides effective control of many annual grasses and broadleaf weeds. The recommended application rate of diuron for weed management in olive orchards is typically around 2-4 kilograms per hectare (Mondino *et al.*, 2014). It is important to follow the manufacturer's instructions regarding application timing and safety precautions. Another herbicide option is oxyfluorfen. Oxyfluorfen is a selective herbicide that targets broadleaf weeds and grasses. The recommended application rate of oxyfluorfen for weed control in olive orchards is approximately 1-2 liters per hectare (Fuentes *et al.*, 2019). It is crucial to apply oxyfluorfen during the pre-emergence or early post-emergence stage of weeds for optimal control.

Flowering, pollination and fruit set

Flowering, pollination, and fruit set are critical stages in the reproductive cycle of olive trees (*Olea europaea*). These processes directly impact the yield and quality of olive fruits. Understanding the timing of flowering, the mechanisms of pollination, and factors influencing fruit set is essential for successful olive fruit production in India.

Flowering

Olive trees in India typically flower during the spring season, usually from April to May (Connor & Fereres, 2005). The flowering stage is characterized by the emergence of small flowers containing stigma, style, and anthers that hold pollen grains.

Pollination

Olive trees require cross-pollination for successful fruit production as they are predominantly self-incompatible. Natural wind pollination plays a significant role in olive pollination, with pollen grains being dispersed by the wind. However, in regions with limited wind activity or a lack of compatible olive varieties nearby, artificial pollination methods may be employed (Díaz-Espejo *et al.*, 2006).

Fruit set

Fruit set occurs when the fertilized flowers develop into young fruit. Factors such as temperature, humidity, and nutrient availability influence fruit set. Insufficient pollination, adverse weather conditions, or inadequate nutrition can lead to reduced fruit set and poor crop yield. Proper management practices, including orchard nutrition and maintenance of optimal soil moisture levels, can enhance fruit set in olive trees (Rapoport & Ortega, 2010).

Physiological disorders and their management

Physiological disorders in olive fruit refer to non-infectious conditions that can affect fruit quality and appearance. Here is a brief explanation of the physiological disorders of olive fruit and their management strategies

Olive fruit pitting

Olive fruit pitting is a physiological disorder characterized by the formation of small, sunken pits or depressions on the fruit surface. It is commonly associated with environmental stress factors such as water stress, extreme temperatures, or nutrient imbalances.

Management

To manage olive fruit pitting, it is important to maintain optimal irrigation practices to prevent water stress. Adequate soil moisture levels and balanced nutrition can help minimize the occurrence of this disorder (Fernández-Escobar *et al.*, 2004). Regular soil and leaf analysis can guide appropriate fertilization practices.

Oleuropein accumulation

Oleuropein is a bitter compound present in olive fruit. High levels of oleuropein can lead to an excessively bitter taste in the fruit, affecting its quality and market value. Environmental conditions, such as temperature fluctuations and water stress, can influence the accumulation of oleuropein.

Management

Management of oleuropein accumulation involves optimizing cultural practices to minimize stress conditions. Proper irrigation scheduling, maintaining stable temperatures, and avoiding extreme fluctuations can help reduce the accumulation of oleuropein (Proietti *et al.*, 2013). Proper harvesting techniques and timing can also influence the bitterness of the fruit.

Fruit abscission

Fruit abscission refers to the premature shedding or dropping of olive fruit before maturity. This physiological disorder can be influenced by various factors, including climatic conditions, inadequate pollination, or imbalances in hormones.

Management

To manage fruit abscission, it is crucial to ensure adequate pollination by maintaining healthy populations of pollinators in the orchard. Optimizing irrigation practices to prevent water stress during critical fruit development stages can also minimize fruit abscission (De la Rosa *et al.*, 2006). Additionally, the application of plant growth regulators, such as gibberellic acid, can help reduce fruit drop.

Major diseases and their management

Anthracnose

Anthracnose is a fungal disease caused by various species of the Colletotrichum genus. It affects olive fruit, causing dark lesions with concentric rings. Infected fruit may exhibit premature ripening, shriveling, and rotting, leading to yield losses and a decline in fruit quality.

Management

Chemical control is commonly used to manage anthracnose in olive fruit production. Fungicides such as azoxystrobin and copper oxychloride have shown effectiveness against Colletotrichum spp. (Serra *et al.*, 2016). Application methods and dosages of fungicides should be based on factors such as disease severity, fruit development stage, and specific fungicide recommendations. Integrated disease management strategies, including cultural practices like orchard sanitation, pruning infected branches, and optimal irrigation, can further enhance disease control (Debbi *et al.*, 2017).

Olive Leaf Spot/Bird's-Eye Spot/Peacock Spot

Olive leaf spot, also known as bird's-eye spot or peacock spot, is caused by the fungus Spilocaea oleagina. It primarily affects olive leaves, causing small, dark lesions with light-colored centers resembling bird's eyes or peacock feathers. Severe infections can lead to defoliation, reduced fruit yield, and quality deterioration.

Management

Chemical control is often employed for managing olive leaf spot. Copper-based fungicides and tebuconazole have demonstrated efficacy against Spilocaea oleagina (Moral *et al.*, 2014). Application methods and dosages should be determined based on disease severity, leaf development stage, and specific fungicide recommendations. Integrated disease management practices, including removal of infected leaves, proper pruning for enhanced air circulation, irrigation management, and balanced nutrition, can complement chemical control and improve disease management (Moral *et al.*, 2016).

Major insect-pests and their management

Black Scale (*Saissetia oleae*)

Black Scale is a common pest that infests olive trees. It is a sap-sucking insect that forms hard, black, waxy scales on the branches and leaves of olive trees. Severe infestations can weaken the tree, reduce vigor, and affect fruit production.

Management

Management of Black Scale typically involves the use of insecticides. Systemic insecticides such as imidacloprid and contact insecticides like pyrethroids have shown efficacy against Black Scale (Daane *et al.*, 2008). Timing of insecticide applications should be based on the life cycle and activity of the scale insect. Regular monitoring and early detection are crucial for effective management.

Parlatoria Scale (*Parlatoria oleae*)

Parlatoria Scale is another scale insect species that infests olive trees. It appears as small, circular, brownish scales on the branches, leaves, and fruit of olive trees. Heavy infestations can lead to leaf yellowing, premature leaf drop, and reduced fruit quality.

Management

Control measures for Parlatoria Scale involve the use of insecticides. Systemic insecticides such as imidacloprid and pyrethroids have demonstrated effectiveness against Parlatoria Scale (Borelli *et al.*, 2012). Proper timing and repeated applications may be necessary to target different life stages of the scale insect.

Apple Weevil (*Phlyctinus callosus*)

Apple Weevil is a pest that can damage olive fruit by creating small, irregular-shaped holes on the surface. The larvae of Apple Weevil feed on the flesh of the fruit, causing further damage and potential fruit drop.

Management

Management of Apple Weevil typically involves cultural practices such as maintaining proper orchard sanitation and implementing pheromone traps to monitor and reduce populations (Department of Primary Industries and Regional Development, 2021). Insecticide applications may be considered if populations exceed economic thresholds.

Garden Weevil (*Otiorhynchus sulcatus*)

Garden Weevil, also known as Black Vine Weevil, can be a pest of olive trees. The adults feed on the leaves, creating characteristic notches along the edges, while the larvae feed on the roots, causing root damage.

Management

Integrated pest management (IPM) strategies are commonly employed to manage Garden Weevil. This may include cultural practices such as removing leaf litter and debris, using physical barriers like sticky traps, and employing biological control agents such as entomopathogenic nematodes (Berkvens *et al.*, 2009). Insecticide applications may be considered if populations exceed economic thresholds.

Rutherglen Bug (*Nysius vinitor*)

Rutherglen Bug is a piercing-sucking insect that can damage olive fruit. It feeds on the developing fruit, causing shriveling, deformities, and blemishes. Infestations can result in reduced yield and marketable fruit.

Management

Management of Rutherglen Bug often involves the use of insecticides. Pyrethroids and neonicotinoids have been used effectively to control Rutherglen Bug in olive orchards (Department of Agriculture and Food Western Australia, 2017). Proper timing and repeated applications may be necessary to target the insect at different life stages.

Olive Lace Bug (*Froggattia olivinia*)

Olive Lace Bug is a pest that can cause significant damage to olive leaves. It feeds on the underside of leaves, causing yellowing, stippling and a silvering effect. Heavy infestations can lead to defoliation and a decline in tree health.

Management

The management of Olive Lace Bug can involve several approaches. Cultural practices such as regular monitoring and maintaining proper orchard hygiene can help detect and prevent infestations. Natural enemies like predatory insects and parasitoids can provide biological control and should be conserved in the orchard (Kapantaidaki *et al.*, 2017). Insecticides, such as pyrethroids and neonicotinoids, can be used when populations exceed the economic threshold or when natural control measures are insufficient.

Maturity indices

Maturity indices for harvesting olive fruits are essential in determining the optimal time to harvest for achieving the desired quality and flavor characteristics. Various parameters are considered to assess the maturity of olive fruits.

Physical maturity indices involve evaluating changes in fruit color, size, and firmness. The color of the fruits changes from green to purple or black as they mature, depending on the cultivar (Beltran *et al.*, 2004). Additionally, the fruit size may increase, and the firmness may slightly decrease as the fruits ripen.

Chemical maturity indices focus on the biochemical composition of the fruits, particularly the oil content and fatty acid profile. The oil content increases as the fruits mature, reaching its peak at full maturity. The fatty acid composition also undergoes changes during maturation, with an increase in monounsaturated fatty acids and a decrease in polyunsaturated fatty acids (Tous *et al.*, 2002).

Sensory maturity indices involve subjective evaluations of taste, aroma, and texture by trained sensory panels. These sensory attributes determine the overall quality and flavor profile of the fruits. The sensory evaluation is usually done in combination with the physical and chemical maturity indices to ensure a comprehensive assessment. It is important to consider multiple maturity indices to make informed decisions about the optimal harvest time. Regular monitoring of these indices, along with expert knowledge and experience, helps determine the ideal maturity stage for harvesting high-quality olive fruits.

Harvesting

Harvesting the olive fruits at the right time is crucial to ensure optimum quality and oil yield. The timing of harvest depends on the maturity of the fruits and the intended use, whether for table olives or olive oil production. In India, the harvesting period for olives typically falls between November and February (Hussain *et al.*, 2018). Harvesting can be done manually by hand-picking or through mechanical methods such as shaking the tree or using mechanical harvesters (Hussain *et al.*, 2018). Care should be taken to handle the fruits gently to avoid damage or bruising. The exact timing may vary depending on the cultivar and the specific region.

Post-harvest management

Post harvest management plays a vital role in maintaining the quality and shelf life of olive fruits. Immediately after harvesting, it is essential to handle the fruits with care to prevent bruising or damage. The fruits should be promptly transported to the processing facility or storage area to minimize any quality deterioration. Sorting and grading of the fruits based on size and quality are typically done to ensure uniformity. Various post-harvest treatments, such as washing, drying, and storage in suitable conditions, are carried out to preserve the freshness and extend the shelf life of the fruits (Di Vaio *et al.*, 2018). Storage conditions such as temperature and humidity should be optimized to maintain fruit quality. Various post-harvest treatments, such as olive oil extraction or processing for table olives, should be conducted to ensure the desired product quality (Di Vaio *et al.*, 2018).

Yield

The yield of olive fruits in India can vary depending on factors such as cultivar, orchard management practices, and environmental conditions. On average, olive trees in India can yield between 5 to 10 kilograms of fruits per tree (Vishwakarma *et al.*, 2020). However, it is important to note that olive orchards may take several years to reach full production potential. Proper orchard management, including irrigation, nutrition, and pest control, can significantly impact the yield.

Future prospects

The advancement of crop improvement techniques holds significant promise for the development of improved olive cultivars in India. Breeding programs and genetic engineering are key areas that can drive crop improvement efforts. Breeding programs involve the selection and hybridization of olive varieties to create new cultivars with desirable traits such as high yield, disease resistance, and tolerance to environmental stresses. These programs can address the specific challenges faced by Indian olive growers, such as adapting to local climatic conditions and pest management. Genetic engineering offers opportunities to introduce specific genes into olive plants, enhancing traits like drought tolerance, disease resistance, and oil quality.

Sustainable and environmentally friendly farming practices also play a crucial role in crop improvement for olive fruit production. Practices such as organic farming promote the use of natural inputs, minimize the use of synthetic chemicals, and enhance soil health. Adopting precision farming techniques that leverage modern technologies like remote sensing, drones, and data analytics can optimize resource management, improve irrigation efficiency, and enhance overall farm productivity.

In addition to genetic and sustainable farming approaches, it is important to address the challenges posed by climate change. Developing olive cultivars that are more resilient to changing climatic conditions, including increased temperatures and water scarcity, is essential. Emphasizing traits like heat and drought tolerance, disease resistance, and adaptability to variable weather patterns will help ensure the long-term viability and sustainability of olive fruit production in India.

References

Barranco, D., Fernandez-Escobar, R., & Rallo, L. (2009). Olive growing. In Handbook of Plant Breeding (pp. 1-64). Springer.

Barranco, D., Fernández-Escobar, R., & Rallo, L. (2013). Olive growing. In The Olive Tree Genome (pp. 377-410). Springer.

Bartolini, G., & Prevost, G. (2002). The olive tree: Olea europaea L. CRC Press.

Beltran, G., Aguilera, M. P., & Del Rio, C. (2004). Biochemical and sensory characteristics of table olives fermented with Lactobacillus plantarum starter cultures. *Food Microbiology*, 21(6), 731-739.

Besnard, G., Garcia-Verdugo, C., Rubio de Casas, R., Treier, U. A., Galland, N., & Vargas, P. (2008). Polyploidy in the olive complex (Olea europaea): Evidence from flow cytometry and nuclear microsatellite analyses. Ann Bot, 101(1), 25-30.

Besnard, G., Khadari, B., Navascués, M., Fernández-Mazuecos, M., El Bakkali, A., Arrigo, N., Baali-Cherif, D., Brunini-Bronzini de Caraffa, V., Santoni, S., Vargas, P., & Savolainen, V. (2007). The complex history of the olive tree: from Late Quaternary diversification of Mediterranean lineages to primary domestication in the northern Levant. Proceedings of the Royal Society B: Biological Sciences, 274(1621), 1547-1554.

Bhardwaj, S. D., & Gupta, M. (2019). Cultivation of Olives. In Olive and Olive Products (pp. 51-72). Springer.

Chadha, K.L. (2019). Hand Book of Horticulture. Second Edition, Volume 1, Indian Council of Agricultural Research, New Delhi. pp. 378-381.

Connor, D. J., & Fereres, E. (2005). The physiology of adaptation and yield expression in olive. *Horticultural Reviews,* 31, 155-229.

Di Vaio, C., Marrese, P. P., Masi, P., & Catalano, P. (2018). Postharvest management of table olives. In Olives and Olive Oil as Functional Foods (pp. 173-191). John Wiley & Sons.

Díaz-Espejo, A., Fernández, J. E., & Cuevas, M. V. (2006). Influence of temperature on olive (Olea europaea L.) pollen performance in relation to relative humidity and genotype. *Environmental and Experimental Botany*, 55(3), 187-195.

Fuentes, M. J., Dominguez, M., De la Rosa, R., Franco, J. A., & Rallo, L. (2019). Evaluation of two herbicides for weed control in olive orchards. *Crop Protection*, 116, 100-105.

Gucci, R., Lodolini, E. M., Rapoport, H. F., & Tattini, M. (2014). Training systems and pruning in olive growing. In Advances in Olive Tree (*Olea europaea* L.) *Crop Science* (pp. 97-128). Springer.

Hussain, S., Mushtaq, S., Khan, I. A., & Mahmood, A. (2018). Olive (Olea europaea L.) cultivation: Present status and future prospects. *Acta Ecologica Sinica*, 38(4), 266-275.

Lavee, S., Serret, M. D., & Rapoport, H. F. (2007). Olive tree training and pruning. In Olive Production Manual (pp. 87-118). University of California Agriculture and Natural Resources.

Mondino, P., Cassano, G., Iannelli, R., Galati, A., De Luca, M., & Santoro, F. (2014). Glyphosate use for weed control in olive groves. *Journal of Agricultural Engineering*, 45(4), 115-118.

Nayak, S., Debnath, M., & Choudhary, P. (2019). Nutrient Management in Olive (Olea europaea L.) Cultivation. In Olive Cultivation in the Mediterranean Region (pp. 61-80). Springer.

Rapoport, H. F., & Ortega, E. E. (2010). Olive flowering: Field and modelling studies for understanding tree behaviour under different climatic conditions. *Spanish Journal of Agricultural Research*, 8(S1), S27-S36.

Sánchez-Rodríguez, F., Dominguez, M , and Ruiz-Sánchez, M. C. (2021). Winter pruning in olive trees: Effects on tree performance, productivity, and fruit quality. *Scientia Horticulturae*, 278, 109847. https://doi.org/10.1016/j.scienta.2020.109847

Tous, J., Romero, A., & Plana, J. (2002). Olive oil: Composition and quality. In Handbook of olive oil: Analysis and properties (pp. 63-86). Springer.

Vishwakarma, K., Jadhav, M. R., & Baig, M. M. V. (2020). Olive Cultivation in India: Current Status and Future Perspectives. Indian Horticulture Journal, 10(2), 157-161.

Vossen, P. (2007). Olive oil: History, production, and characteristics of the world's classic oils. *HortScience*, 42(5), 1093-1100.

Berkvens, N., Moens, R., & Verhaeghe, A. (2009). Occurrence and control of Otiorhynchus sulcatus (Fabricius) (Coleoptera: Curculionidae) in strawberries and ornamentals: a review. *Communications in Agricultural and Applied Biological Sciences*, 74(3), 569-579.

Borelli, S., Di Leo, V., & Russo, A. (2012). Parlatoria oleae: an unusual location on olive fruit and management perspectives. *Bulletin of Insectology*, 65(2), 211-215.

Daane, K. M., Yokota, G. Y., Krugner, R., & Vossen, P. M. (2008). Monitoring Saissetia oleae populations in California olive orchards to improve treatment timing decisions. *Journal of Economic Entomology*, 101(2), 349-358.

Department of Agriculture and Food Western Australia. (2017). Rutherglen Bug. Retrieved from https://www.agric.wa.gov.au/pest-insects/rutherglen-bug

Department of Primary Industries and Regional Development. (2021). Apple weevil. Retrieved from https://www.agric.wa.gov.au/pest-insects/apple-weevil

Kapantaidaki, D. E., Roditakis, N. E., Raptopoulos, D., Papaioannou-Souliotis, P., & Stathas, G. J. (2017). Olive lace bug, Froggattia olivinia (Costa) (Hemiptera: Tingidae): biology, ecology, and management. Insects, 8(4), 128. https://doi.org/10.3390/insects8040128

Debbi, A., Triky-Dotan, S., & Zvieli, Y. (2017). Effect of Foliar and Soil Applications of Copper Products on Olive Anthracnose. *Plant Disease*, 101(9), 1637-1644. https://doi.org/10.1094/PDIS-12-16-1781-RE

Moral, J., Trapero, A., Mercado-Blanco, J., & López-Escudero, F. J. (2014). Control of olive leaf spot caused by Spilocaea oleagina with prochloraz. *Crop Protection*, 63, 33-38. https://doi.org/10.1016/j.cropro.2014.04.011

Moral, J., de Lapeña, R., & Roca, L. F. (2016). Olive leaf spot: progress in the development of integrated control. *Phytoparasitica*, 44(5), 517-533. https://doi.org/10.1007/s12600-016-0556-x

Serra, I., Streito, J. C., Larignon, P., & Guerin-Dubrana, L. (2016). Efficacy of Azoxystrobin and Difenoconazole for Controlling Olive Anthracnose. *Plant Disease*, 100(11), 2215-2221. https://doi.org/10.1094/PDIS-01-16-0034-RE

De la Rosa, R., Rallo, L., León, L., Guerrero, A., & Ruiz-Sánchez, M. C. (2006). Water relations and growth of olive trees under different irrigation regimes. *Agricultural Water Management*, 84(1-2), 274-282. https://doi.org/10.1016/j.agwat.2006.02.013

Fernández-Escobar, R., Belaj, A., Aguilera, M. P., & Jiménez, A. (2004). Effect of different water regimes on fatty acid composition and oil quality parameters in olive (Olea europaea L.). *European Journal of Lipid Science and Technology*, 106(3), 158-163. https://doi.org/10.1002/ejlt.200300907

Proietti, P., Nasini, L., Businelli, D., Esposto, S., Veneziani, G., Taticchi, A., Selvaggini, R., & Urbani, S. (2013). Effect of fruit ripening on "Cornicabra" virgin olive oil quality: oxidative stability and phenolic compounds. Journal of Agricultural and Food Chemistry, 61(9), 2004-2012. https://doi.org/10.1021/jf304414r

13

Advances Production Technology of Kiwifruit

Afiya R S

PG Scholar, Department of Horticulture, School of Life Sciences, Central University of Tamil Nadu, Thiruvarur – 610005
Email: afiarajan@gmail.com

1.0 Introduction

Kiwi or kiwifruit (*Actinidia deliciosa*) belongs to the Actinidiaceae family that is a fruiting vine originated from Yangtze Valley of China. Kiwifruit was named because it resembles a flightless bird, kiwi which is New Zealand's national symbol. It is also referred as **"New Zealand's horticultural wonder**". In china kiwi was named as Yang Tao and also as **"China's miracle fruit"**. Over the past century, kiwis have also acquired a variety of strange names, including **"melonette**", **"wood berry", "strange fruit", "monkey peach, "macaque pear", "mihoutau", "Chinese gooseberry"** (Tyagi *et al.*, 2015) **"vine pear", "sun peach"** and the particularly enticing moniker **"hairy bush fruit**" (Hugh Cambbell and Julia Haggerty., 2008). Kiwifruit is a nutrient dense fruit which is exceptionally known for its flavour and vitamin C content (Guroo *et al.*, 2017). Digestive potential, antioxidant property, healthy metabolism, immune function and protection from coronary artery disease are the therapeutic benefits of kiwi (Feruguson., 2003; Boeing *et al.*, 2012; Singeltary., 2012). It reduces the risk against cancer (Puri and Manjula., 2018), arteriosclerosis, cardiovascular diseases and irritable bowel syndrome (Chang *et al.,* 2012) as it is a rich source of ascorbic acid and polyphenols. Due to palatability, aromatic compounds present in kiwifruit attract many consumers. Unique flavour and aroma is increased due to the presence of aldehyde E-2-hexenal, methyl benzoate, ethyl butanoate and esters. China tops in kiwifruit production followed by Italy. Two biggest exporters of Kiwi in world are Italy and New Zealand (Testolin

Fig. 1.1: Kiwifruit

& Ferguson, 2009). Kiwifruit has excellent pharmacological profile and it helps in development of nutritional supplements for welfare of mankind. Farmers can earn high return per unit area as it bears heavily every year and has no crop failure. Thus, kiwifruit has bright prospects in our country.

1.1.0 Origin and distribution

The kiwi is a fruiting vine that naturally grows as a deciduous plant in the mountain ranges of South Western China, where it is native to Central China. In the early 20th century, kiwi fruit arrived in New Zealand from China. But New Zealanders, who control more than 70% of global trade, fully realised the fruit's economic potential (Gyeltshen *et al.*, 2011). The first cultivated Kiwi fruit's were reportedly first Europeans and North Americans by early 1900's, (Janick and Paul., 2008). After 1960, its cultivation in other nations began to pick up steam, and today it is grown commercially in the USA, Italy, China, Japan, France, Germany, and Australia (Himelrick and Powell, 2002). It was first introduced to India at Lal Bagh Garden in Banglore, but the plant did not bear fruit because it did not receive the necessary chilling hours during the winter for its bud break. It was later introduced to Shimla Hills in 1963, where the plant began to bear fruit in 1969 (Pandey and Joshi, 1997).

A total of 2,47,109 hectares of land are used to grow kiwifruit globally, with a 40,22,650 tonne annual production (FAOSTAT, 2018). Arunachal Pradesh in India has approximately 3379 hectares of land under kiwifruit, producing about 6047 tonnes of fresh kiwi fruits, (Economic Survey of Arunachal Pradesh report, 2016–17). More than half of the world's kiwi production comes from Arunachal Pradesh, followed by Nagaland (Mani *et al.*, 2018). According to a 2015–16 report, despite being the state with the highest kiwifruit production, Arunachal Pradesh overall kiwifruit productivity is very low (1.2 MT/ha) when compared to other states like Nagaland (12 MT/ha) and Mizoram (3.4 MT/ha) (Yadav and Shalendra, 2018).

1.2.0 Composition

Kiwi contains a number of minerals, including calcium, iron, potassium, magnesium, manganese, copper, phosphorus, zinc, and selenium, which are essential for the body's various metabolic processes and have varying levels of bioavailability depending on how quickly each mineral is absorbed (Chawla *et al.*, 2019, 2020). It significantly contributes to the major flavonoid and phenolic contents of kiwi fruit by providing phyto-chemicals such as caffeic acid, gallic acid, syringic acid, salicylic acid, ferulic acid, and protocatechuic acid (Satpal *et al.*, 2021). Kiwifruit also showcase multiple biological activities including anti-oxidant, anti-diabetic, anti-inflammatory, anti-hypertensive, anti-carcinogenic, antifungal, antiviral, anti-asthmatic, hepatoprotective, anti-platelet, anti-nociceptive, anti-HIV, anti-microbial, anticonstipation, cytotoxic, anti-tumor and anti-thrombin (Chawla *et al.*, 2016). (Chawla *et al.*, 2016). Recently, a brand-new tocopherol analogue (tocomonoenol) in kiwifruit was found and studied. Its contribution to the kiwifruits' overall antioxidant activity has been identified (Fiorentino *et al.*, 2009). The delicate scent of kiwifruit is brought on by ethyl butanoate, hexanal, and trans hex-2-enal (Pandey and Tripathi., 2014). Up to 185 mg of vitamin C per 100 g of fresh weight (FW) of these fruits, this slightly decreases during cold storage. Additionally identified were dietary fibre (2%–3%) and the proteolytic enzyme actinidin, which is related to papain (Yamanaka *et al.*, 2004).

Table 1: Nutritional value of two commercially important kiwifruit cultivars (USDA Nutrient *Source*)

Nutritional value per 100 g	**Fuzzy kiwifruit (*Actinidia delicosa*)**	**Golden kiwifruit (*Actinidia chinensis*)**
Energy	255 kJ (61 kcal)	251 kJ (60 kcal)
Carbohydrates	14.66 g	14.23 g
Sugars	8.99 g	10.98 g
Dietary fiber	3g	2g
Fat	0.52g	0.56g
Protein	1.14g	1.23g
Vitamin		
Vitamin A	122 µg	114 µg
Thiamine (B1)	0.027 mg	0.024 mg
Riboflavin (B2)	0.025 mg	0.046 mg
Niacin (B3)	0.341 mg	0.28 mg
Panthothenic acid (B5)	0.183mg	0.5mg
Vitamin B6	0.063mg	0.057mg
Folate (B9)	25 µg	34 µg
Choline	7.8mg	5mg
Vitamin C	92.7mg	105.4mg
Vitamin E	1.46mg	1.49mg
Vitamin K	40.3 µg	5.5 µg
Minerals		
Calcium	34mg	20mg
Iron	0.31mg	0.29
Magnesium	17mg	14mg
Manganese	0.098mg	0.058mg
Phosphorous	34mg	29mg
Potassium	312mg	316mg
Sodium	3mg	3mg
Zinc	0.14mg	0.10mg

1.3.0 Uses

Because it has a sedative effect, the stem and roots are used to treat rheumatoid arthritis, liver and esophageal cancer, as well as urinary tract stones (Ferguson and Bollard. 1990, Shastri *et al.*, 2012). The ability of seed is to thin blood due to the presence of vitamin E and omega-3 fatty acids (Chawla *et al.*, 2016). Due to the high phenolic content, the peel extract has been shown to have anti-microbial and anti-viral properties. It also has anti-oxidant and anti-cancer properties (Motohashi *et al.*, 2001; Chawla *et al.*, 2016; Alim *et al.*, 2019). Due to its extensive pharmacological profile, the kiwi fruit has a number of health advantages. Cancer, diabetes, asthma, HIV/

AIDS, and cardiovascular diseases are all prevented by it. It significantly contributes to the reduction of metabolic disorders like dyslipidemia, low-density lipoprotein, triglycerides, hypertension, abnormal glucose metabolism, vascular inflammation, and homeostatic disorder (Stonehouse *et al.*, 2013).

Against cancerous cells that are malignant, kiwis act as a cytotoxic agent. Due to its high fibre content, it lowers the risk of colon cancer (Motohashi *et al.*, 2002). It has actinidin, a proteolytic enzyme that affects how proteins are digested in the stomach and the intestine. It also promotes digestive health as a result of its laxative properties, which aid in the treatment of constipation (Stonehouse *et al.*, 2013). It serves as a good option for treating type 2 diabetes mellitus because it has a low glycemic impact (Mishra *et al.*, 2017, Wilson *et al.*, 2018). It has a significant amount of carotenoids, ascorbic acid, and citric acid, which improves a person's iron status (Beck *et al.*, 2011, Diaz *et al.*, 2017). Since kiwi is a good source of folic acid, it is advantageous for women who are expecting because it promotes the growth of the foetus (Tyagi *et al.*, 2015).

2.0 Botany

2.1.0 Morphology

All *Actinidia* species are perennial, scrambling, climbing plants that are primarily deciduous but can also be evergreen in some cases. Kiwis are the fruits of nine-meter-tall trees, shrubs, or woody vines (Morton. 1987). The plant is distinguished by fleshy roots that are highly branched and have a propensity to disperse on the upper surface of soil. It has flexible stems that are tender and twinning at the top and woody at the bottom. The tendrils on the stems are used to hook onto supports (Strik and Cahn., 1996, Saquet and Brackmann., 1995). Shoots are two varieties: terminating and non-terminating. Kiwi lacks tendrils; instead, non-terminating shoots serve as their replacement (Meena and Singh. 2021). The kiwifruit leaves are simple, alternate, short or long petiolate, oval to nearly circular, cordate at the base, and deciduous. Young leaves have red hairs covering them; mature leaves have dark-green upper surfaces that are hairless (Morton., 1987, Saquet and Brackmann., 1995).

The kiwi plant is dioecious, or bisexual; male and female flowers grow on separate plants, and both sexes must be planted close together to produce fruit. The fragrant kiwifruit flowers are produced singly or in groups of three in the leaf axils. They have 5 to 6 petals, are 2.5 to 5 centimetres wide, and have central tufts of numerous stamens on both sexes. However, the female flowers lack pollen that is viable (Morton. 1987). Even though flowers are produced in few-flowered cymes, the majority of lateral flowers do not grow, resulting in only one flower per axil. The ovary is surrounded by a whorl of many stamens. They are smaller and non-viable pollen in pistillate flowers. Male flowers are typically larger than female flowers. Male flowers have a pronounced whorl of many stamens that are longer than in flowers as well as in rudimentary ovaries and reduced styles. These flowers only set very small fruit on rare cases, suggesting that a self-fruitful, hermaphroditic cultivar may be cultivated in the future (Meena and Singh. 2021). According to the cultivar, the fruit is a berry that grows in clusters that can be ovoid, spherical, or elongated. It has typically soft, brownish-green hair on the outside. Small black seeds are arranged in a circle around the centre of its green flesh (Morton., 1987, Saquet and Brackmann., 1995).

2.2.0 Flowering and fruit set

In general, kiwifruit vines have a brief juvenile period, and flowering begins after 2-3 years of planting. However, it takes longer time for own-rooted plants to bear fruit and yield sizeable and profitable harvests after the fourth or fifth year. Mixed buds, from which both leaves and flowers emerge, give rise to the flowering shoots. The basal four or five lateral buds of a current season growth that develops from year-old wood bear the flowers. Small dichasia or pseudocymes of flowers appear in the leaf axils distal to those basal buds at the base of flowering shoots. Flowers never form on the shoot apex, which remains vegetative and can be found on both terminating and non-terminating shoots. The flowering in Allison, Bruno, Monty, and Abbott begins in the third week of April and ends in the first week of May under the mid-hill conditions of Himachal Pradesh (Solan), while in cv. Hayward; the flowering begins in the first week of May and ends in the second week of May. However, in Shimla conditions, flowering starts in the last week of April and ends in the third week of May at higher elevations (Meena and Singh. 2021).

3.0 Climate and Soil

A deciduous vine known as the kiwi can grow in a variety of climates (Meena and Singh. 2021). They require soils with excellent drainage as well as enough moisture. With 600 to 800 chilling hours to break dormancy, it can be grown successfully in warm temperate to subtropical regions between 3000 and 5500 feet MSL. Contrary to other cultivars, the Hayward cultivar requires 800 hours of winter rest between 0 and 7 degrees Celsius. Frost and low temperatures (-2.5°C or lower) in the spring and fall are very harmful because they kill young shoots and fruit buds. They are vulnerable to late spring or early autumn frosts because they require a lengthy frost-free period of 270–300 days from budburst to commercial harvest (Feruguson and Seal. 2008). In the summer, high temperatures (above 38°C) combined with high solar radiation (loo) and low humidity may result in scorching of the leaves, sunburned fruits, and even plant death. For healthy growth and development, 120–150 cm of rain, evenly distributed throughout the growing season (Jindal *et al.*, 2021). Because of its large, robust leaves, vigorous vegetative growth, and vine-like habit, the plant cannot withstand strong winds throughout the year. Therefore, windy areas should be avoided or windbreaks should be built before planting (Meena and Singh. 2021).

Its cultivation requires deep, well-drained sandy-loam soil with a good amount of organic matter. The ideal pH range for fruit production and vine growth is 5.5 to 6.5. (Jindal *et al.*, 2021). Their optimum pH is around 7.0, which is neutral. Heavy or poorly drained soils, as well as salinity, are not tolerated by the plants. Heavy soils that can become flooded are completely inappropriate. The texture of the soil is improved, and soil moisture is preserved, by a thick layer of humus (Yadav *et al.*, 2017). The highest yield is obtained when the soil pH is less than 6.9, but Mn deficiency makes higher pH values, up to or greater than 7.3, detrimental. In heavy wet soils, plants cannot endure wet feet for an extended period of time. The highest yield is obtained when the soil pH is less than 6.9, but Mn deficiency makes higher pH values, up to or greater than 7.3, detrimental. In heavy wet soils, plants cannot endure wet feet for an extended period of time.

4.0 Species and cultivars

4.1.0 Species

Actinidia is a genus with 76 species and about 125 taxa that are recognised globally (Huang and Ferguson. 2007). The family Actinidiaceae, order Ericales, class Magnoliopsida, and division Magnoliophyta are all home to the genus Actinidia (Feruguson and Huang. 2007). *Actinidia chinensis* Planch. And *A. deliciosa* (A. Chev.) are two closely related species from which the kiwifruit used in international trade are selections (Feruguson and Seal. 2008). The Fuzzy Kiwifruit, which belongs to the species *A. deliciosa*, is the most popular variety of the kiwifruit. Other species also produce edible fruit, including the Golden Kiwifruit (*A. chinensis*), Chinese Egg Gooseberry (*A. coriacea*), Baby Kiwifruit (*A. arguta*), Arctic Kiwifruit (*A. kolomikta*), Red Kiwifruit (*A. melanandra*), Silver Vine (*A. polygama*), and Purple Kiwifruit (*A. purpurea*) (Jindal *et al.*,2021). *A. deliciosa* and *A. chinensis* are the most produced species, with 37% and 31% of total world production, respectively, while *A. kolomitka* and other inter-specific hybrids have the lowest production, with 8% and 7%, respectively. In fact, *A. arguta* already accounts for 17% of all kiwis produced globally (Fideghelli *et al.*, 1980).

The fuzzy kiwifruit, or *Actinidia deliciosa*, has a fuzzy, dull-brown skin and bright-green flesh. It produces more and produces larger fruits with lower respiration rate and ethylene sensitivity. Therefore, fuzzy kiwi has the longest storage (Guroo *et al.*, 2017). The fuzzy kiwifruit has heart-shaped leaves (Liang and Ferguson., 1984). Actinidia chinensis is a species with smooth, bronze skin and a beak-shaped stem attachment. The flavour of this species is sweeter and more aromatic. When ripe, the yellow fruited kiwi is less hairy than the fuzzy kiwifruit and tastes better eaten unpeeled. Because of its limited storage life, commercialization is limited (Guroo *et al.*, 2017). Hort 16A was marketed in 1992 in New Zealand under the trade name ZESPRITM GOLD (Feruguson., 1999). Hort16A experienced significant losses as a result of the PSA (*Pseudomonas syringae* pv. *actinidae*) bacterium. It has been discovered that a new golden kiwifruit variety called "Gold3," sold by Zespri under the trade name "SunGold," is more disease-resistant but is not quite as sweet as the Hort16A, with a hint of tanginess (Guroo *et al.*,2017).

Actinidia arguta berries exhibit rapid softening during storage, followed by significant dehydration. The dioecious climbing plant *Actinidia melanandra* produces clusters of large, reddish-green berries that resemble grapes and are sweeter and more flavorful than the fuzzy kiwifruit, *Actinidia deliciosa*. Berries from *A. melanandra* can be eaten whole and have a smooth texture. This abundantly reproducing species has excellent winter hardiness and can endure low temperatures of –20 °C (Kaya *et al.*, 2016). By breeding this species with various *A. arguta* genotypes, new cultivars have been created. *A. arguta* x *A. melanandra* 'Ken's Red' is a popular hybrid cultivar. *A. melanandra's* berries store a variety of biologically active substances (Guroo *et al.*, 2017). These berries are known for having incredibly high antioxidant activity. However, little is known about this species, and little research has been done on it. Additionally, it cannot be produced commercially due to its short shelf life.

4.2.0 Cultivars

As a dioecious plant, kiwifruit produces flowers on separate plants that are staminate and pistillate. After extensive research throughout New Zealand, various names for types that showed promise for commercial production were published in 1958.

A. Pistillate varieties

Hayward: This cultivar bears Hayward Wright's name because he chose it from a random seedling. It is a widely used cultivar due to its large, lovely oval-shaped fruit with excellent keeping qualities. The fruit is broad and flat, and it is much wider than it is long. It is regarded as having superior flavour and contains a lot of sugar and ascorbic acid. Compared to other cultivars, the columella, or central core, is larger. At the nodes, the fruits are borne singly. This cultivar matures slowly and bears fruit in comparably small quantities. For commercial production, it needs late flowering pollinators and close planting. Due to the longer chilling requirements of the vine, it is better suited to high hills.

Allison: Although it resembles the cultivar Abbott, its fruits are a little bit wider than they are long. The medium-sized fruits have a slight taper on both ends. It has a higher sweetness than other cultivars. It is a heavy bearer and appropriate for foothill and mid-hill regions. The flower's petals are overlapping and crinkled at the edges. Low acidity and ascorbic acid.

Abbott: Fruit is oblong in shape, medium in size and hairy all over. This cultivar has a sweet flavour, low ascorbic acid content, medium titrable acidity, heavy yielding potential, and regular bearing potential. Early flowering results in a type that matures quickly. It can be grown successfully in mid- and foothill regions because it has a low chilling requirement.

Monty: Its wide bearing range and late maturation make it a good choice for lower elevations. The fruits are similar to those of Abbott and Allison and are oblong, but they are slightly wider near the blossom end. To get medium-sized fruits, a prolific bearer is needed, and sometimes thinning is also necessary. It has a medium sugar blend and a higher acidity.

Bruno: The largest fruits of all the cultivars have a length-to-breadth ratio of roughly 1.9:1, a weight range of 40–70g per fruit, and a slight taper toward the stem end. Fruits contain ascorbic acid in large quantities. Early flowering and maturing variety that is suitable for lower hills and valleys and has a high yield.

Jintao: At the Wuhan Institute of Botany in China, a cultivar of Actinidia chinensis with yellow flesh was chosen. With respect to yield, bud density, and bud fertility, cultivar Hayward excels. Small-sized fruit (86g per fruit). It develops prior to Hayward. Fruit quality and TSS are higher, but fruit firmness is lower than in Hayward.

Chuhong: It is a brand-new cultivar with red flesh that matures quickly and was developed from *Actinidia chinensis var. rufopulpa* wild plants. The fruit has an appealing green-red-yellow pattern from outer to inner parts, and its central core has red flesh. Fruit that is ripe has an acid content of 1.47 percent, an average soluble solid content of 16.5%, and a maximum soluble solid content of 21%.

(B) Staminate varieties

Allison: Due to its consistent, abundant flowering, which coincides with pistillate kiwifruit cvs, it is considered to be the best pollinator and blooms earlier than Tomuri. Monty, Abbott, Allison, Bruno, and Monty.

Tomuri: It is a great pollinator for Hayward cultivars because it blooms late and in clusters of 5 (1-7) flowers. Flowers are borne in groups and at the end of the season, and peduncle hair is short. It is effective at pollinating the cultivar Hayward.

Matua: Flowers are borne in groups and early in the season, and peduncle hair is short. For early and midseason pistillate cultivars, the cultivar is an effective pollinator.

Table 3: Varieties under various species (Pinto and Vilela, 2018)

A. deliciosa		***A. chinensis***	***A. arguta***	***A. kolomikta***
Summer Kiwi Bruno Abbott Monty Saanichton 12 Blake	Jing Gold Hayward Allison, Elmwood Greensil Gracie Vicent	A19 (Enza Gold) JB Gold (Kiwi Kiss) Sun gold Sorel	Ananasnaja Meader Larger Jumbo Issai Transcarpathia	Santyabraskaya Szymanowski

5.0 Propagation

5.1.0 Sexual Propagation

In the past, the kiwifruit was grown from seed, giving rise to a close relative of plants with a good variety of traits. The pulp is passed through a fine sieve to extract the seed, and then washed. When the seed has been extracted, it is dried in the shade and kept in sealed polythene bags at 4-5°C until it is sown. The seed should be stratified (30–40 days) at a low temperature (3–4°C) before sowing. With a GA treatment at a concentration 2000 to 3500 ppm, seed germination can also be enhanced (Meena and Singh. 2021). The seedlings are transplanted into well-prepared nursery beds when the monsoon arrives, spacing them 15 cm apart and arranging them in rows 20–25 cm apart. The seedlings grow to graftable size in a year. By stratification (4°C for 37 days) and 2500 ppm GA3 (24-hour treatment), seeds germinate at a rate of about 94% in 3 weeks in Bruno and Abbott (Beutel, 1981). Similar to this, in Himachal Pradesh, the highest seed germination (76%) was achieved by stratifying cv. Bruno seeds for 4 weeks at 4°C and treating them for 24 hours with 2500 ppm GA3. Due to the high and uniform germination rates of seeds from this cultivar, Bruno seedlings are frequently used as rootstock in India (Chadha, 2019).

5.2.0 Vegetative Propagation

5.2.1.0 Grafting

Kiwi plant propagation also involves grafting. Despite the fact that grafting or budding a nursery plant into a seedling takes almost two years to complete. This approach is the most straightforward and economical. The best method of grafting in whip and tongue grafting (Zuccherelli., 1979). Kiwi plants can be created by tongue grafting kiwi seedlings in January and February. For dormant season grafting, the scion wood is removed from the shoot that is one year old. For dormant season grafting, the scion wood is taken from the one-year-old shoot, whereas for summer budding, it is taken from the current season's growth. With a 98% success rate, tongue grafting has been found to be the most effective of the various grafting techniques. Under Indian conditions, tongue grafting performed during the first week of February had a 90% success rate for bud-take. The highest mean bud-take rate, sprouting rate, shoot diameter, and shoot length were observed with whip and splice grafting carried out in the month of February, according to Zenginbal *et al.*(2006).

5.2.1.1.0 Rootstocks

In kiwifruit, seedlings Bruno and Abbott, are frequently used as a rootstock. Seeds of these cultivars are preferred due to their excellent germination and robust seedling vigour. Despite the fact that grafting is the main method of kiwifruit propagation in the hills of India, specialised rootstock cultivars are not frequently used there. According to some recent research, specialised rootstock varieties can increase kiwifruit's tolerance to cold temperatures and help it grow in less-than-ideal soil conditions (Oliveira and Fraser 2005). However, these rootstocks are not frequently used in India. Due to its strong vigour and Psa tolerance, Bruno is preferred. Hayward, Kaimai, Hort16A, and Bounty are additional rootstocks. The plant that was grafted onto the Hayward rootstock was less vigorous but produced more inconsistent fruit. Hort16A and Kaimai are both extremely vulnerable to Psa. Bounty, a more recently developed rootstock, is resistant to Psa and seems to be able to withstand both dry and wet foot conditions. Because Bounty is less vigorous than Bruno, full-canopy establishment must occur more quickly.

5.2.2.0 Budding

The scion wood is collected from current seasons growth for summer budding. When the stem diameter is between 6 and 8 mm at the end of the first season, seedlings are typically ready for budding. The expansion of the current season serves as the foundation for future buds. T-budding is used to place one or two buds on each stem 10 cm above the ground. Over time, chip budding causes up to ninety-five percent of the buds to be taken. The bud is firmly fixed into place by tagging with polyethylene tape. Following bud formation, during the subsequent dormant season, the highest of the vascular plant is cut higher than the union. The developing young shoot needs to be shielded from the bud, which is incredibly fragile and just breaks.

5.2.3.0 Cutting

This approach is the quickest and simplest. To grow one's own rooted plants, softwood, semi-hard, and hard wood cuttings are used. The ideal cutting is 10–15 cm long, 0.5–1.0 cm thick, with internodes that are relatively short and at least 4-5 buds. For semi-hard cuttings, the middle of the growth shoot from the current season should be used in June–July, and for hard wood cuttings, the shoot from the previous year should be used in January–February. The cuttings are first prepared, dipped in a 4000-5000 ppm IBA solution for 10–15 seconds, and then planted for rooting in a mist propagation chamber or an open field (Beutel, 1981). IBA 4000 ppm application had the greatest effect on increasing the callus percentage and the length of the longest root (Miri-Nargesi and Sedaghathoor, 2015). After 90 days, it was determined that the cuttings of different Kiwi cultivars differed in their rates of budding and rooting and that the Allison cultivar's cuttings responded well in terms of percentages of rooting and sprouting (Pio *et al.*, 2010). At the end of July, woody cuttings of the cultivars Hayward, Monty, Abbott, Allison, Tomuri, and Bruno were collected; they were 20 cm long and had a 0.8 cm basal diameter (Pio *et al.*, 2010). At the cutting's peak and base, they each made a straight cut. At the base of the cutting, incisions were made, and IBA (indol butyric acid) at 8000 mg dm-3 was applied for 10 seconds. Cuttings that were 15-20 cm long, 0.5-1.0 cm thick, and had internodes that were relatively short exhibited 70% rooting under low-cost construction. After 90 days of planting, jute bags should be removed (Bordaloi *et al.*, 2017).

5.2.4.0 Tissue Culture

For *Actinidia chinensis*, meristem tip culture was initially started. These three steps are organized as follows: Place the meristem tip, which should be between 0.2 and 0.5 mm long, in a container with the media composition specified below for each stage. Place the cultures in the dark for the entire previous night before moving them to a growth room that is 24 1°C. 20–40 days later, rosettes measuring 5–10 mm and having 4–10 leaves will be produced. Rosettes should be transferred to the culture vessel containing the proliferation medium. Additionally, keep the cultures at 24°C with a 16-hour photoperiod. Transfer shoots to the rooting media when they are 3-3.5 cm tall and have 2-4 leaves.

According to a study by Shen *et al.*, single node explant cultures can be initiated on MS media with 2 uM BA supplementation. Then, explants were transferred to MS media supplemented with 1 mg/L 1 BA and 0.1 mg/L IAA in order to achieve shoot multiplication. According to a study by Kamenicka and Rypak(1989), using MS medium containing 2.mg/L GA3 and 1.mg/L BA, Kiwi stem nodes can multiply. The cultures were kept at 24/28°C day/night temperatures with a 16-hour photoperiod. As the shoots mature, they are treated for 5 minutes with 10 mg/L or 20 mg/L of IBA. Transferring the shoots to a rooting medium containing 1 mg/L IBA plus 0.1% activated charcoal can also promote rooting. The alternative method proposed by Wiyaporn *et al.*involved growing Kiwi using explants of stem segments, nodes, or petioles in MS medium supplemented with BA and 2iP at 1–7 mg/L.

Kiwi (*A. chinensis*) can be successfully cultured using petiole segments as an explants on MS medium supplemented with sucrose @ 20 g/L, zeatin @ 1 mg/L, and IAA @ 0.025 mg/L (Pais et al, 1987). Chiariotti *et al.*(1991) used an A. deliciosa root explant to regenerate shoots and roots. The explant was cultured on Ms medium with 2 mg/L . The regeneration of whole plants using root explants on Ms media supplemented with 2 mg/L BA and 0.1 mg/L IBA was reported in a different study by Kumar and Sharma. Gui *et al.*in 1992 used endosperm as an explants on a basic MS medium supplemented with 3.mg/L zeatin, 0.5 mg/L 2,4-D, and 400 mg/L cycloheximide, the callus was in this case induced from the endosperm of Kiwi.

6.0 Cultivation

6.1.0 Planting

Kiwifruit can be planted on flat or gently sloping land. Depending on the cultivars being grown and the chosen training system, plant spacing varies. Planting is typically done with a 6 m × 4 m in varieties like Allison, Abbott, and Monty trained on T-bar Trellis system. While Hayward is planted at 5 m × 5 m spacing, it is less vigorous. For better fruit production in the pergola system of training, a spacing of 6 m × 6 m is advised. During the dormant season, planting is done (December to January).Every female plant is placed next to a male during planting, and the male plants are dispersed throughout the block. The male to female ratio must be 1:8 or 1:9 (Jindal *et al.*, 2021).

6.2.0 Canopy management

6.2.1.0 Training

Kiwifruit is a fruit that grows on a vine, much like grapes, so it needs a similar training structure. While the vines are growing, training the kiwifruit vine is a crucial component that requires constant attention. Similar to grapes, kiwifruit vines need structural support to be produced commercially. It is best to build the supporting frames right away after or even before planting. A vine can be trained using a variety of structures, including the Kniffin, T-bar, and pergola systems, but the latter two are the most well-known and frequently used.

T-bar trellis

In a T-bar trellis system, iron and concrete pillars that are about 1.8 metres tall above the ground are built in a straight line, 6 metres apart from one another. Each pole has a cross arm (1.5 m) fixed to it that carries five outrigger wires spaced 45 cm apart. Vines are trained up to the centre wire as a single stem before two leaders in the opposing directions are chosen or developed. Temporary fruiting arms are chosen at a right angle along both sides of each of these permanent leaders, spaced 25–30 cm apart (Figure 1.2).

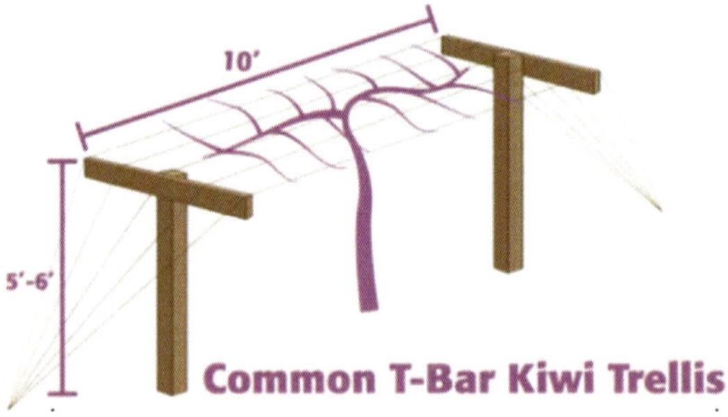

Fig. 1.2: T-bar trellis

Training on pergolas

The pergola system is similar to a T-bar for training vines. On the built-up pillars, a flat-topped network of crisscross wire is prepared. Although this system is expensive to prepare, it produces higher yields when vines are trained on it

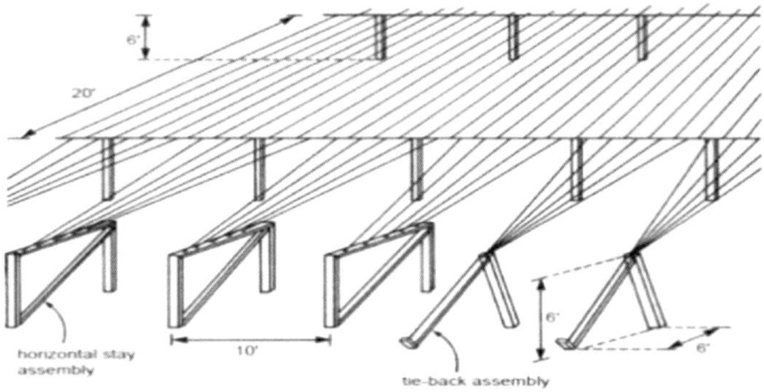

Fig. 1.3: Pergola system for Kiwifruit (Strik and Cahn, 2000)

6.2.2.0 Pruning

The primary goal of pruning is to increase fruit yield and control excessive vegetative growth. When pruning, the following guidelines should be kept in mind. The fruit grows on growth from a shoot that was planted one year ago. On current season growth, the basal buds of nodes 4–12 are the only productive nodes. Vines grow 2-3 m each year, and if they aren't pruned in the summer and winter, they become overcrowded. The shoots created by heading back on older wood not bear fruit normally in the first year.

In accordance with these guidelines, vines are pruned so that fruiting areas are accessible each year, necessitating the use of young wood. A lateral replacement system lasting three to four years achieves this. The fruiting lateral is pruned during dormancy to 2 vegetative buds after the **last fruit**. These vegetative buds produce the fruiting shoots in the second year, which are then pruned once more. Lateral shoots' arms are pruned and allowed to bear fruit for three to four years. After that, the lateral is cut off from the main branches, and other laterals are chosen and pruned appropriately to maintain a balance between vegetative and reproductive growth for the continued production of fruit. In the summer, the shoot is pruned beyond 6 to 8 buds from the final fruit in June or July. According to Biassi *et al.*, 1986 pruning at three different lengths in Hayward cultivar found that higher number of buds left on the canes the higher were the number of bearing shoots.

Zero-leaf pruning

This technique involves cutting off specific fruiting shoots on the lateral just above the final (distal) fruit so that there are no axillary buds from which secondary re-growth can arise. The fruiting canopy's light levels are kept at a good level by this pruning, which controls excessive vine vigour. It stops the vine from growing vigorously, and the zero leaf pruned shoot won't grow back much. The main goal of this technique is to reduce the price of pruning.

6.0 Pollination and role of pollinators

Given that the plants are technically dioecious, pollination is extremely important to the kiwifruit crop. One male plant is planted for every nine late pistil plants in order to ensure effective pollination. Pollination is primarily accomplished by wind and insects, but honeybees have been introduced into an orchard to further increase fruit set and size. For efficient pollination, kiwifruit orchards require 8–9 honeybee colonies per hectare (Palmer-Jones & Clinch 1974). In addition to insect pollination, hand pollination is necessary to produce fruits that are larger and of higher quality. Pollen is also known to be carried by flies and wind and to be deposited on the stigma, but these pollinators are less important than bees (Jay and Jay 1984; Craig *et al.* 1988). However, because kiwifruit flowers are dioecious and require a large amount of pollen to be transported, it is challenging to fully pollinate them. Female flowers need to receive roughly 1300-6000 pollen grains (1000 black seeds).

7.0 Intercultural Operation

7.1.0 Intercropping

Intercrops like strawberry, peas, beans, cowpeas, and vegetable crops like tomato, ginger, etc. can also be grown in the open space between the trees during the first two years.

7.2.0 Mulching

Kiwifruit orchards should be kept clean and the tree basin area should be mulched with 15 cm-thick hay grass or black polythene.

7.3.0 Nutrient management

The vines produce a lot of fruit and have a lot of vegetative growth, so they need enough manure and fertiliser to grow normally and produce fruit. The amount of fertilisers to be used depends on the age of the vine and the fertility of the soil. Each year of age, a basal dose of 20 kg FYM and 0.5 kg of an NPK fertiliser mixture containing 15% N should be applied. Every year, 60-80 kg FYM, 800g N, 560g P2O5, and 1200g K2O/vine are applied to fully bearing vines (8 years and older). FYM is applied in December and January along with full doses of P2O5 and K2O, and N fertiliser is applied in two equal dressings, half before bud-burst and the remaining full dose at the beginning of monsoon season, or in July. It is advised to use drip irrigation to apply the recommended NPK fertilisers in eight equal split doses starting at bud bursting and spaced 10 days apart. This method results in a 25% fertiliser cost savings and increases fruit yield by quality over the conventional method. According to Rathore., 1991 coupling the application of basal dose of 2-30kg FYM with 0.5 kg NPK up to 5years of age is effective.

7.3.1.0 Role of micronutrients

Calcium foliar spray caused fruit firmness, which is useful for improving fruit quality after harvest. The highest impact on yield per tree, fruit firmness, fruit pH, and percentage of the fruit's dry matter was seen when 350 g urea, 500 g potassium sulphate, 500 g superphosphate, and 80 g iron sulphate were consumed along with foliar sprays of 5 p.t. zinc chelate and iron chelate and 0.5% foliar spray of calcium at various stages for each tree. 24kg N, 3.5kg P, 4.8 kg K, 4.7kg Ca and 2kg Mg were utilized by Hayward cutivar (Ferguson and Eiseman., 1983).

7.4.0 Thinning

For the greatest amount of exportable yield, thinning can be done several times over the course of the growing season. The quantity and quality of the crop are both affected by heavy crop lead on the vine. To maintain an ideal crop load for better quality, thinning is crucial. This procedure can be started at any point in the development of the bud, even before it turns into a flower. The removal of lateral flowers or fruits is known as hand thinning. In order to increase fruit size and quality, growth regulators like CPPU (N-(2-Chloro-4-pyridyl)-N'-phenylurea) must be applied exogenously. Fruit size, weight, and yield were all significantly increased by spraying 10–20 ppm CPPU two weeks after full bloom. For better fruit size, yield, and quality fruits in Indian conditions, commercial advice has been to dip kiwifruit in 5–10 ppm CPPU.

7.5.0 Girdling

Increased dry matter, fruit weight, and flower quantity the following season are all benefits of trunk girdling. The taste of kiwifruit has been known to be enhanced by trunk and cane girdling. In order to prevent roots from competing with fruits for the carbohydrates produced by the leaves, the process isolates roots while the girdle is open. This facilitates the transfer of vital plant sugars from the roots to the developing fruits, increasing fruit size and improving fruit flavour.

8.0 Irrigation Management

Because of their vigorous vegetative growth and larger leaf surface area, kiwifruit have very high water needs. For a 16–17 m^2 canopy area, fully grown vines typically need 80–100L of water per day for total transpiration during the summer. Supplemental irrigation is crucial for increasing fruit productivity because rainfall in the majority of kiwifruit growing regions is unevenly distributed. While bearing vines should be irrigated at 20% depletion of soil moisture from field capacity (5–6 days intervals) during the summer to produce better-sized fruits, young vines should be irrigated every two to three days. In kiwifruit, drip irrigation has been found to be extremely beneficial because it increases yield by 20% and saves water by 25% compared to surface irrigation. According to Miller *et al.*(1998), moderate water stress experienced late in the growing season enhanced fruit quality by increasing soluble solids and hastening the accumulation of sucrose. *Actinidia chinensis* Planch cv, which is six months old, and the effects of prolonged flooding on its growth. Under controlled circumstances for 28 days, Abbot plants and some effects on stomatal behaviour and leaf water relations were studied (Save and Serrano, 1986). Stomatal closure and reductions in transpiration rate, xylem water potential, osmotic potential, and turgor potential were brought on by flooding. The chlorosis and necrosis of leaves, the production of hypertrophied lenticels, and the appearance of a small number of adventitious roots on the submerged portions of the stems were all also affected by flooding.

9.0 Major Insect-pest and their management

Leaf Roller

The most prevalent leaf roller is the omnivorous leaf roller (*Platynota stultana*). When they feed, it directly damages fruit by scarring the surface. Any systemic insecticides from the bud burst to pea stage applied at intervals of 15-20 days can lessen the infestation. To get rid of the eggs and caterpillars, the orchard's remaining pruned branches need to be dried out and burned. To create a shelter to live in, they often roll the leaves into a tube or web together leaf edges or leaves and fruit. According to Folias *et al.*, 1994 spraying of Lufenuran @ 5g/100 litres at an interval of 28 days will give significant control.

Greedy Scale

The leaves, canes, trunk, and fruits are all found to be infested by greedy scale (*Hemiberlesia rapax*). Because of their colour resemblance to bark, insects on vines are frequently difficult to see. It is advised to spray oil during the dormant season after spraying an organophosphate insecticide during the growing season. Blank *et al.*, 1994 recommended the spraying of organophosphate insecticide during growing season and during dormant season application of mineral oil to control greedy scale.

Nematodes

Most of the nematodes connected to kiwifruit roots are root knot nematodes (*Meloidogyne icognita*). The nematode known as root knot causes the kiwifruit root system to exhibit distinctive distortion, knotting, and galling. Roots that have been severely infected are much shorter and have fewer branches than usual. Making sure that planting material is nematode-free is crucial. By treating the bare root of plant in fenamiphos 0.1% a.i. with aqueous dip for 1hour gave a moderate control of root knot nematodes (Dale and Van der Mespel., 1972). Appication of fenamphos and oxamyl@ 1kg a.i./ha to bare root in well established kiwifruit orchard followed by irrigation will reduce nematode population (Grandison, 1983; Grandison and Spencer., 1978).

10.0 Major Diseases and their management

Crown rot, collar rot and root rot (*Phytophthora coctorum, P. cinnamoni, P. citricola, P. lateralis, P. megasperma var. megasperma*)

Disease symptoms include slowed growth of the shoots and Small and chlorotic leaves. Vine productivity may decline gradually over several seasons or suddenly in some cases. Roots and root crowns have a reddish-brown discoloration that is visible when the root is cut in half. The pathogen persists on the soil's infected plant tissues and on hosts that are weeds as oospores. Disease development is favoured by atmospheric conditions of relative humidity of 90% and temperature in the range of 10–20 °C. In 1987, Cacioppo reported that for the control of Phytpphthora spp. and Armillaria mellea trunk injection of Fosetyl aluminium had failed. Soil drenching with systemic fungicides like ridomyl are effective.

Bacterial Leaf Spot and Blossom Blight

Disease symptoms include angular spots with a halo, though not all leaf spots clearly show the halo, brown discoloration of buds, and, in more severe cases of infection, the leakage of red-rusty gum are some of the symptoms. Not all symptoms manifest at once. Strong winds, torrential rain, and pollen on the wind all contribute to the spread of disease. It is also thought that footwear, vehicles, orchard tools, animals, and people can spread. Cool temperatures, consistent rain, and high humidity are favourable conditions. It is sensitive to temperature and functional between 10 and 20°C.

Fungal Leaf Spot

The cortical tissue has a dark discoloration and white mycelial strands are present. Root-like rhizomorphs extend from roots into the soil. White mycelial mats may be present under bark close to the soil line. The fungus-pathogen lives for a long time on ill roots and wood underground. Continually wet soil encourages the spread of diseases

Storage rot

Shriveled fruit that may have grey fungal growth mostly at the stem end, occasionally around the sepals, or over the entire surface of the fruit are signs of decay and the pathogen. Internal fruit tissues that are ill appear dark green and wet. On the infected fruit, sclerotia of the fungus with a diameter of up to 0.2 inches (5 mm) may develop in the later stages of the disease. When it rains during bloom or especially at harvest, this disease is worse. For the pathogen's spores to germinate and infect, moisture is necessary.

Rhizoctonia Stem Rot

A reddish brown dry cortical root rot that may spread to the stem base is brought on by the pathogen. Later in the growing season, infections at the plant's base (cortical rot) may cause plants to break off in strong winds. Yellowing or wilting of the leaves is foliar symptoms. Infected crop debris and soil serve as the primary inoculum sources for the pathogen. The disease is more common in the wetter months. The disease develops as a result of the high humidity, moderate temperature, and moist soil.

Sclerotinia rot (*Sclerotinia sclerotiorum*)

Fruit and stems that are infected initially appear to be wet. On the area that is infected, fluffy, cottony fungal growth is visible. Sclerotia, which are small, hard, black fungi, eventually form embedded in the cottony mould. Fruits are frequently infected through the blossom end and become rotted. These rotted fruits may contain watery sclerotia. The pathogen can survive in the form of mycelium in living or dead plants, sclerotia in infected plant parts, or on the soil's surface or in contaminated seed. The development of the disease is favoured by high humidity (90-95%), the average temperature (18–25°C), and wind current. To protect fruitlet from disease, application of fungicides at late blossom and petal fall stage followed by pruning of vine to remove senescent male flowers will be effective.

11.0 Physiological Disorders

Hail Damage

Depending on the hailstorm's intensity and the time of year it occurs, kiwi plants can sustain varying degrees of damage. Fruit with pits and shredded leaves are typical signs of hail damage. Canes can be damaged, endangering the crop the following year. In severe situations, vines may become completely defoliated and lose their entire crop.

Frost Damage

The degree of damage to kiwifruit plants depends on the intensity, type, and timing of the frost as well as how long it lasts below freezing. Even slight frosts in the spring can harm the developing shoots, buds, and flowers. Flowers may fail and foliage may burn. Frost in the fall can harm fruit and cause leaf drop. The trunks and shoots of dormant vines can sustain significant damage from severe frost during the winter. On trunks, this typically manifests as splitting damage.

12.0 Harvesting and Yield

As the skin, flesh, or size of the fruit do not change at fruit maturity in kiwifruit, determining the ideal fruit maturity is challenging. For fruit harvest, a maturity index of 6.2°Brix TSS is deemed adequate (Burdon, 2015) The best index to determine the ideal harvesting time was discovered to be Days from Full Bloom to Harvest along with TSS (Burdon, 2015). In lower elevations, the number of days from full bloom to harvest ranges from 190+4 to 205+4 to 223+4 in higher elevations (Pandey and Tripathi., 2014). In addition, at maturity, the hairs on the fruit skin can be easily removed and used to determine when the fruit is ready for harvest or maturity. Fruits are gently twisted during manual harvesting, and then picked without a stalk.

13.0 Grading and Packing

In India, there is no set standard for packing and grading, but the following grades are listed: "A" grade >70g, "B" grade (50-70g), and "C" grade 50g (Rymbai *et al.*,2022). However, the international grading scale uses a "A" grade of above 100g and a "B" grade of 70–100g. 70g is rated a C (Rymbai *et al.*,2022). Kiwifruit is packaged in cardboard boxes weighing 3 to 5 kg in India. Kiwifruit, however, are packaged in tray packs in other nations (Thakur *et al.*, 2017). An out case of cardboard or plastics, a performed plastic packet tray-pack, a plastic liner corrugated strawboard, and a cardboard lid make up a standard tray. A standard tray can fit up to 33 fruits. Fruits can be kept at room temperature for 6-7 weeks, but refrigeration is required for effective commercial storage. It can be kept in cold storage for 4-6 months at 0°C and 90% relative humidity (Thakur *et al.*,2017).

14.0 Post Harvest Management

Kiwifruits are primarily consumed fresh, but they can also be processed into these products. *Actinidia delicosa* is green kiwifruit which is not typically processed because doing so damages the chlorophyll that gives it its appealing green colour (Cassano *et al.*, 2007). Additionally, the distinctive flavour of green kiwifruit is lost. The Golden Kiwifruit, *Actinidia chinensis*, has emerged as a substitute for food processors. The 'Jintao' variety has produced good results when processed; the fruit's yellow colour holds up well in things like jams and juices (Jiodeani *et al.*, 2017). However, some kiwifruits are also processed into juices, fortified drinks, purees, candies, frozen, dehydrated, and lyophilized products, kiwifruit leathers, distilled spirits, and kiwifruit preserved in syrup (Torregiani *et al.*, 1994).

Fresh-cut Kiwi fruit that has undergone minimal processing (peeling, slicing) is physically harmed or injured, which increases respiration rates and ethylene production and shortens shelf life (Abe *et al.*, 1991). It has been reported that handling and storing minimally processed kiwifruit slices at refrigeration temperatures (90% RH) associated with calcium chloride or calcium lactate treatment can extend their shelf life by up to 9–12 days (Agar *et al.*, 1999). Fresh-cut kiwifruit's dehydration, microbial spoilage, and respiratory activity were all delayed by MAP (Modified Atmospheric Package) when combined with alginate-based coatings (Mastromatteo *et al.*, 2011).

Alginate reportedly prevented fruit from drying out and reduced respiratory activity in slices of peeled kiwifruit (Mastromatteo., 2011). It has been discovered that coating red kiwifruit (*Actinidia melanandra*) with chitosan preserves higher concentrations of phenolics and ascorbic acid. Additionally, it reduced the rates of decomposition (Kaya *et al.*, 2016). In freshcut kiwifruit, it has been discovered that an aloe vera coating preserved the fruit's firmness and stopped ascorbic acid losses and ripening-related yellowing. Additionally, it decreased the growth of microorganisms (Benitez *et al.*, 2015). The mucilage made from the cladodes of the cactus pear (*Opuntia ficus-indica*) has a highly branched, intricate polymeric structure that is carbohydrate in nature. *Opuntia ficus-indica* has been shown to maintain the firmness, ascorbic acid, and pectin content of kiwifruit slices (Allegra *et al.*, 2016).

15.0 Future Thrust

In our nation, kiwifruit has a promising future and is considered to be one of the significant future commercial fruits. The economic revolution brought about by the apple for growers in high hills has been a boon for farmers in mid and low hills who grow low-value stone fruits. There is a tonne of room for its cultivation even though it is only recently being grown in our country and the total production and productivity are very low.

References

Abe K, Watada AE. (1991). Ethylene absorbent to maintain quality of lightly processed fruits and vegetables. *J Food Sci* 56: 1589-1592.

Agar IT, Massantini R, Hess-Pierce B, Kader AA.(1999). Postharvest CO2 and ethylene production and quality maintenance of fresh-cut kiwifruit slices. *J Food Sci* 64: 433-440.

Alim, A., Li, T., Nisar, T., Ren, D., Zhai, X., Pang, Y., & Yang, X.(2019). Antioxidant, antimicrobial, and antiproliferative activity-based comparative study of peel and flesh polyphenols from *Actinidia chinensis*. *Food & Nutrition Research*, 63, 1577.

Allegra A, Inglese P, Sortino G, Settanni L, Todaro A. (2016). The influence of *opuntia ficus-indica* mucilage edible coating on the quality of Hayward kiwifruit slices. *Postharvest Biol Technol* 120: 45-51.

Anonymous (2012). Production of Kiwi (fruit) by countries. *UN Food and Agriculture Organization.*

Beck, K., Conlon, C. A., Kruger, R., Coad, J., & Stonehouse, W.(2011). Gold kiwifruit consumed with an iron-fortified breakfast cereal meal improves iron status in women with low iron stores: A 16-week randomised controlled trial. *British Journal of Nutrition*, 105(1), 101–109.

Benitez S, Achaerandio I, Pujola M. (2015). *Aloe vera* as an alternative to traditional edible coatings used in fresh-cut fruits: A case of study with kiwifruit slices. *LWT - Food Sci Technol*. 61:184–193.

Beutel, J.(1981). Kiwifruit propagation. Extension Pomologist, U.C. Davis. pp 1-2.

Biassi, R. Costa, G. and Succi, F. (1986). *Proc. 22nd Int. Hort. Congress*, California, No. 304 (Abs).

Blank, R.H. Olson, M.H. Tomkins, A.R. Greaves, A.J. Waller, J.E. and Pulford, W.M. (1994). *New Zealand. J. Crop & Hort. Sci*. 22(2): 195-202.

Boeing H, Bechthold A, Bub A.(2012). Critical review: vegetables and fruit in the prevention of chronic diseases. *Eur J Nutr*. 51(6):637–663.

Bordoloi, R., Deka, B.C., Singha, A.K., Kumar, B., Jat, P.C., Sarma, C.K. and Borgohain, R. (2017). Technology inventory for Northeast India. Published by ICAR Agricultural Technology Application Research Institute, Umiam, Meghalaya – 793103. pp 163-218.

Boudreau MD, Beland FA .(2006). An evaluation of the biological and toxicological properties of Aloe Barbadensis (Miller), *Aloe vera*. *J Env Sci Health-part C* 24: 103-154. 43.

Benitez S, Achaerandio I, Pujola M, Sepulcre F .(2013). *Aloe vera* based edible coatings improve the quality of minimally processed 'Hayward' kiwifruit. *Postharvest Biol Technol* 81: 29-36.

Burdon, J. (2015). *Acta Hortic*., 1096: 257-266.

Cassano A, Donato L, Drioli E.(2007). Ultrafiltration of kiwifruit juice: Operating parameters, juice quality and membrane fouling. *J Food Eng* 79: 613-621.

Chadha, L.K. (2019). In: Handbook of Horticulture (vol.1), Indian Council of Agricultural Research. Directorate of Knowledge Management in Agriculture, New Delhi, pp. 328-335.

Chang CC, Lin YT, Lu YT. (2010). Kiwifruit improves bowel function in patients with irritable bowel syndrome with constipation. *Asia Pac J Clin Nutr*. 19(4):451–457.

Chawla, H., Parle, M., & Yadav, M. (2016). Medicinal potential and phytopharmacology of Actnidia deleciosa. *International Journal of Experimental Pharmacology*, 6(1), 20–25.

Chawla, P., Kumar, N., Kaushik, R., & Dhull, S. B. (2019). Synthesis, characterization and cellular mineral absorption of nanoemulsions of Rhododendron arboreum flower extracts stabilized with gum arabic. *Journal of Food Science and Technology*, 56(12), 5194–5203.

Chawla, P., Kumar, V., Bains, A., Singh, R., Sadh, P. K., Kaushik, R., & Kumar, N. (2020). Improvement of mineral absorption and nutritional properties of Citrullus vulgaris seeds using solid-state fermentation. *Journal of the American College of Nutrition*, 39(7), 628–635

Chesoniene L, Daubaras R, Viskelis P .(2004). Biochemical composition of berries of some kolomicta kiwi (*Actinidia kolomicta*) cultivars and detection of harvest maturity. *Acta Hortic* 663: 305-308.

Chiariotti, A., Caboni, E. and Frattarelli, A. (1991). Shoot regeneration from in vitro roots of kiwi. *Acta Horticulturae,* 289, 97±9.

Collins AR, Harrington V, Drew J, Melvin R. (2003). Nutritional modulation of DNA repair in a human intervention study. *Carcino* 24: 511-515.

Correa-Betanzo J, Jacob JK, Perez-Perez C, Paliyath G. (2011). Effect of a sodium caseinate edible coating on berry cactus fruit (*Myrtillocactus geometrizans*) phytochemicals. *Food Res Int* 44: 1897-1904.

Craig, J.L., Stewart, A.M., Pomeroy, N., Heath, A.C.G. and Goodwin, R.M. (1998). A review of kiwifruit pollination: Where to next? *New Zealand Journal of Experimental Agriculture,* 16:4, 385-399.

Dale, P.S. and Van der Mespei, G.I. (1972). *Plant Diseases Reporter* 58: 850-851.

Debersaques, F. and Mekers, O. (2010). Growth and production of kiwifruit and kiwiberry. *Soils, plant growth and crop production, 2*, p.8.

Díaz, G. A., Latorre, B. A., Lolas, M., Ferrada, E., Naranjo, P., & Zoffoli, J. P. (2017).

Identification and characterization of *Diaporthe ambigua, D. australafricana, D. novem,* and *D. rudis* causing a postharvest fruit rot in kiwifruit. *Plant Disease*, 101(8), 1402–1410.

FAO. FAOstat database. (2018).

FAOSTAT. (2018). www.faostat.org

Ferguson A, Huang H. (2007). Genetic resources of kiwifruit: domestication and breeding. *Horticultural Reviews*. 33:1–121.

Ferguson AR, Ferguson LR. (2003). Are kiwifruit really good for you? *Acta Horticulturae*. 610:131–135.

Ferguson AR, Seal AG. (2008). Kiwifruit. In: Temperate Fruit Crop Breeding, Hancock, Jim F (ed) Germplasm to Genomics, East Lansing, MI, USA.

Ferguson AR. (1999). New temperate fruits: Actinidia chinensis and Actinidia deliciosa. In: J Janick, editor. Perspectives on new crops and new uses. Alexandria, VA: ASHS Press. p. 342–347.

Ferguson, A.R. and Eiseman, J.A. (1983). *New Zesland J. Agric. R*es. 26: 115-117.

Ferguson, A.R. and Ferguson, L. R. (2003). *Acta Hortic*., 610: 131-138.

Fideghelli C. (1980). Cultivar di kiwi introdotte nel mondo dal. Kiwi Informa. 2012;7–9:20–29.

Fiorentino A, Mastellone C, D'Abrosca B, Pacifico S, Scognamiglio M, Cefarelli G, Caputo R, Monaco P (2009) δ-Tocomonoenol: a new vitamin E from kiwi (Actinidia chinensis) fruits. *Food Chem.* 115(1):187–192

Fiorentino A, Mastellone C, D'Abrosca B, Pacifico S, Scognamiglio M. (2009) Tocomonoenol: A new vitamin E from kiwi (Actinidia chinensis) fruits. *Food Chem* 115: 187-192.

Folias, G. Weish, R. Bland. S. Patterson, T.M. (1994). Proc. 47th New Zealand Plant Protectopn Soc. pp. 341-345.

Gammon CS, Kruger R, Minihane AM, Conlon CA, Von Hurst PR. (2012). Kiwi fruit consumption favourably affects plasma lipids in a randomized controlled trial in hyper-cholesterolemic men. *Br J Nutrs*14: 1-11.

García JC, García G, Ciordia M. (2014). Variedades de kiwi. *Tecnología agroalimentaria*. 14:2–7.

Gonzalez, M. V., Rey, M. and Rodriguez, R. (1995). Plantlet regeneration from petioles of kiwifruit microshoots. *HortScience*, 30, 1302±3

Grandison, G.S. (1983). *Plant Diseases*, 67(8): 899-900.

Grandison, G.S. and Spencer, M.M. (1978). Proc. 31st New Zealand Weed and Pest Control Conference pp. 147-150.

Gui, Y. L. (1979). Induction of callus and regeneration of plantlets in stem segment culture of Chinese gooseberry. *Acta Botanica Sinica*, 21, 339±44

Guroo I, Wani SA, Wani SM, Ahmad M, Mir SA. (2017) A Review of Production and Processing of Kiwifruit. *J Food Process Technol*. 8: 699.

Gyeltshen, S., Tshering, K., Phuntsho, L., Khampa., Penjor, T., Tshering, G. and Tomiyasu, Y. (2011) *J. Rene. Nat. Res.* Bhutan., 7(1): 11-20.

Himelrick, D.G. and Powell, A. (2002) Kiwifruit production guide, Alabama Cooperative Extension System, www. aces. edu. pp. 1-7.

Huang HW, Ferguson AR. (2007). Actinidia in China: Natural diversity, phylogeographical evolution, interspecific gene flow and kiwifruit cultivar improvement. Acta Hortic 753: 31-40.

Hugh Campbell and Julia Haggerty, 'Kiwifruit - Early history, names and varieties', Te Ara – the Encyclopedia of New Zealand, http://www.TeAra.govt.nz/en/kiwifruit/page-2 (accessed

31 December 2022) Story by Hugh Campbell and Julia Haggerty, published 24 Nov 2008

Janick, J., & Paull, R. E. (Eds.). (2008). *The encyclopedia of fruit and nuts*. CABI.

Jay, D; Jay, C. (1984). Observations of honey bees on Chinese gooseberries ('kiwifruit') in New Zealand. Bee world 65: 155-166.

Jindal K. K., Sharma D P, Laxuman Sharma. (2021). Hi-Tech Kiwi Fruit Production Techniques in North West and North East Himalayan States Potential for Transformation from Subsistence Farming to Sustainable Horticulture Towards Improving Livelihoods of Farmers. *International Journal of Tropical Agriculture*, 39(3).

Kamenicka, A. and Rypak, M. (1989). The regeneration of Actinidia chinensis Planch cultured in vitro. *Polnohospodarstvo*, 35, 811±8.

Kaya M, Cesoniene L, Daubaris R, Leskauskaite D, Zabulione D. (2016). Chitosan coating of red kiwifruit (Actinidia melanandra) for extending of the shelf life. I*nt J Biol Macromol.* 85: 355-360.

Lintas, C., Adorisio, S., Cappelloni, M. and Monastra, E., (1991). Composition and nutritional evaluation of kiwifruit grown in Italy. *New Zealand journal of crop and horticultural science*, *19*(4), pp.341-344.

Mani, G., Kundra, A. and Haque, A. (2018) *Agric. Econ. Res. Rev*., 31(1): 123-130.

Mastromatteo M, Mastromatteo M, Conte A, Del Nobile MA. (2011). Combined effect of active coating and MAP to prolong the shelf life of minimally processed kiwifruit (*Actinidia deliciosa* cv. Hayward). *Food Res Int* 44: 1224-1230.

Meena, R.K. and Singh, K.K., 2021. 16. Advanced Production Technology of Kiwifruit Under Temperate Region. In: Akshat Uniyal & Isha Sharma, (eds) Trends in Agriculture: Traditional And Modern Approaches. pp. 303-312

Miri-Nargesi, S.M. and Sedaghathoor, S. (2015) South-West J. *Hort. Biol. Environ*., 6(1): 33-42.

Mishra, S., Edwards, H., Hedderley, D., Podd, J., & Monro, J. (2017). Kiwifruit non-sugar components reduce glycaemic response to co-ingested cereal in humans. *Nutrients,* 9(11), 1195.

Morton J. Kiwifruit (*Actinidia deliciosa*). (1987). In: Julia F. Morton, Miami, editors. Fruits of warm climates. *FL.* p. 293–300.

Motohashi, N., Shirataki, Y., Kawase, M., Tani, S., Sakagami, H., Satoh, K., Kurihara, T., Nakashima, H., Mucsi, I., Varga, A., & Molnár, J. (2002). Cancer prevention and therapy with kiwifruit in Chinese folklore medicine: A study of kiwifruit extracts. Journal of Ethnopharmacology, 81(3), 357–364. https://doi.org/10.1016/S0378-8741(02)00125-3

Motohashi, N., Shirataki, Y., Kawase, M., Tani, S., Sakagami, H., Satoh, K., Kurihara, T., Nakashima, H., Wolfard, K., Miskolci, C., & Molnár, J. (2001). Biological activity of kiwifruit peel extracts. *Phytotherapy Research*, 15(4), 337–343.

Oliveira. M. M. and L. G. Fraser. (2005). Chapter 1.1 Actinidia spp. Kiwifruit. in *Biotechnology of Fruit and Nut Crops*. R. E. Litz editor. CAB International, Oxfordshire, England.

Oms-Oliu G, Rojas-Grau MA, Gonzalez A, Varela P, Soliva-Fortuny G. (2010) Recent approaches using chemical treatments to preserve quality of fresh-cut fruit: A review. *Postharvest Bio Technol* 57: 139-148.

Pais, M. S. S., Oliveira, M. M. and Barroso, J. (1987). Use of petiole segments of Actinidia chinensis (kiwi) for plant differentiation and production of friable calli for protoplast isolation. *Acta Horticulturae*, 212, 687±90.

Palmer-Jones, T.; Clinch, P.G. (1974): Observations on the pollination of Chinese gooseberries variety 'Hayward'. *New Zealand journal of experimental agriculture* 2: 455-458.

Pandey, G. and Joshi, B.D. (1997). Kiwi - A future fruit for Indian hills, NBPGR, New Delhi-12.

Pandey, G. and Tripathi, A.N. (2014) Kiwifruit A boon for Arunachal Pradesh. Programme Coordinator, Krishi Vigyan Kendra, Yachuli (ICAR), Govt. of Arunachal Pradesh, Lower Subansiri District - 791120 (Arunachal Pradesh). pp. 1-17

Patel RK, Yadav DS and Yadav RM. (2005). Physico-Chemical Studies of Kiwi Cultivars under mid hill situation of Meghalaya. *Indian J. Hill farming*, 18 (1&2):165-166.

Pinto T, Vilela A. (2018). Kiwifruit, a botany, chemical and sensory approach a review. *Adv Plants Agric Res.* 8(6):383–390.

Pio, R., Costa, F.C., Curi, P.N. and Moura, P.H.A. (2010) *Sci. Agraria.*, 11: 271-274.

Puri DK, Manjula SN. (2018). Nutritional Content and Therapeutic Potential of Kiwifruit. *World Journal of Pharmacy and Pharmaceutical Sciences.* 7(6):536–565.

Rana, V.S. (2013). In: Fruit production in India (Ed. W.S. Dhillon), Narendra Publishing House, Delhi, pp. 319-328.

Rathore, D.S. (1991). In: Temperate fruit, eds. S.K. Mitra, D.S. Rathore and T.K. Bose, *Horticulture and Allied Publishers*, 27/3, Chakraberia Lane, Calcutta.

Revilla, M. A., Rey, M. A., Gonzalez-Rio, F., Gonzalez, M. V., Diaz-Sala, C., & Rodriguez, R. (1992). Micropropagation of Kiwi (*Actinidia spp.*). *High-Tech and Micropropagation* II.

Rico D, Martin-Diana AB, Barat JM, Barry-Ryan C. (2007). Extending and measuring the quality of fresh-cut fruit and vegetables: A review. *Trend Food Sci Technol* 18: 373-386.

Saquet AA, Brackmann A.(1995). Kiwifruit Production. Ciência Rural. 25(1):177–182.

Satpal D, Kaur J, Bhadariya V, Sharma K. (2021). Actinidia deliciosa (Kiwi fruit): A comprehensive review on the nutritional composition, health benefits, traditional utilization, and commercialization. *J Food Process Preserv.*

Shen, X. S., Wan, J. Z. and Luo, W. Y. (1990). Propagation in vitro of Chinese gooseberry (Actinidia chinensis) through the development of axillary buds. *Scientia Horticulturae*, 42, 45±54.

Singletary K. (2012). Kiwifruit: Overview of Potential Health Benefits. *Nutrition Today.* 47(3):133–147

Stonehouse, W., Gammon, C. S., Beck, K. L., Conlon, C. A., von Hurst, P. R., & Kruger, R. (2013). Kiwifruit: Our daily prescription for health. *Canadian Journal of Physiology and Pharmacology*, 91(6), 442–447.

Strik B, Cahn H. Growing Kiwifruit. (1996). Corvallis: Oregon State University.

Tahir, J., Hoyte, S.and Bassett, H. (2019). *Hort. Res.*, 6:

Testolin, R. and Fergusson, A.R. (2009). Kiwifruit (Actinidia spp.): Production and Marketing in Italy. *New Zealand J. Crop Hortic. Sc.,* 37: 1-32.

Torreggiani D, Forni E, Pelliccioni L. (1994). Modification of glass transition temperature by osmotic dehydration and kiwi color frozen stability: *Research and innovation in the food industry.* pp: 621-630.

Torres JA. (1994). Edible film and coating from protein: Protein functionality in food systems. Marcel Dekker press.

Tyagi, S., Nanher, A. H., Kumar, V., Nishad, S. K., Ahmad, M., & Bhamini, K. (2015). Kiwifruit: Health benefits and medicinal importance. *Rashtriya Krishi*, 10(2), 98–100.

Watada AE, Abe K, Yamuchi N. (1990). Physiological activities of partially processed fruits and vegetables. *Food Techno*l 44: 116−122.

Wilson, R., Willis, J., Gearry, R., Hughes, A., Lawley, B., Skidmore, P., Frampton, C., Fleming, E., Anderson, A., Jones, L., Tannock, G., & Carr, A. (2018). SunGold kiwifruit supplementation of individuals with prediabetes alters gut microbiota and improves vitamin C status. Anthropometric and Clinical Markers. Nutrients, 10(7), 895.

Wiyaporn, S., Subhadrabandhu, S. and Sahavacharin, O. (1990). In vitro vegetative multiplication of kiwifruit. *Acta Horticulturae*, 279, 447±59.

Workman, D. (2020). Kiwifruit exports by country, Worldstopexports.com/kiwi-fruit exports-by-country

Yadav, H. and Shalendra. (2018). Kiwi value chain analysis and market assessment for Lower Subansiri district, Arunachal Pradesh: Report submitted to Mission for Integrated Development of Horticulture, Ministry of Agriculture and Farmer's Welfare, Govt. of India.

Yamanaka M, Oota T, Fukuda T, Nishiyama I. (2004). Varietal differences in actinidin concentration and protease activity in fruit juice of Actinidia species. *J Japanese Soc Food Sci Technol* 51: 491-494.

Zenginbal, H., Ozcan, M. and Demir, T. (2006*). Int. J. Agric*. 1: 597-602

Zhang L, Li ZZ, Wang YC, Jiang ZW, Wang SM. (2010). Vitamin C, flower color and ploidy variation of hybrids from a ploidy unbalanced *Actinidi*a interspecifi cross and SSR characterization. *Euphytica* 175: 133-143

Zuccherelli, G. (1959*). Frutticoltura*, 41 (3/4): 39-43.

14

Advances Production Technology of Strawberry

Amit Kumar*[1], Ramdeen Kumar[2], Shalini Singh[1], Sajeel Ahamad[3] and Vibhu Pandey[1]

[1]*Ph.D. Research Scholar, Department of Fruit Science, College of Horticulture, Sardar Vallabhbhai Patel University of Agriculture and Technology, Meerut-250110, Uttar Pradesh, India.*

[2]*Ph.D. Research Scholar*, Department of Horticulture, Institute of Agriculture Science, Banaras Hindu University, Varanasi-221005 Uttar Pradesh, India

[3]*Ph.D. Research Scholar, Division of Food Science & Post-harvest Technology, Indian Agricultural Research Institute, New Delhi-110012 India*

**Email: amitworld701@gmail.com*

Introduction

Strawberry (*Fragaria x ananassa* Duch.) is a perennial herb that belongs to the Rosaceae family (Bayat *et al.*2019). Strawberry is sweet and has a distinct fragrance. Strawberries are a popular fruit that are known for their sweet taste, bright red color, and high nutritional value. It is abundant in several vitamins and nutrients, including carotene and anthocyanins. It is popularly regarded as the important table fruit of millions of people around the globe and well-liked by consumers. Strawberry, which has 98% edible component, is regarded as a whole fruit. typically eaten fresh and raw. They are tasty, healthful, and low in calories. Strawberries grow well in India in conditions that are often temperate in the north, subtropical in the plains, or tropical at high altitudes. The primary reason strawberry production is still in its infancy in India, despite its handsome fruits and palatable flavour, is that it is a perishable

Fig. 1.1: Strawberries

crop (Pramanick *et al.* 2012). The strawberry is a fruit that grows all over the world. The Roman poets Virgil and Ovid wrote about it in the first century, and gardeners have been growing strawberries in England since the sixteenth century (Boriss *et al.* 2006).

Importance, composition and uses

Strawberries hold immense importance both in terms of economic value and nutritional benefits. They are high-value crop and have a strong demand in both domestic and international markets (Gupta *et al.*, 2019). The increasing popularity of strawberries in India is mainly due to their unique taste, attractive appearance, and versatility in culinary applications. Additionally, strawberries are rich in essential nutrients, antioxidants, and phytochemicals, making them a valuable addition to a healthy diet (Bhardwaj *et al.*, 2020). The growing awareness of health-conscious consumers has further boosted the demand for strawberries in the country.

The composition of strawberries contributes to their unique taste, aroma, and health benefits. Strawberries are rich in various essential nutrients, including vitamin C, folate, potassium, and dietary fiber (Savant *et al.*, 2017). They are also a good source of antioxidants, such as anthocyanins, flavonoids, and ellagic acid, which possess numerous health-promoting properties (Bhardwaj *et al.*, 2020). These bioactive compounds in strawberries have been associated with potential anti-inflammatory, anticancer, and cardioprotective effects.

Strawberries find extensive usage in various sectors, ranging from culinary applications to the pharmaceutical and cosmetic industries. In the culinary field, strawberries are widely used in the production of jams, jellies, preserves, ice creams, desserts, and beverages (Narzary *et al.*, 2018). Their vibrant color and sweet flavor make them a popular choice for garnishing and enhancing the visual appeal of dishes. Furthermore, strawberries are increasingly being incorporated into health foods, smoothies, and functional beverages due to their nutritional value.

Nutrient composition

According to the USDA, the nutritive value per 100g of the edible portion of strawberries is as follows:

Nutritive value of apple fruit/100 g edible portion

Contents	Amount with Unit	Contents	Amount with Unit
Energy	32 kcal	Magnesium	13mg
Protein	0.67g	Phosphorus	30mg
Total Lipid (fat)	0.30g	Potassium	153mg
Carbohydrates	7.68g	Sodium	1mg
Fiber	1.5g	Vitamin	C 58.8mg
Sugars	4.89g	Vitamin	K 24.7μg
Calcium	21mg	Folate	23μg
Iron	0.41mg		

Source: USDA Nutrient Database: kcal = kilocalorie, g = grams, μg = micrograms; mg milligrams

Origin and distribution

The strawberry is a fruit that originated in France, and its widespread popularity can be seen by the fact that global output and area have both grown significantly in the last ten years. 50 and 30 percent, respectively, of the overall production, are accounted for by Europe and North America. France, Italy, and Poland are the top three strawberry-producing nations in Europe. Strawberries are now grown in many parts of the world, including North America, Europe, Asia, and South America. The largest producers of strawberries include the United States, Spain, Italy, and Mexico. Strawberries are also grown in many other countries, including Japan, Australia, and South Africa. They are also grown in greenhouses in colder climates. The wild strawberry, or Fragaria vesca, is believed to be the ancestor of the modern strawberry. This wild strawberry can still be found growing in the wild in many parts of Europe and North America. Over time, various cultivars of strawberries have been developed through selective breeding. These cultivars have been developed to have larger and sweeter fruit, as well as to be more resistant to disease and pests. Today, there are many different cultivars of strawberries that are grown around the world.

Area and production

The major strawberry-producing nations are Turkey, Spain, Egypt, and Mexico. The majority of strawberries produced are eaten fresh. Its cultivation was previously limited to mountainous regions in India, including Himachal Pradesh, Uttaranchal, regions of Uttar Pradesh, and the Kashmir valley. The Hilly Tracts of Himachal Pradesh, Uttaranchal, sections of Uttar Pradesh, and Kashmir are the only places in India where strawberry is grown (Lyngdoh 2014). However, that's because of the advent of day-neutral cultivars, it has expanded to tropical and subtropical regions as well. Like Solan and Kullu in Himachal Pradesh, Jylikot, Dehradun, and Nainital in Uttarakhand, Srinagar in Jammu and Kashmir, Darjeeling in West Bengal, Hissar, and Gurgaon (Haryana). Strawberry production is exclusive to Muzzafar Nagar, Saharanpur, and Meerut (U.P.), Abohar, Alwar, Jalandhar, and Gurdaspur (Punjab), Mahabaleshwar and Panchgani (Pune), and Banglore (Karnataka). Strawberries cover 3,000 hectares area in India, producing 20,000 MT in total production (Anon, 2020).

Taxonomy

The history of strawberries dates back more than 2200 years. It is believed that producers' custom of covering their plants with a layer of straw when the berries started to emerge is where the name "strawberry" originated. The 600 strawberry types that are currently accessible are descendants of five or six original wild species (Fragaria spp.), which are a rose family plant. *Fragaria chiloensis* Duch. and *Fragaria virginiana* Duch., two predominantly dioecious octoploid species, were crossed to develop the cultivated strawberry (*Fragaria X ananassa* Duch.). *Fragaria virginiana* and *Fragaria chiloensis* Duch. are thought to have spontaneously bred in Europe at the beginning of the seventeenth century. All cultivated strawberry types are octoploid (2n=56) with an X = 7 basic chromosomal number and are members of the Rosaceae family. There are at least 17 other species in the genus *Fragaria* in addition to *F. x ananassa*. Four species—*F. chiloensis, F. daltoniana, F. niigerrensis*, and *F. vesca*- out of the 17 species have been recorded to be grown in various regions of the world.

Strawberries are classified into the following categories-

Rank	Scientific Taxonomical Details
Kingdom	*Plantae* - Plants
Subkingdom	*Tracheobionta* - Vascular plants
Superdivision	*Spermatophyta* - Seed plants
Division	*Magnoliophyta* - Flowering plants
Class	*Magnoliopsida* - Dicotyledons
Subclass	*Rosidae*
Order	*Rosales*
Family	*Rosaceae* Juss. - Rose family
Genus	*Fragaria* L. - strawberry
Species	*ananassa* Duch.

Botany

Strawberries are characterized by their low-growing habit, accompanied by a cluster of leaves at the plant's base and long, slender runners that facilitate propagation and spreading. The compound leaves typically consist of three leaflets with toothed margins. The plant produces small white or pink flowers that eventually develop into juicy, red, heart-shaped fruits known as strawberries. Strawberries exhibit adaptability, thriving in a broad range of climatic conditions from temperate to subtropical regions. They prefer well-drained soil enriched with organic matter. Strawberries have a unique reproductive system, relying on cross-pollination by insects, particularly bees. The flowers contain both male and female reproductive organs, and intriguingly, the seeds are located on the exterior of the fruit, making strawberries an "accessory fruit" in botanical classification. The botany of strawberries highlights their ability to flourish in diverse environments and their fascinating reproductive mechanisms, contributing to the growth of these delightful and vibrant fruits enjoyed worldwide.

Strawberry plant parts

The strawberry plant has several distinct parts:

Leaves: The leaves of the strawberry plant are green, glossy and have a toothed edge. They are arranged alternately on the stem, and are used for photosynthesis.

Flowers: The strawberry plant produces small, white flowers that are pollinated by bees and other insects. The flowers have five petals, and are located at the tips of the runners or on short stems near the base of the plant.

Fruit: The strawberry plant produces a red, juicy fruit that is an aggregate accessory fruit. Each fruit is composed of many small drupelets that are fused together. The fruit is a popular food item and is used in a variety of desserts and other culinary applications.

Runner: The strawberry plant produces runners, or stolons, which are stem-like structures that grow horizontally above the ground. These runners eventually produce new plants at their tips.

Roots: The strawberry plant has a shallow root system, consisting of a taproot and many fibrous roots. These roots absorb water and nutrients from the soil.

Crown: The crown of the strawberry plant is the area where the leaves and roots meet. It is the most important part of the plant as it allows the plant to survive through winter and produce new leaves and runners.

Stolons give rise to a primary stem called a crown, in which leaves, roots, and stolons are generated in a repeating pattern. The inflorescence forms the end of the crown's varying number of internodes. A specialised true stem called a stolon, which differs significantly from short internode crowns in having long internodes and vascular structures Runners and stolons are prostrate growths that are produced by the axillary buds of the crown during the growing season. In the crown, inflorescence grows terminally. Dichasial cymes- a term that describes the branch structures of strawberry inflorescence—are highly variable. Two lateral branches and a central flower emerge from the principal peduncle, the inflorescence's axis (secondary flowers). Similar to this, two secondary branches give rise to tertiary ramifications, each of which ends in a tertiary bloom. The timing and quality of fruit production are greatly influenced by the inflorescence structure. It is known as a pseudocarp in botany and is an aggregate fruit.

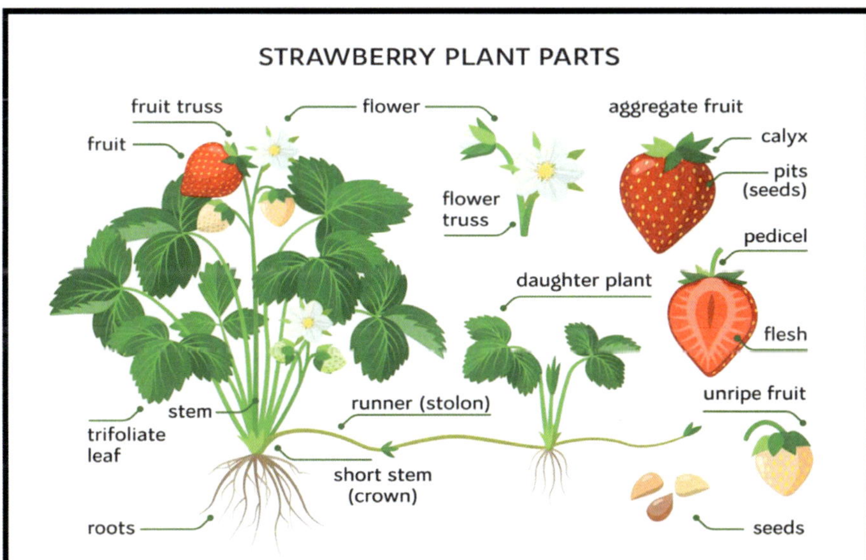

Fig. 1.2: Strawberry Plant Parts

Flowering, pollination and fruit set

Flowering

Strawberry blossoms emerge in groups. The inflorescence is a raceme with dichotomous branching. Primary, secondary, tertiary, and quaternary flowers are found in a typical inflorescence. In farmed strawberries, there are two types of flowers: hermaphrodite and pistillate.

Pollination and fruit set

The strawberry blossoms grow in groups. The raceme-shaped inflorescence has dichotomous branching. The primary, secondary, tertiary, and quaternary flowers are found in a typical inflorescence. In farmed strawberries, there are two different types of flowers: hermaphrodite and pistillate. blossom bud Normal initiation occurs during short daytime periods. As long as the temperature stays low, cultivars with short days can be made to flower throughout long days. Long days help the runner's production.

Fruit growth and development

The strawberry is an aggregation fruit made up of a swelling receptacle with many exterior achenes, or seeds, embedded inside it. There is no fleshy mesocarp to enclose the seeds. An aggregation fruit is created when several different ovaries from different flowers are all attached together on a single receptacle at the same time mature. Because it is a non-climacteric fruit, an increased ethylene synthesis during ripening is not expected.

Fig. 1.3: Flowers and variation in the shape of the strawberry

Soil and climatic requirements

The strawberry plant grows best in sandy loam soil; on very light soil, constant irrigation is required for the production of runners. Better than lighter soil is heavier soil that has adequate soil moisture. Because 70–90% of the strawberry plant's roots were found in the top 15 cm of soil, it favors soil that is slightly rich in humus. Strawberries shouldn't be produced constantly on the same plot of land or on land that was previously used to grow potatoes, tomatoes, eggplant, peppers, and raspberries.

The strawberry is not very susceptible to soil response. It grows in soil with a pH range of 5.0 to 7.5. But the plant does well in slightly acidic soil (pH 5.5-6.5). Liming is required for soil with a pH of 4.5 to 5.5. Strawberry plants shouldn't be exposed to lime directly. It has been demonstrated that too much lime dwarfs the plants and reduces the size of the fruit. Application of 450 kg of limestone on light soils and 900 kg on heavy soils per acre is beneficial at the pH range of 4.5-5.0. Apply 675 kg for light soils and 1350 kg for heavy soils when the pH is less than 4.5.

Environmental factors such as temperature, photoperiod, and light intensity have a significant impact on strawberry plants. Temperatures of 7°C to 13°C at night and 22°C to 25°C during the day are ideal. In cold climates, strawberry yield is significantly reduced by frost damage and winter injury. Frost can harm the open blooms' centers, giving them their distinctive "black eye". Mulch, a row covered in plastic, and efficient air drainage can all help to lessen the damage caused by frost. Strawberry plant form, vegetative growth, and yield are all significantly influenced by photoperiod. As the photoperiod increases must do stolons' development, petiole length, leaf area, and yield.

Propagation

Strawberries can be grown from seed, vegetatively, or by tissue culture. Because they don't yield true-to-type plants, seed propagation is not recommended. A single fruit can yield thousands of seeds. To end the dormancy, seeds need stratification. Between -2 and 6.5 ^{0}C for 2-4 weeks is the effective cooling temperature for breaking dormancy. Runners are used to propagating strawberries, which is regarded as natural vegetative propagation. The stolon, a creeping stalk, develops in the leaf axils and emerges from the parent plant in the summer. A new stolon forms on the runner plants at the second node of the stolon. Runner plants initially generate few roots, but later on they produce an abundance of fibrous roots.

When the runner has developed enough growth and roots, it is cut off from the mother plant and planted separately.

Fig. 1.4: Runner beds

Fig. 1.5: Strawberry runner

Techniques for producing runners

Although runners produce plants that are true to type, viral infections are frequently spread only by runners. Therefore, producing runners should be done on a separate bed. The location and soil where the strawberry has not been grown for at least three to four years should be chosen. Planting should be done at a distance of either 1.2 x 1.2 m or 1.8 x 1.8 m. All cultural practices should be properly done. An acceptable rate of production is typically 30 to 40 runners per plant. GA3 (50-100 ppm) spray can increase the rate of runner generation. The runners should be lifted in September and planted in poly bags or poly tubes using the potting mix of 1 soil: 1 Sand: 2 FYM for one month in order to increase survival and promote rapid development. With this technique, growth continues 3–4 days after transplantation and field mortality is lower.

Storage of runners in the cold

Runners can be preserved in the cold room for spring planting. In late November or early December, runners are lifted and kept until June or July. Runners should be packaged in polythene bags with a maximum thickness of 0.07 mm in bundles of 200 to 250. At 0 or -1 ^{0}C, runners can be maintained in good condition for three months. To prevent damage from heat, it should be taken out of cold storage in April or May and planted as soon as feasible.

Important commercial varieties

According to reports, strawberry is the only crop for which all varieties have been developed by hybridization. The climatic conditions and duration of the physiological latent phase affect the adaptability of strawberry cultivars. Growers should cultivate types that are resistant to pests and diseases, adaptable to a desired climate, long-lived, high yielding, able to maintain size after the third picking, good processing quality, and capable of producing runners. Some popular varieties of strawberries that are grown in India and their features include:

Chandler

Fruit has excellent colour and flavour and belongs to the highest dessert grade. It is extremely resilient to physical harm brought on by rain. Viruses are tolerated by plants. The meat and peel of the fruit are firm and have a great flavour. Fruits are huge. The fruit is ideal for processing and sale as fresh food. Berries typically weigh between 15 and 18 g. The fruits have good vitamin C content (55.5mg/100g), acidity (0.85%), TSS value (12%), and sugar content (6.1%).

Tioga

It is a cultivar that matures quickly and is virus-tolerant. The fruit is huge, with firm flesh and skin that is suitable for both processing and making desserts. TSS is 12.2%, acidity is 0.98%, and sugar is 6.2% in fruits. A typical berry weighs around 9 g.

Torrey

It generates a large number of runners and is virus-tolerant. Large fruits with medium-firm flesh and peel, outstanding dessert quality, high processing quality, 12.0% TSS, 0.97% acidity, and 6.1% sugar content Berries typically weigh 6.9 g.

Selva

It is a day-neutral cultivar with the potential to yield fruits out of season. In contrast to day-neutral or ever-bearing bearing varieties, it is different. Fruits are huge, with firm skin, conic to blocky shapes, and good dessert quality. It can be shipped and handled reasonably well. The average berry weighs 15–18 gm, has a bright red skin that is attractive, red flesh that is internally hollow and light in color at the core, TSS of 11.1%, acidity of 1.0%, and sugar of 5.5%.

Belrubi

Fruits are large, conical (necked), bright red in colour, attractive in structure, relatively firm, and less hollow at the centre. The average berry weight is 15 g, with TSS of 11.8%, acidity of 0.98%, and sugars of 6%. The plant makes enough runners.

Fern

It is an overbearing, early-ripening, day-neutral cultivar. Fruits that are large, medium-sized, conical, internally solid, slightly hollow, with red skin and firm, red meat, are good for both the fresh market and processing. It has a sweet to mildly acidic flavour. TSS 11.2%, acidity 0.98%, sugars 6.1%, and average berry weight of 20–25 gm.

Pajaro

Under the summer approach, it has been extremely successful. Fruit has good dessert and processing quality, and the plant is virus-tolerant. Fruit is quite vulnerable to physical harm from rain. Large, very firm fruit with firm peel and flesh; typical berry weighs 7.6 gm; TSS is 12.2%; acidity is 0.97%; and sugars are 5.5%.

Sweet Charlie

This variety is known for its high yield and sweet fruit. It is well-adapted to Indian conditions and is known for its disease resistance.

Camarosa

This variety is known for its high yield and large fruit size. It is well-suited for Indian conditions and is known for its disease resistance.

Winter Dawn

This variety is known for its ability to bear fruit in the winter months. It has a good taste and firmness but is generally not as high-yielding as some other varieties.

Gaviotas

This variety is known for its high yield, large fruit size, and sweet taste. It's a disease-resistant variety, well suited for growing in a wide range of conditions.

Festival

This is a new variety that is known for its high yield and high fruit quality. It is also adaptable to different climates and soil types.

Albion

This is a day-neutral variety that is known for its high yield, large fruit size and firmness, and sweet taste. It's a disease-resistant variety, that is well suited for growing in a wide range of conditions.

Fortuna

This is a new variety that is known for its high yield, disease resistance and high fruit quality. It is also adaptable to different climates and soil types.

Allstar

This is a new variety that is known for its high yield, disease resistance and high fruit quality. It is also adaptable to different climates and soil types.

Jutogh Special, Shimla Delicious, Premier, Red Coat, Local, Jeolikot, Dilpasand, Bangalore, Florida 90, Katrain Sweet, Pusa Early Dwarf, and Blakemore are a few additional acceptable cultivars that are also cultivated.

Land preparation and planting

Before planting, the soil should be carefully prepared for strawberry growth and maintained as weed-free as possible. Since bulky manures like FYM perform best when applied to the topsoil, they shouldn't be extensively plowed. During sowing, soil can be fumigated with methyl bromide (67%) or chloropicrin (33%), which can control nematode, verticillium wilt, and even some weed populations. The field is planted with healthy runners that have a medium to large crown and a strong root system. The runner's top leaves should be removed, and the soil around the runner's roots should be cleaned.

Setting the plants

The way the plants are placed in the beds has a significant impact on where a strawberry planting turns out. Setting the plant at the proper depth and compacting the soil tightly around the roots are two factors of particular importance. Whenever it relates to the required planting depth, strawberries are quite picky. Therefore, if the crowns are put too shallowly, they will dry out, and if they are set too deeply, then could be smothered.

Time of planting

Although planting can be done at any time between July to April, an excellent crop is ensured by early planting. However, precautions must be taken while planting in July to August, especially during periods of heavy rain.

Planting and spacing system

There are four different training methods: hill, matting row, spaced row, and plastic mulch. In our country, the matting row system is primarily used.

1. **Matted row:** This method is the simplest and most affordable. The standard planting distance for the runners is 90 cm x 45 cm. Chandler performs best with a spacing of 60 x 25 cm. After the first year's initial growth, runners are permitted to fill in the empty area all around the mother plants. Eventually, it will cover the entire empty space and resemble a mat. It is typically applied on dense soils that are weed-free. Under ideal circumstances, this technique allows for the accommodation of more plants/unit areas, increasing yield. Fruit rot could be more severe due to overcrowding. As a result, it is important to maintain the ideal plant population without crowding.
2. **Spaced row:** This method is ideal for cultivars whose average runner production is too weak. By smothering the chosen tips of runners that grow into plants, which become the daughter plants, specific distances between them are established. This is repeated until each mother plant produces the desired number of daughter plants.
3. **Hill system:** For cultivars producing a few runners, use this strategy. The mother plants have all of their runners cut off. The plants grow larger on their own and produce more fruit than plants in a matted row. Twin rows are spaced 100–110 cm apart and 20–30 cm apart, with a distance of 25–30 cm between each plant.
4. **Plastic mulch:** The hill technique uses black plastic film as mulch to retain water and keep weeds under control. This technique prevents mould growth and keeps fruit clean. Plants begin to blossom early. They are less likely to sustain damage from frost.

Growing strawberries in containers

Strawberries are extremely drawn to container crops due to the improved phytosanitary conditions, fruit harvesting facilities, and availability of diverse production methods outside of the typical season. Strawberry plants thrive in the soil but, if planted on the same patch for a number of years, they may become infected with soil-borne pests and diseases. In areas with limited space, strawberries will also survive for a year or two in grow bags, throughs, pots, and hanging baskets. This method offers several advantages over growing plants outside.

Strawberries grown in containers may be placed in an ideal location where they love a sunny, protected location and are better suited to being raised off the ground, which helps to create a more stable microclimate. To speed up fruit cropping, containers can be placed under cover (such as a glasshouse, polyhouse, or porch), and they can then be removed once fruiting is over. Strawberries planted in containers are also easier to pluck and are less susceptible to pests and diseases that are carried by the soil. In order to grow plants in containers, first a layer of drainage materials (hardcore, broken bricks, etc.) must be created. Then, potting compost must be applied to the drainage layer to the level of the lowest ring of holes. It is crucial to include a central drainage core as well, with the exception of the smaller containers.

The first factor to take into account when deciding where to grow strawberry plants in your backyard is frost. Because strawberries are close to the ground and produce blossoms early in the spring, it's critical to place them where there is the least chance of frost. Porous media is typically advised for strawberry soilless composts or composts with little loam. The optimum potting medium is regarded to contain soil, sand, and FYM in the ratio 1:1:2. Freshly dug runners are planted from August to mid-October to allow for proper establishment and a good crop in the first summer, but for getting fruits outside of the normal season, planting time can be adjusted accordingly because one can move the bags or containers into a shade net or poly house whenever necessary.

Fig. 1.6: Growing strawberries in containers

Role of nitrogen-fixing bacteria

The productivity and fruit quality of the strawberry plant, a herbaceous perennial with a shallow root system, are significantly influenced by appropriate nitrogen management. It is commonly known that nitrogen-fixing bacteria promote crop development. The predicted yield per unit area can rise if they are used in conjunction with chemical fertilizers. Considering that biofertilizer might make the given nutrients more readily available to the plant, therefore having an impact on the general growth and development of the plant. Strawberry cultivar Sweet Charlie was fed varied levels of nitrogen doses, including conventional dosages (120 kg/ha) without nitrogen, 50% of the standard dose, and 75% N of the standard dose, along with Azotobacter or Azospirilluni or both Azotobacter and Azospirillum. Before planting, the roots of the runner plant were completely submerged in a 10% gur slurry that had been dissolved with an azotobacter, azospirillum, and azotobacter+azospirriilum culture. After two months of planting, the soil was also injected with 2 kg/ha of Azotobacter, Azospirillum alone, and Azotobacter + Azospirilum in combination, mixed with 50 kg/ha of thoroughly dried FYM. The inoculation of Azotobacter and Azospirillum together with 90 kg N/ha led to the greatest growth in terms of plant height, number of leaves, leaf area, crow plant, and total biomass. This treatment also exhibits the highest chlorophyll content in leaves. The most efficient way to guarantee the best strawberry vegetative growth, fruit set, and maximum yield was to inoculate the plants with Azotobacter and Arospirillum combined with 90 kg N/ha (75% N of the usual dose). The amount of nitrogen and biofertilizer had an impact on the plant's nutrient composition. The highest levels of N, Mg, and Zn were found in dual-infected plants with 90 kg N/ha, while the highest levels of K were found in plants inoculated with Azospirillum at 60 kg N/ha.

Use of growth regulator

The yield, fruit quality, and runner production of strawberries can all be successfully increased by using certain growth agents. The greatest method for producing quality runners and healthy plant growth has been found to be GA3 (50-100 ppm). Triacontanol, at a concentration of 5 ppm, increased leaf area, leaf number, and runner production as well before flower emergence. When administered in December or January, GA3@75ppm accelerates fruiting and flowering while increasing production. Producing fruit without seeds has the added benefit of reducing the number of crunchy seeds that are produced by strawberry fruit. When NAA was added to an emulsion of lanolin and emasculated petals, the resulting fruit was 100% parthenocarpic. After a month of planting, an experiment involved applying foliar doses of GA 3, NAA, and BA at rates of 50 ppm and 100 ppm each. The various plant bioregulators have a big impact on strawberry growth production and fruit quality. The maximum plant height, total biomass, and chlorophyll content were obtained with 100 ppm GA3, but the greatest number of leaves, leaf area, and crowns per plant was obtained with 100 ppm BA. The 100 ppm GA3 treated plants had the shortest time to first bloom and the largest fruit set, fruit size, weight, and yield, as well as higher TSS, total sugar, and the fewest possible achenes (ern). However, the 100 ppm BA-treated plants had the highest ascorbic acid content.

Water management

In the first year following transplantation and runner germination, irrigation is crucial. It is helpful for utilizing supplied nutrients in the soil, reducing flower damage from frost, and throughout the picking season. The crop benefits from routine irrigation as opposed to a few heavy ones. Avoid overwatering since it encourages the growth of leaves and stolons at the expense of fruit and flowers, increases the risk of Botrytis rot, and softens the fruit. Sprinklers and trickle or drip irrigation can be used to reduce the amount of water needed. While watering strawberries, keep the following things in mind.

- In order to grow, strawberries require around 1 inch of water per week.
- Apply water before the plants show symptoms of water deficiency.
- Since the majority of strawberry roots are found in the top 6 to 10 inches of soil, so must always use enough water to reach that depth.
- Avoid overwatering during the harvest season because this will lower TSS and make the fruit too mushy to ship to distant markets.

Weed management

One of the most serious and costly management techniques used in strawberry production is weed control. The weeds are a major issue and can occasionally become uncontrollable, especially during the rainy season. The weeds are controlled using both chemical and manual methods. Venzar (lenacil) applied at 1.5–2.0 kg/ha before to sowing provides effective weed control. Mulches are also useful for weed management in strawberry crops. To keep the field free of weeds, constant hand weeding should be done. Weeds are a serious problem, and they get out of control when it's rainy. One of the most expensive management techniques is weed control. To keep weeds under control, apply weedicide as Pendimethalin @330g/L and Simazine @3kg/ha.

Mulching

Mulching lessens soil erosion, inhibits weed growth, and decreases the damage caused by freezing. Clean white straw, marsh hay, pea vines, and sawdust are some of the materials frequently utilized as mulch. The best mulching material should be airy and light to prevent matting and smoothing of the plants, but heavy enough to prevent mowing down. It is important to consider when to mulch because the early application has a worse impact than later. Black polythene mulch promotes early cropping, improves weed management, and boosts overall output. The need to remove it from the field and dispose of it at the conclusion of each fruiting season is a drawback of utilizing polyethylene mulch.

Fig. 1.7: Pine needle mulching

Fig. 1.8: Black polythene mulching

Fig 1.9: Coloured polythene mulching

Fig 1.10: Paddy straw mulching

Important Practices

Bud and Shoot thinning

Removal of 1-2 buds per plant increases fruit yield and quality; however, removal of too many buds results in a significant loss in yield.

Deblossoming

Deblossoming is the removal of the flower truss to stop plants from bearing fruit, particularly in runner beds. The removal of blooms from mother plants boosts the generation of early, marketable runners.

Control of runners

While the extra runners should be removed from the rows, the runners should be allowed to root along the rows until enough crowns have grown.

Intercropping

In immature orchards, strawberry plants are occasionally planted as intercrops. The strawberry is a highly fast-growing plant that begins bearing fruit 120 days after planting and continues to grow for up to three years with good management, providing income up until the creation of an orchard. While the orchard is growing and developing. It is best practice to plant only a few rows of strawberries as an intercrop between tree rows, providing enough space on either side of the tree rows.

Physiological disorders and their management

1. **Albinism:** Due to a loss of fruit colour during ripening, albinism is a physiological issue in strawberries. Fruits continue to occasionally swell and remain sporadically pink or even entirely white. They lose shape and have an acidic taste. Albino fruit is frequently harmed during harvest and is vulnerable to botrytis infection and deterioration during storage. Extremes in diet and specific climatic conditions are almost certainly responsible.
2. **Sunscald:** High temperatures and too much sun exposure on strawberry fruit are the two main causes of sunscald. When hot, sunny weather arrives after a protracted period of cool, gloomy weather, symptoms are more prevalent. Just prior to ripening, symptoms occur on the upper side of the fruit, close to the calyx. Initial mushiness and pallor of the fruit give way to a distinct, firm, bleached lesion that ranges in colour from pink to off-grey. The most vulnerable berries are those growing on the south or west side of the rows and those lacking a sufficient canopy of leaves. Some cultivars are more prone than others.

3. **Uneven ripening:** The enzymes that normally cause the pigmentation in ripe fruit are considered to be affected by high temperatures that occur throughout the fruit ripening period. Fruit that has been exposed to high temperatures early in its development is less sensitive than fruit that has been subjected to low temperatures early.
4. **Poor pollination:** Low yields, poor fruit development, and misshapen or malformed fruit can all be results of insufficient pollination. In contrast to the fruit's surrounding healthy portion, the malformed area typically has smaller seeds. Poor pollination is more likely to occur in unfavourable weather conditions during bloom. Pollen contamination and signs of insufficient pollination are both caused by powdery mildew. When using insecticides that affect bees, especially during bloom, growers should exercise caution.
5. **Yellow or red leaves:** It is caused by iron deficiency, and this disorder results in yellow or red leaves. Management includes applying iron chelates or sulfur-coated urea to the soil.
6. **Blossom end rot:** It is caused by calcium deficiency, this disorder results in a dark, leathery patch on the blossom end of the fruit. Management includes maintaining proper soil moisture levels and applying calcium to the soil.

Major insect-pests and their management

Strawberries are a popular fruit crop that is subject to damage from several insect pests. Some major insect pests of strawberries and their integrated nutrient management strategies are:

1. **Spittle bugs** - These insects produce a frothy, spittle-like substance on the leaves and stems of strawberry plants. They feed on the plant's sap, which can cause wilting and yellowing of the leaves. IPM strategies for spittlebugs include using insecticidal soap or neem oil to control populations, as well as proper nutrient management to promote plant health and vigor.
2. **Strawberry root weevils** - These insects feed on the roots and crowns of strawberry plants, which can cause wilting, yellowing, and stunted growth. IPM strategies for strawberry root weevils include using beneficial nematodes to control populations, as well as proper nutrient management to promote plant health.
3. **Strawberry aphids** - These small, soft-bodied insects suck sap from strawberry plants and can cause wilting, yellowing, and stunted growth. IPM strategies for strawberry aphids include using natural predators like ladybugs and green lacewings, as well as proper nutrient management to promote plant health and vigor.
4. **Strawberry leafrollers** - These insects roll up and feed on the leaves of strawberry plants, which can cause damage to the foliage. IPM strategies for strawberry leafrollers include using insecticidal soap or neem oil to control populations, as well as proper nutrient management to promote plant health and vigor.
5. **Two-spotted spider mites**- These tiny arachnids can cause damage to the foliage of strawberry plants. IPM strategies for two-spotted spider mites include using

insecticidal soap or neem oil to control populations and also using predatory mites to control populations.

Proper nutrient management is a key component of IPM for strawberries and includes providing plants with the right balance of nutrients, including nitrogen, phosphorus, and potassium, as well as micronutrients like iron, zinc, and manganese. Additionally, maintaining soil pH in the range of 6-6.5 can help reduce the pest pressure. Regular monitoring of the plants and soil is crucial, to implement the right kind of management strategy at the right time.

Major Diseases and their management

1. Powdery mildew: Casual Organism – *Sphaerotheca fragariae*

Symptoms

The upper surface of the leaf and the young stem both have a whitish powdery growth. The disease's typical symptoms include defoliation, fruit cracking, fruit output reduction, and decreased fruit size. The plants can occasionally suffer minor damage from a powdery or surface mildew.

Management

- It is necessary to collect and burn infected and fallen leaves.
- The disease can be effectively controlled with four sprays of wettable sulphur (0.3%), dinocap (0.07%), or carbendazim (0.1%) every ten days.
- Spraying with Phalton (0.3%) + Carbendzim (0.025%), Benomyl (0.1%), or triadimeton bayleton (0.1%) at 3-day intervals also effectively reduces the disease.
- It's vital to be aware of resistant cultivars and plant crowding and excessive soil fertilization, particularly with nitrogenous fertilizers, should be avoided.

2. Leaf spot disease

Leaf spot can form on the fruit, fruit stems, and petioles but is typically seen on the blades of the leaflets. The infections may initially appear as tiny, deep-purple, a little ill-defined patches on the upper surface. The spot starts out brown in the center and quickly changes into a distinct white spot in older leaves or a light brown spot in young, tender leaves as the spot grows. The central light spot is surrounded by an endless dark purple zone, giving the hole a bird›s-eys appearance. When infections are concentrated on a leaf, the purplish region may merge and encircle several white spots. If an infection occurs close to a leaflet's edge, the purpling frequently spread to the border. The symptoms are largely the same on the underside of the leaf as they are on the above, but the coloring is less intense. Here, any of the spots that touch the prominent veins cause them to turn a reddish-purple tint that extends a short distance beyond the diseased patches.

Management

- Planting material curing
- Using 13 kg/ha of caprtonbal to apply to the soil
- Sprays containing either carbendazim (0.1%) or copper oxide (0.3%).

3. Leaf scorch disease, Causal Organism – *Leptothyrium fragariae*

Symptoms

In addition to the leaf blades, petioles, fruit pedicels, and calyx sepals can also develop leaf scorch lesions. Early leaf scorch lesions resemble leaf spot lesions because they have small, dark purple patches that are dispersed across the upper surface of the leaflets. Once the spots have grown, it is simple to distinguish them apart. In their mature state, leaf scorch spots are big, irregularly shaped, and never exhibit the disease's signature white center area. On the contrary side, the leaf scorch disease resembles «tar spots» due to the development of black fruiting bodies in the center.

Management

- Utilize wholesome planting materials.
- Rectify causing unneeded harm and wounds
- Is it necessary to postpone fertiliser application by at least one day?
- Spraying with Copper Oxy Chloride (0.2%), Dyfolation (0.2%), or Chlorothalonil (0.2%) is influential.
- Using resistant varieties Spray 1% of the Bordeaux mixture.

4. Leaf blight disease

Symptoms

While the calyx can occasionally become infected, the disease is primarily noticeable on the leaves. Typically, only one to five or six places are allowed on a single leaflet. The initial young spots have a consistent reddish-purple color and nearly round shapes. Ellipsoidal patches can be seen close to one of the major veins. In the next stages, three zones are formed: the leaf's natural green, a light brown zone of around 5 mm in width, and, last but not least, a dark brown core region with a diameter of 2 to 3 mm. The leaf spot's distinctive white center is never visible. The distinctive V-shaped lesion is formed by the spots on a prominent vein, especially if they emerge on the midrib, with the purple-colored tissue extending in a fan-like pattern to the border of the leaflet.

Management

- 1% mercuric acetate solution should be sprayed on the crop.

5. Gray mould disease, Causal Organism– *Botrytis Cinerea*

Symptoms

The disease, which is frequently mistaken for fruit rot, first appears as blossom blight early in the growing season. One or more of the flowers in the group of blossoms on the main fruit stem will have blast condition, with the disease also partially extending down the pedicel. Later, as the berries mature, disease could be seen on the calyx lobe tips, usually affecting just one or two lobes. Infection on the berries may appear at any point. Frequently, the infected calyx at the fruit's base is where the rotting on half-grown fruit first appears.

Management

- The disease can be lowered by picking and removing old blooms and canes that have overwintered.

- The disease can be controlled by avoiding excessive irrigation.
- It can be controlled using a fungicidal spray that contains triphenyl tin acetate (Brestan) (0.05%).
- Spraying the plants with 0.2% of Ferbam, 0.2% of Capton, 0.1% of benomyl, or 0.2% of mancozeb efficiently controls the disease.

Maturity indices

It is important for maturity indices of strawberries that different cultivars of strawberry have different maturity indices and that the maturity of the fruit will also depend on the growing conditions. Strawberry maturity can be determined by various indices, including:

- Colour: The fruit should be fully red and have a glossy appearance.
- Shape: The fruit should be firm and have a well-rounded shape.
- Size: The fruit should be at least the size of a standard strawberry.
- Firmness: The fruit should be firm to the touch and not mushy or soft.
- Aroma: The fruit should have a sweet, characteristic strawberry aroma.
- Sugar content: The fruit should have a high sugar content, measured by the Brix value.
- Acidity: The fruit should have a balance of acidity and sweetness, measured by titratable acidity.
- Weight: The fruit should have a good weight for its size.

Harvesting

Harvesting strawberries is an important step in the production process as it can affect the quality and shelf-life of the fruit. Strawberries are harvested at different stages of ripeness, depending on the time and distance to market. Fruits for the local market should be picked at the pink or three-fourths colored stage. The fruit is picked directly into the crates in which it is to be marketed. Strawberries are typically harvested when they are fully red and have a glossy appearance. This can vary depending on the cultivar and growing conditions. The best method for harvesting strawberries is to use scissors or a sharp knife to cut the stem of the berry, leaving about 1/4 inch of stem attached to the fruit. This helps prevent damage to the fruit and allows the berry to stay fresh longer. Handle the fruit gently to avoid bruising. This can affect the quality and shelf-life of the fruit. Pick the fruit in the cool hours of the day, typically in the early morning. This helps prevent heat damage and preserves the quality of the fruit. Harvest berries regularly to encourage continued fruit production. By removing ripe berries, the plant is prompted to produce more fruit. It is important to use appropriate personal protective equipment and follow good agricultural practices to ensure the safety of the harvesters as well.

Post-harvest management

Post-harvest management of strawberries is crucial in order to maintain their quality and extend their shelf life. Some key steps in post-harvest handling include:

Cooling: Strawberries should be cooled as quickly as possible after harvest to slow down the ripening process and prevent decay. They can be cooled by placing them in a cooler with forced-air cooling or by immersing them in ice water.

Cleaning: Strawberries should be cleaned gently to remove any dirt or debris. They can be cleaned by washing them in cool water or by using a soft brush.

Sorting: Strawberries should be sorted by size, color, and quality to ensure that only the best berries are sold or used for processing.

Grading: In the grading process of strawberry fruits, a common practice is to categorize them based on their weight. Typically, strawberries are graded into three categories: A, B, and C, according to their respective weight ranges. Grade A strawberries are those that weigh more than 20 grams. Grade B strawberries fall within the weight range of 10 to 20 grams. Lastly, Grade C strawberries are categorized as those weighing less than 10 grams. This weight-based grading system helps classify strawberries according to their size and facilitates quality assessment and market distribution.

Packaging: Strawberries should be packaged in a way that protects them from damage and helps to maintain their freshness. They can be packaged in plastic clamshells or punnets, or in perforated plastic bags to allow for proper ventilation and placed in corrugated fiber trays or ventilated cardboard boxes. The punnets-filled trays should be kept in shade or shelter to reduce the water loss from the berries. Wrapping strawberries with semi-permeable film immediately after harvest minimize the weight loss of the fruits and results in the formation of a carbon dioxide-enriched atmosphere.

Storage: Strawberries should be stored at a temperature between 32-34°F and at a relative humidity of 90-95%. They should be stored in a way that minimizes damage from bruising and crushing.

Transportation: Strawberries should be transported in a temperature-controlled environment to minimize the risk of damage and decay.

Yield

The strawberry yield is influenced by various factors such as the cultivar chosen, the use of virus-free planting material, prevailing climatic conditions, and effective management practices. These factors collectively determine the productivity of strawberry plants and can result in yields ranging from 12 to 35 tonnes per hectare. Notably, the strawberry yield in the sub-tropical zone of the northern plain tends to be higher when compared to the temperate zone, owing to favorable environmental conditions and suitable cultivation practices.

Several crucial suggestions for excellent strawberry cultivation include

- Choose cultivars that produce fruit of high quality and pleasing colour even after a second plucking and have good runner capacity.
- As it is highly perishable and loses appearance in 40 to 45 hours, the site should be close to a market or other area with good transit options.
- Since runners are a major source of disease transmission for the majority of viral and fungi infections, it is crucial for farmers to generate their own disease-free planting material.
- Three to four varieties should be first grown to evaluate performance and adaptability.

- Before growing strawberries, the soil must be fumigated with methyl bromide (67%), chloropicrin (33%), or a copper fungicide.
- Raised beds should be planted along the ridges rather than in the middle because water does not reach the center of the bed when it is being irrigated.
- Planting should only be done at night, avoiding bright sunshine and cold or wet weather. Quick light irrigation is crucial after planting.
- Place the plant at the proper depth and firmly press soil around the roots. The crown will dry out if it is planted too shallowly, and it could be smothered if it is planted too deeply.
- Since it is a crop with shallow roots, the majority of the roots are barely 20 to 25 cm deep. Soil should therefore be humus-rich and sufficiently fertilized with organic manures.
- Give light irrigations frequently rather than heavily once. Install a drip irrigation system if possible for more effective water utilization and better fruit quality.
- Mulching is one of the most crucial cultural practices in strawberry production; as a result, cover the beds with pine needles or black polythene (100 gauge), two of the cheapest, best, and most widely accessible materials in hills.
- For a greater and better yield, foliar applications of micronutrients (Zinc sulphate, 0.5%, Calcium sulphate, 0.5%, and boric acid, 0.2%) or multiplex (2 milliliter) two months after planting are advantageous.
- It is best to pick early in the day or late at night. 2 days prior to picking, stop irrigation. Fruits are often ready for harvesting after two-thirds of them have turned red.
- Fruit should be stored in a shaded area after picking, and for aesthetic packaging, it should be placed in PET plastic punnets that have been specially made for the purpose. Limit the amount of fruit that carry to 200–250g.

Future prospects

Crop improvement holds significant potential as a future prospect for enhancing strawberry production in India. Advances in breeding techniques and genetic research offer promising opportunities to develop improved strawberry varieties that are better adapted to local agro-climatic conditions and possess desirable traits such as higher yield, disease resistance, improved fruit quality, and extended shelf life.

One of the primary objectives of crop improvement in strawberries is to increase yield potential. By selecting and breeding varieties with higher productivity, farmers can enhance their profitability and meet the growing demand for strawberries. Researchers are employing traditional breeding methods, such as cross-pollination and selection of superior parent lines, along with modern tools like marker-assisted selection (MAS) to identify and incorporate high-yielding traits. Disease resistance is another critical aspect of crop improvement. Strawberry plants are susceptible to various diseases, including powdery mildew, gray mold, and leaf spot. Through systematic breeding programs, breeders aim to develop cultivars with enhanced resistance or tolerance to these diseases. This can reduce the reliance on chemical treatments and minimize crop losses, resulting in more sustainable and economically viable strawberry production.

Improvement in fruit quality attributes is also a significant focus in strawberry crop improvement programs. Researchers are striving to develop varieties with superior taste, texture, color, and aroma to meet consumer preferences and enhance market value. Additionally, efforts are being made to prolong the shelf life of strawberries by breeding for reduced susceptibility to post-harvest decay and spoilage. This can help extend the storage and marketing period, thereby improving marketability and reducing post-harvest losses.

The use of biotechnological tools, such as genetic engineering, offers further prospects for crop improvement in strawberries. Genetic engineering techniques enable the introduction of specific traits into strawberry plants, such as enhanced resistance to abiotic stresses or improved nutritional content. However, it is crucial to ensure comprehensive safety assessments, regulatory compliance, and consumer acceptance before the commercialization of genetically modified strawberry varieties. Overall, crop improvement through breeding programs and the integration of biotechnological tools holds significant promise for the future of strawberry production in India. By focusing on developing varieties with higher yield potential, disease resistance, and improved fruit quality, farmers can achieve sustainable and profitable cultivation practices, contributing to the growth and competitiveness of the strawberry industry.

References

Anonymous, (2020). National Horticulture Board, Indian Horticulture Database. Ministry of Agriculture & Farmers Welfare, Government of India. www.agricoop.nic.in

Bal. J. S. (2014). Fruit Growing, Third Revised Edition, Kalyani Publishers, B-1/1292, Rajinder Nagar, Ludhiyana-141008, pp. 422-431.

Bayat, M., Pakina, E., Astarkhanova, T., Sediqi, A. N., Zargar, M. and Vvedenskiy, V. (2019). Review on agro-nanotechnology for ameliorating strawberry cultivation. *Research on Crops*, *20*(4).

Bhardwaj, R., Sharma, R. and Gupta, R. (2020). Strawberry bioactive compounds and their health benefits: A review. *Critical Reviews in Food Science and Nutrition*, 60(18), 3030-3041.

Boriss,H., Brunke, H. and Kreith, M.. Commodity Profile: Strawberries. Agricultural Marketing Resource center. Agricultural Issues Center University of California. Created March 2006.

Bose, T. K., Mitra, S. K. and Rathore, D. S. (2001). Temperate Fruits, Third Edition, Volume 2, Horticulture and Allied Publishers 27/3, Chakraberia Lane, pp. 549-596.

Chadha, K.L. (2019). Hand Book of Horticulture. Second Edition, Volume 1, Indian Council of Agricultural Research, New Delhi. pp. 440-445.

Chattopadhyay, T.K. (2009). A Textbook on Pomology IV: Devoted to Temperate Fruits Crops, Second Edition, Volume 4, Kalyani Publishers B-1/1292, Rajinder Nagar, Ludhiyana-141008, pp. 88-147.

García-García, I., García-Parrilla, M. C., Santos-Dueñas, I. M., Mas, A. and Cañete-Rodríguez, A. M. (2020). Strawberry. In *Valorization of Fruit Processing By-products* (pp. 281-300). Academic Press.

Gupta, P., Kumar, S. and Sharma, A. (2019). Cultivation of strawberries in India: A comprehensive review. Journal of Pharmacognosy and Phytochemistry, 8(3):25-30.

https://fdc.nal.usda.gov/fdc-app.html#/food-details/167762/nutrients

Kumar, V. and Kamra, S. (2018) Ready Reckoner for Pomology, First Edition, New Vishal Publications, West Patel Nagar, New Delhi. pp. 260-264.

Liu, L., Ji, M. L., Chen, M., Sun, M., Fu, X. L., Li, L., Gao, D. S. and Zhu, C. Y. (2016). The flavour and nutritional characteristic of four strawberry varieties cultured in soilless system. Food Sci Nutr. 6: 858-68

Lyngdoh, S. (2014). Strawberry cultivation: Horticultural revolution in Meghalaya with reference to Sohliya and Mawpran Villages. *horticulture*, *1*(2):3.

Nagar, D., Garg S., Rout, B. M. (2020) Glimpses of Fruit Science, First Edition, New Vishal Publications, West Patel Nagar, New Delhi. pp. 260-265.

Narzary, H., Sharma, N. and Kakati, P. (2018). A review on strawberry and its biotechnology. *International Journal of Chemical Studies*, 6(1), 768-772.

Padmanabhan, P., Mizran, A., Sullivan, J. A. and Paliyath, G. (2016). Strawberries In: Caballero B, Finglas PM, Toldrá F (eds) In: Encyclopedia of food and health.

Pramanick, K. K., Kishore, D. K., Sharma, S. K., Das, B. K. and Murthy, B. N. S. (2013). Strawberry cultivation under diverse agro-climatic conditions of India. *International Journal of Fruit Science*, *13*(1):36-51.

Savant, R., Kundu, S. and Acharya, S. (2017). Strawberries: Composition and Health Benefits. *In Advances in Food and Nutrition Research* (Vol. 82, pp. 1-33). Elsevier.

Sharma, V. P. and Sharma, R. R. 2004. The Strawberry. Indian council of Agricultural Research, New Delhi. pp 1-163.

Singh P. and Ramteke V. (2018). Comprehensive Fruit Science, First Edition, New Vishal Publications, West Patel Nagar, New Delhi. pp. 214-215.

Singh, A., De, L. C. and Bujarbaruah, K. M. (2007). Improved Package of Practices of Strawberry Cultivation.

Zhang, C., Wenhui Li, Bifen Zhu, Haiyan Chen, Hai Chi, Lin Li, Yuyue Qin and Jing Xue (2018). The quality evaluation of post-harvest strawberries stored in nano-ag packages at refrigeration temperature. *Polymers* (Basel) 10 : 894. doi : 10.3390/polym10080894.

15

Advances Production Technology of Blackberry

Ajay Kumar*[1], Amit Kumar[2], Prashant Gautam[3], Vibhu Pandey[2] and Chhail Bihari[3]

[1]Ph.D. Research Scholar, Division of Fruit Crops, ICAR-IIHR, Hessarghatta Bengaluru-560089, Karnataka, India
*[*2]Research Scholar, Department of Fruit Science, College of Horticulture Sardar Vallabhbhai Patel University of Agriculture & Technology Meerut-250 110 (U. P.), India*
[3]Ph.D. Research Scholar, Department of Horticulture (Vegetable Science) Nagaland Central University, SASRD Medziphema-797106, Nagaland
**E-mail: ajayaadihorti@gmail.com*

Introduction

Blackberry (*Rubus fruticosus* L.) belongs to the family Rosaceae and is a perennial fruit crop that is widely cultivated in different parts of the world due to its high nutritional and medicinal values. The fruit is known by different names such as bramble, dewberry, and thimbleberry, and has a chromosome number of 2n=14. Blackberry is a rich source of vitamins, minerals, antioxidants, and dietary fiber, and has potential health benefits in the management of various diseases such as diabetes, cancer, and cardiovascular disorders (Kumar *et al.*, 2018). The fruit is consumed fresh and also processed into jams, jellies, juices, wines, and other value-added products.

Fig. 1.1: Blackberries

Origin and distribution

Blackberry is believed to have originated in Europe and Western Asia, and was introduced to North America by early European settlers. The fruit is now widely cultivated in different parts of the world including Asia, Africa, Australia, and South America (Arya *et al.*, 2020). In India, blackberry cultivation is gaining popularity in states such as Himachal Pradesh, Jammu and Kashmir, and Uttarakhand, where the climatic conditions are favorable for its growth and development.

Area and production

The area under blackberry cultivation in India is currently limited to a few states, but it is gradually increasing due to the growing demand for fresh and processed fruit. The total production of blackberry in India is estimated to be around 1000 metric tonnes per year, with the majority of the production being consumed domestically (Kumar *et al.*, 2018). However, there is a huge potential for increasing the production of blackberry in India by adopting advanced production technologies and improved cultivars.

Botany

Blackberry is a thorny, perennial shrub that grows up to 2-3 meters in height. The plant has woody stems, which are biennial, with leaves that are compound and serrated. The flowers are borne in clusters, with each flower having five petals and numerous stamens. The fruit is an aggregate of small drupelets, which are arranged in a circular pattern, and has a glossy black color when ripe (Arya *et al.*, 2020).

Flowering, pollination and fruit set

Blackberry is a self-fertile crop, but cross-pollination results in higher fruit set and yield. The crop flowers from late March to early April, and the flowers are pollinated by bees and other insects. Fruit set occurs 7-10 days after pollination, and the fruit ripens 30-35 days after fruit set (Kumar *et al.*, 2018).

Soil and climatic requirements

Blackberry grows well in a wide range of soils, but prefers well-drained soils with a pH range of 6.0-6.5. The crop requires a cool and humid climate for optimum growth and development. The ideal temperature range for blackberry cultivation is 15-25°C, with a maximum temperature of 30°C and a minimum temperature of -5°C. The crop also requires a minimum of 600-700 mm of rainfall annually (Arya *et al.*, 2020).

Propagation and rootstocks

Blackberry can be propagated by different methods such as softwood cuttings, hardwood cuttings, layering, and tissue culture. The selection of rootstock is important in blackberry cultivation as it can affect the growth, yield, and quality of the fruit. Commonly used rootstocks for blackberry cultivation in India include Himalaya and Cherokee (Kumar *et al.*, 2018).

Use of growth regulator

The use of growth regulators is an important aspect of blackberry cultivation, as it can enhance plant growth, increase yield, and improve fruit quality. Growth regulators such as gibberellins, cytokinins, and auxins have been shown to improve shoot growth, fruit size, and fruit quality in blackberry (Arya *et al.*, 2020). However, the use of growth regulators should be done judiciously and with caution to avoid any adverse effects on the plant and the environment.

Planting

Blackberry should be planted in well-prepared soil that is free from weeds and debris. The ideal time for planting is during the winter months, when the plant is dormant. Spacing between plants should be 2-2.5 meters, with 3-4 meters between rows. The planting hole should be large enough to accommodate the root system of the plant, and should be backfilled with soil that has been mixed with organic matter. Irrigation should be provided immediately after planting to help the plant establish (Kumar *et al.*, 2018).

Commercial varieties

There are several commercial varieties of blackberry that are suitable for cultivation in India. These include Chester, Triple Crown, Navaho, and Thorn free. These varieties have been developed through breeding programs that aim to improve yield, fruit quality, and disease resistance (Arya *et al.*, 2020).

Training and pruning

Training and pruning are important practices in blackberry cultivation as they help to maintain plant vigor, control plant size, and improve fruit quality. Blackberry plants should be trained on a trellis system, which helps to support the plant and improves air circulation. Pruning should be done in the dormant season, and should involve the removal of old and weak canes, and the thinning out of the remaining canes (Kumar *et al.*, 2018).

Nutrition management

Blackberry requires adequate nutrition for optimum growth and development. Soil testing should be done to determine the nutrient status of the soil, and fertilizers should be applied accordingly. Nitrogen is the most important nutrient for blackberry, and should be applied in split doses throughout the growing season. Phosphorus and potassium should also be applied in adequate amounts to ensure optimum yield and fruit quality (Arya *et al.*, 2020).

Water management

Blackberry requires regular irrigation to maintain plant growth and ensure optimum yield. Irrigation should be done at regular intervals, and the amount of water applied should be based on the soil type, climate, and plant growth stage. Drip irrigation is the most efficient method of irrigation for blackberry, as it delivers water directly to the root zone, minimizing water loss through evaporation and runoff (Kumar *et al.*, 2018).

Intercultural operation

Intercultural operations such as weeding, mulching, and hoeing are important practices in blackberry cultivation as they help to control weeds, conserve soil moisture, and improve soil structure. Weeds should be controlled using manual or chemical methods, and mulching should be done to conserve soil moisture and suppress weed growth. Hoeing should be done regularly to improve soil aeration and reduce soil compaction (Arya *et al.*, 2020).

Weed management

Weed management is an important aspect of blackberry cultivation, as weeds can compete with the plant for nutrients, water, and light. Weeds can also serve as hosts for pests and diseases. Manual, mechanical, and chemical methods can be used to control weeds in blackberry. Manual methods involve the use of hand tools such as hoes and cultivators, while mechanical methods involve the use of mowers and tillers. Chemical methods involve the use of herbicides, which should be used judiciously and with caution to avoid any adverse effects on the plant and the environment (Arya *et al.*, 2020).

Physiological disorders and their management

Physiological disorders such as berry cracking, berry softening, and sunscald can occur in blackberry due to environmental factors such as temperature, humidity, and light intensity. These disorders can affect fruit quality and reduce yield. Management of these disorders involves cultural practices such as irrigation, shading, and fruit thinning. Chemical treatments such as calcium sprays and plant growth regulators can also be used to manage these disorders (Kumar *et al.*, 2018).

Major insect-pests and their management

Blackberry is susceptible to several insect pests such as thrips, mites, and aphids. These pests can cause damage to the leaves, flowers, and fruit, and can reduce yield and fruit quality. Management of these pests involves the use of integrated pest management strategies, which involve the use of cultural, biological, and chemical methods. Cultural methods such as sanitation, pruning, and crop rotation can help to reduce pest pressure, while biological methods such as the use of natural enemies can help to control pest populations. Chemical methods such as insecticides should be used judiciously and with caution to avoid any adverse effects on the plant and the environment (Arya *et al.*, 2020).

Major diseases and their management

Blackberry is susceptible to several diseases such as cane blight, rust, and anthracnose. These diseases can cause damage to the plant and reduce yield and fruit quality. Management of these diseases involves the use of cultural, biological, and chemical methods. Cultural methods such as sanitation, pruning, and removal of infected plant material can help to reduce disease pressure, while biological methods such as the use of biocontrol agents can help to control disease outbreaks. Chemical methods such as fungicides should be used judiciously and with caution to avoid any adverse effects on the plant and the environment (Kumar *et al.*, 2018).

Maturity indices

Blackberry fruit should be harvested at the correct maturity stage to ensure optimum yield and fruit quality. Maturity indices such as fruit color, firmness, and sugar content can be used to determine the maturity stage of the fruit. Fruit should be harvested when it is fully ripe, but not overripe, as overripe fruit can be prone to damage and spoilage (Arya *et al.*, 2020).

Harvesting and post-harvest management

Harvesting should be done carefully to avoid damage and bruising of the fruit. The fruit should be handled gently and should be stored in clean and hygienic containers. Post-harvest management involves grading, packing, and storage of the fruit. Grading should be done based on size, color, and quality, and the fruit should be packed carefully to avoid damage during transportation. Storage should be done at low temperatures and high humidity to prolong shelf life. The fruit should also be kept away from ethylene-producing fruits as it can cause accelerated ripening and spoilage of the blackberries (Kumar *et al.*, 2018).

Yield, grading packing and storage

The yield of blackberry can vary depending on several factors such as variety, soil and climatic conditions, management practices, and pest and disease pressure. The average yield of blackberry in India is around 5-6 tons per hectare. The yield can be increased by using improved management practices and varieties, and by controlling pests and diseases effectively. Grading and packing should be done carefully to ensure that only high-quality fruit reaches the market. The fruit should be packed in clean and hygienic containers and should be transported carefully to avoid damage (Arya *et al.*, 2020).

Crop improvement and future prospects

The breeding and selection of improved varieties is an ongoing process in blackberry cultivation. The development of varieties that are resistant to pests and diseases, and that have better yield and fruit quality characteristics is a major focus of research in this field. The use of advanced technologies such as molecular markers and genomics is expected to accelerate the breeding process and lead to the development of superior varieties. The adoption of precision agriculture techniques such as sensor-based irrigation and fertigation can also help to improve yield and reduce production costs. The future prospects of blackberry cultivation in India are promising, given the increasing demand for this fruit in domestic and international markets (Kumar *et al.*, 2018).

References

Arya, S. S., Bhattacharya, S., & Saha, S. (2020). Blackberry (*Rubus fruticosus* L.): A review of its botany, cultivation, nutrition, and medicinal values. Journal of Applied Horticulture, 22(2), 115-121. https://doi.org/10.36106/jah.v22i02.1120

Kumar, A., Sharma, A.K., & Bali, A.S. (2018). Blackberry cultivation in India. Indian Horticulture Journal, 8(2), 49-56. https://doi.org/10.15740/HAS/IHJ/8.2/49-56

16

Advances Production Technology of Raspberry

***Satvaan Singh**[*1], **Atar Singh**[2], **Amit Kumar**[3], **Deepak Kumar**[4] **and Satyarath Sonkar**[5]*

[1]*Ph.D. Research Scholar, Department of Floriculture and Landscaping Architecture*
[2]*Assistant Professor, Department of Genetics and Plant Breeding*
[3]*Ph.D. Research Scholar, Department of Fruit Science,*
[4]*Ph.D. Research Scholar, Department of Vegetable Science, Sardar Vallabhbhai Patel University of Agriculture and Technology Meerut-250110, Uttar Pradesh, India*
[5] Ph.D. Research Scholar, *Department of Fruit Science, Chandra Shekhar Azad University of Agriculture & Technology, Kanpur, Uttar Pradesh (208002), India*
**Email: satvaansinghflori@gmail.com*

Introduction

Raspberry, scientifically known as *Rubus idaeus* L., belongs to the family Rosaceae. It is also known as "red raspberry" or "European raspberry," and its chromosome number is 2n=14. Raspberry is an important and popular fruit crop due to its rich flavor, high nutritional value, and various medicinal properties. Raspberry is rich in vitamins, minerals, dietary fiber, and antioxidants, making it a valuable component of a healthy diet (Rahman *et al.*, 2020). Raspberry is widely used in the food industry for jam, jelly, juice, ice cream, and other processed products.

Fig. 1.1: Raspberries

Origin and Distribution

Raspberry is believed to have originated in Eastern Europe and Western Asia (Farrar, 2016). Raspberry was introduced to India during the British era and was initially cultivated on a small scale in the hilly regions of the country.

Today, raspberry cultivation has spread to various parts of India, including Himachal Pradesh, Jammu and Kashmir, Uttarakhand, and Sikkim. The major raspberry-growing states in India are Himachal Pradesh and Jammu and Kashmir.

Area and Production

Raspberry cultivation in India is still in its nascent stage, with a very small area under cultivation. According to the National Horticulture Board, the total area under raspberry cultivation in India was 360 hectares in 2019-20, with a production of 1,323 tons. Himachal Pradesh is the leading raspberry-producing state in India, accounting for more than 90% of the total production (National Horticulture Board, 2020).

Botany

Raspberry is a perennial shrub that can grow up to 2 meters in height. It has compound leaves, thorny stems, and produces pink to white flowers, which give way to the fruit. Raspberry fruit is an aggregate of drupelets, which are small, juicy, and red in color.

Flowering, Pollination and Fruit Set

Raspberry plants are self-incompatible, which means they require cross-pollination for fruit set. The flowers are pollinated by bees and other insects. In India, the raspberry flowering season varies depending on the region, with the flowering period starting from February to March in Himachal Pradesh and from March to April in Jammu and Kashmir. Optimal pollination can be achieved by maintaining a good population of honeybees in the orchard.

Soil and Climatic Requirements

Raspberry plants grow best in well-drained soils that are rich in organic matter. The ideal pH range for raspberry cultivation is 5.5 to 6.5. Raspberry is a temperate crop that requires a cool climate for optimum growth and fruiting. In India, raspberry cultivation is primarily confined to the hilly regions where the climate is cool and moist.

Propagation and Rootstocks

Raspberry can be propagated by seeds or vegetatively through stem cuttings or layering. In India, raspberry propagation is mainly done by stem cuttings. The use of rootstocks is not very common in raspberry cultivation in India.

Use of Growth Regulator

The use of growth regulators in raspberry cultivation is limited in India. However, some studies have shown that the application of gibberellic acid (GA3) can increase the fruit size and yield of raspberry (Chen *et al.*, 2019).

Planting and Orchard Establishment

Raspberry plants are usually planted in the fall or early spring. The spacing between the plants varies depending on the variety and the trellising system used. In India, the most commonly used trellising system is the T-trellis. The orchard should be well-irrigated and fertilized to ensure good growth and development of the plants.

Commercial varieties

There are several commercially important raspberry varieties grown in India, each with its unique characteristics and adaptability to different growing conditions (Singh *et al.*2019; Kumar and Singh 2019; Prasad & Kamble 2018; Sood & Nagarajan 2017; Maiti & Saha 2015). Each of these varieties has different growing requirements and yields, and growers may choose a variety based on their specific needs and growing conditions. Some of the popular varieties are:

Heritage: This is a red raspberry variety that is highly productive and widely grown in India. It is resistant to many common diseases and pests and produces large, firm berries with a sweet, tart flavor.

Nova: This is a popular black raspberry variety grown in India. It is highly productive and produces large, firm berries with a rich, sweet flavor.

Prelude: This is a popular early ripening raspberry variety that is highly productive and adapted to Indian growing conditions. It produces large, firm berries with a sweet, tangy flavor.

Tulameen: This is a popular red raspberry variety grown in India. It is highly productive and produces large, firm berries with a sweet, juicy flavor.

Meeker: This is another popular red raspberry variety grown in India. It is highly productive and produces large, sweet, and juicy berries.

Willamette: This is a popular red raspberry variety grown in India. It is highly productive and produces large, firm berries with a sweet, juicy flavor.

Fallgold: This is a popular yellow raspberry variety grown in India. It is highly productive and produces large, firm berries with a sweet, mild flavor.

Training and Pruning

Raspberries require proper training and pruning to get good yield and quality fruits. In India, raspberries are generally trained to a two-wire trellis system. The training of the plant starts in the first year itself when the plant is about 20-30 cm high. The leader shoot should be tipped at this height to promote lateral branching. The lateral shoots should be spaced at about 20-25 cm apart and trained to the trellis wire. In the subsequent years, the shoots should be tied to the trellis wire and pruned back to a height of 120-150 cm. Pruning should be done in such a way that there are about 6-8 fruiting canes per meter length of the row.

Nutrition Management

Proper nutrition management is important for good yield and quality fruits. Raspberries require a well-balanced supply of nutrients, especially nitrogen, phosphorus, potassium, calcium, magnesium, and sulfur. The soil should be tested before planting to determine the nutrient status and nutrient requirement of the crop. Nutrient management should be done through the use of organic manures and inorganic fertilizers as per the recommendations of the soil test report. The fertilizer should be applied in split doses during the growing season. Foliar application of nutrients can also be done to correct any nutrient deficiency.

Water Management

Raspberries require adequate and uniform water supply for good growth and yield. Irrigation should be done at regular intervals to maintain soil moisture levels. Drip irrigation is the most efficient method for irrigating raspberries in India. The irrigation water should be of good quality and free from contaminants like salt and chlorine. Excessive moisture should be avoided as it can lead to waterlogging and root diseases.

Intercultural operation

Weed control is important for good growth and yield of raspberries. Weeds can be controlled by mechanical, manual or chemical methods. Mulching with organic materials like straw, leaves, or grass can also help in weed control and conserving soil moisture. In addition to weed control, intercultural operations like hoeing, earthing up, and pruning can help in managing the growth and development of the plant.

Major Insect-Pests and their management

Raspberries are susceptible to a number of insect pests like aphids, thrips, fruit flies, mites, etc. The pests can be managed by the use of insecticides, biopesticides, or cultural practices like pruning and sanitation. The insecticides should be used judiciously and as per the recommendations of the manufacturer. The use of biopesticides like neem oil, plant extracts, and microbial agents is a safer alternative to chemical insecticides.

Major Diseases and their management

Raspberries are also prone to a number of diseases like root rot, crown gall, powdery mildew, anthracnose, etc. The diseases can be managed by the use of fungicides, biofungicides, or cultural practices like sanitation and pruning. The fungicides should be used as per the recommendations of the manufacturer. The use of biofungicides like Trichoderma, Pseudomonas, and Bacillus is a safer alternative to chemical fungicides.

Maturity Indices

Raspberries are harvested when they are fully ripe, which is indicated by the color, flavor, and firmness of the fruit. The fruit should be picked carefully and gently to avoid damage. The fruit can be stored for a few days at low temperatures and high humidity.

Harvesting and Post-harvest Management:

Raspberries are hand-harvested when the fruit reaches the desired maturity stage, usually every two to three days, to ensure that the fruit is of the highest quality. Ripe raspberries are tender and easily damaged, so they should be handled carefully during harvest. The harvested fruits should be sorted, graded, and packed into small baskets to prevent damage. They should be handled gently to avoid crushing and damage to the fruit.

Post-harvest handling plays a significant role in maintaining the quality of the fruit during transportation and storage. Raspberries have a short shelf life and are highly perishable. They should be cooled immediately after harvest to remove the field heat and reduce the respiration rate. Cooling should be done as quickly as possible to minimize the loss of fruit quality.

The best method for cooling raspberries is forced-air cooling, which quickly lowers the temperature of the fruit and removes the heat produced during respiration. The optimum temperature for storing raspberries is between 0°C and 2°C, with a relative humidity of 90-95%. Fruit should be stored in a well-ventilated cold storage facility and should be inspected frequently to remove any damaged or overripe fruit (Behn 2017).

Yield

The yield of raspberries depends on many factors, including cultivar, plant density, nutrition, irrigation, and pruning. Under optimal growing conditions, raspberry yields can range from 3-6 tonnes per hectare, depending on the cultivar and the management practices employed.

Future Prospects

Research in the area of raspberry cultivation is ongoing, and several new developments have emerged in recent years. One area of research is the development of new raspberry cultivars with improved traits, such as disease resistance, fruit quality, and yield potential. Another area of research is the development of more efficient and sustainable cultivation methods, such as precision agriculture and organic production systems.

Researchers are also working to improve post-harvest handling techniques to extend the shelf life of raspberries and reduce food waste. Additionally, advancements in molecular biology and biotechnology are opening up new possibilities for the genetic improvement of raspberries, such as the development of new varieties with enhanced nutritional content or improved disease resistance.

References

Anonymous, (2020). National Horticulture Board, Indian Horticulture Database. Ministry of Agriculture & Farmers Welfare, Government of India. www.agricoop.nic.in.

Behn, B. (2017). Raspberry production in California. University of California, Division of Agriculture and Natural Resources. Retrieved from https://anrcatalog.ucanr.edu/pdf/7246.pdf.

Chen, Z., Chen, L., Wang, X., Guo, L., & Gao, H. (2019). Effects of gibberellic acid on the fruit size, quality and yield of raspberry. Acta Horticulturae, (1230), 259-264. doi: 10.17660/ActaHortic.2019.1230.37

Farrar, J. (2016). The complete book of raspberries, blackberries, and brambles. Timber Press.

Kumar, A., & Singh, R. (2019). Advances in small fruit crops cultivation in India: A review. Journal of Applied and Natural Science, 11(2), 612-619.

Maiti, S., & Saha, S. (2015). Small fruits production and its status in India. African Journal of Agricultural Research, 10(12), 1335-1344.

Prasad, N. N., & Kamble, C. K. (2018). Present status and future scope of small fruit cultivation in India. Current Journal of Applied Science and Technology, 29(2), 1-16.

Rahman, M. S., Biswas, S. K., & Hossain, M. A. (2020). Nutritional and therapeutic perspectives of raspberry. Heliyon, 6(9), e04923. doi: 10.1016/j.heliyon.2020.e04923

Singh, G., Singh, P., & Singh, M. (2019). Raspberry production technology in India. In Fruit and Vegetable Production (pp. 271-290). Springer, Singapore.

Sood, S., & Nagarajan, S. (2017). Cultivation of raspberry in India. Journal of Horticultural Sciences, 12(1), 13-18.

17

Advances Production Technology of Cranberry

Satvaan Singh*[1], Atar Singh[2], Amit Kumar[3], Sajeel Ahamad[4] and Imran Ali[5]

[1]Ph.D. Research Scholar, Department of Floriculture and Landscape Architecture,
[2]Assistant Professor, Department of Genetics and Plant Breeding
[3]Ph.D. Research Scholar, Department of Fruit Science
[5]M.Sc. Fruit Science, Department of Fruit Science, College of Horticulture, Sardar Vallabhbhai Patel University of Agriculture and Technology Meerut-250110, Uttar Pradesh, India
[4]Ph.D. Research Scholar, Division of Food Science & Post-harvest Technology, Indian Agricultural Research Institute, New Delhi-110012, India
**Email: satvaansinghflori@gmail.com*

Introduction

Cranberry (*Vaccinium macrocarpon* Aiton) belongs to the Ericaceae family and is native to North America. Cranberries are also known by various other names, including American cranberry, bearberry, and lingonberry. The plant has a diploid chromosome number of 24 and grows in acidic, nutrient-poor soils in wetlands and bogs. It is an important commercial crop worldwide due to its high nutritional value, antioxidant properties, and health benefits. Cranberries are primarily consumed as juice, dried fruit, or in the form of sauces and jellies. They are a rich source of vitamins C and E, dietary fiber, anthocyanins and essential minerals such as potassium, manganese, and copper. The fruit is also used in the pharmaceutical industries due to its anti-inflammatory, antimicrobial, and anti-aging properties (Santos *et al.*, 2017; Wang *et al.*, 2017).

Fig. 1.1: Cranberries

The composition of cranberries is quite remarkable, with high levels of antioxidants, including phenolic compounds such as flavonols, anthocyanins, and proanthocyanidins. These compounds are responsible for the fruit's vibrant red color and are thought to provide many of the health benefits associated with cranberries. Cranberries are an important crop in the United States, Canada, and Chile, with significant economic value. They are primarily consumed as juice or in processed forms, such as sauces and jams. Cranberries are also used in the production of dietary supplements and cosmetics.

Cranberries are an important crop worldwide due to their versatility and numerous health benefits. They are widely used in the food and beverage industry, particularly in the production of juices, sauces, and jams. The unique tart flavor of cranberries also makes them a popular ingredient in a variety of recipes, including baked goods and desserts. Additionally, cranberries have been found to have medicinal properties and are used to prevent and treat various ailments, including urinary tract infections, cardiovascular disease, and cancer.

Origin and Distribution

Cranberry is native to North America and was traditionally used by Native Americans for its medicinal and nutritional properties. The cultivation of cranberries started in the 1800s, and today, the USA is the largest producer of cranberries, followed by Canada and Chile (Lee *et al.*, 2020). In recent years, cranberry cultivation has also spread to other parts of the world, including Europe, Asia, and Australia.

Cranberries are not native to India, but in recent years, there has been an effort to introduce the crop and develop advanced production technologies to grow them in India.

The Indian Council of Agricultural Research (ICAR) and the Horticulture Department of the Government of Jammu and Kashmir have been working on the introduction of cranberry cultivation in India. The efforts began in 2011 when the government of Jammu and Kashmir signed a Memorandum of Understanding (MoU) with the University of Massachusetts, Amherst, USA, to develop cranberry cultivation in the state. The main areas of focus for the development of cranberry cultivation in India include the selection of suitable varieties, identification of optimal growing conditions, and the use of advanced production technologies. Currently, the cultivation of cranberries is being carried out on a small scale in Jammu and Kashmir.

Area and Production

According to FAO STAT (2021), the global production of cranberries was 1.1 million tons in 2019, with the USA accounting for 64% of the total production. Other major producers include Canada, Chile, and Belarus. Cranberry cultivation in other countries, such as the Netherlands and Poland, is also increasing. India is not a major producer of cranberries. In fact, cranberries are not commercially grown in India due to unsuitable climatic conditions. The majority of cranberry cultivation takes place in North America, specifically in the United States and Canada. However, there are some companies in India that import cranberries from other countries for consumption and use in various products.

Botany

The cranberry plant is a perennial shrub that typically grows up to 2 meters in length and has woody stems, small evergreen leaves, and delicate flowers that bloom in the spring. The leaves are dark green and glossy, and the flowers are pink or white, which bloom in June or July. The fruit is a berry, which is initially green and turns red when it matures (Martin *et al.*, 2019).

Flowering, Pollination and Fruit Set

Cranberries are self-pollinating, and their flowers are hermaphrodite, meaning that they have both male and female reproductive organs. The pollination process is mainly done by bees and other insects. The fruit set of cranberries is affected by various environmental factors, such as temperature, light, and water availability (Ehlenfeldt *et al.*, 2017).

Soil and Climatic Requirements

Cranberries grow well in acidic, well-drained soils that are rich in organic matter. The ideal pH range for cranberry cultivation is between 4.0 and 5.5. Cranberries require a cool and humid climate with a minimum annual rainfall of 500-600 mm. They are adapted to frost-prone regions and can withstand temperatures as low as -20°C (Vorsa, 2018).

Propagation and Rootstocks

Cranberries can be propagated vegetatively through stem cuttings or by layering. Rooted cuttings are transplanted into the field, and they develop into a full-sized plant in 2-3 years. Cranberry cultivars are commonly grafted onto rootstocks to improve their performance and resistance to pests and diseases (Poling *et al.*, 2018).

Use of Growth Regulator

Growth regulators are commonly used in cranberry cultivation to improve fruit yield and quality. Gibberellins are the most commonly used growth regulators in cranberry production, which promotes shoot growth and flower bud formation. Ethephon is used to induce fruit ripening and improve fruit color (Poling *et al.*, 2018).

Planting

Planting is an essential component for the successful production of cranberry. Before planting, soil testing is important to assess soil fertility and pH. Cranberries prefer acidic soils with pH levels between 4.0 and 5.5. The site should be well-drained, with access to irrigation. Planting methods include using rooted cuttings or vines. The use of tissue culture plantlets has also been studied as a viable option (Fajardo *et al.*, 2019). Planting should be done in the spring or fall. Beds can be constructed using a variety of methods including a floating board system or a dry bed system. Bed preparation should include the application of fertilizers, herbicides, and pre-emergent weed control methods (Savage, 2015).

Commercial Varieties

Cranberry cultivars are generally classified into two groups: American and hybrid. American cultivars have been traditionally used for fresh fruit, while hybrids are used primarily for processing. There are several commercial varieties of cranberry that are widely grown, including 'Stevens', 'Ben Lear', and 'Pilgrim'. These varieties differ in their fruit quality, yield potential, and disease resistance. 'Stevens' is a widely planted variety known for its high yield potential and good fruit quality. 'Ben Lear' is a later maturing variety with good disease resistance, and 'Pilgrim' is an early maturing variety with high yield potential (Liu *et al.*, 2020).

Training and Pruning

Training and pruning are important management practices to ensure good fruit quality and yield. Young cranberry plants should be trained to produce upright shoots, which will form the main fruiting branches. Pruning should be done in the late winter or early spring to remove dead or diseased wood and to thin out the canopy. This will increase air circulation and sunlight penetration, reducing the incidence of disease and improving fruit quality (Larsen *et al.*, 2021).

Nutrition Management

Proper nutrition management is essential for the production of high-quality cranberries. Fertilizer applications should be based on soil test results and the crop's nutritional requirements. Nitrogen is the most critical nutrient for cranberry production, and applications should be made based on the crop's growth stage. In addition to nitrogen, phosphorus, potassium, and micronutrients are also important for cranberry growth (Eck *et al.*, 2017).

Water Management

Cranberries require consistent soil moisture throughout the growing season, and adequate water management is essential for their production. Irrigation can be achieved through overhead sprinkler systems or flood irrigation. Overhead sprinkler systems are more commonly used in the western United States, while flood irrigation is used primarily in the eastern United States. In addition to irrigation, drainage is also important to prevent waterlogging and root damage (Savage, 2015).

Intercultural Operations

Intercultural operations are essential for cranberry production, and these include practices such as weed control, pest management, and disease control. The use of herbicides is a common practice for weed control, and pre-emergent herbicides can be used to control weeds before they emerge. In addition to herbicides, other cultural practices, such as hand weeding and mechanical cultivation, can also be used for weed control (Liu *et al.*, 2020).

Important Practices

Several important management practices can improve cranberry production. These include the use of frost protection, bird management, and pollinator management. Frost protection can be achieved through the use of sprinkler systems or wind machines. Bird damage can be controlled through the use of netting or the use of auditory repellents.

Pollination can be improved through the use of managed bee colonies (Eck *et al.*, 2017).

Weed Management

Weed management is a significant challenge in cranberry production. A wide variety of weeds can compete with cranberry plants for nutrients, water, and light, and can reduce yield and quality. The use of herbicides is a common practice for weed control in cranberry production. Pre-emergent herbicides can be used to control weeds before they emerge, and post-emergent herbicides can be used to control weeds after they have emerged. Hand weeding and mechanical cultivation can also be used for weed control (Liu *et al.*, 2020).

Physiological Disorders and their Management

Cranberry plants can be affected by several physiological disorders that can reduce yield and quality. One such disorder is "false blossom," which can result in a reduction in fruit size and yield. False blossom is caused by a phytoplasma, and management includes the removal of infected plants and the use of insecticides to control leafhopper populations, which can spread the disease (Stafne *et al.*, 2017). Another physiological disorder is "internal breakdown," which can occur during storage and result in the breakdown of the fruit's internal tissues. Management includes the use of proper storage conditions and post-harvest handling practices (Stafne *et al.*, 2017).

Major Insect-Pests and their Management

Cranberry production is also affected by several insect pests, which can reduce yield and quality. One such pest is the cranberry fruit worm, which can cause damage to developing fruit. Management includes the use of insecticides and the use of pheromone traps to monitor and control the pest population (Stafne *et al.*, 2017). Another pest is the cranberry weevil, which can cause damage to the buds and flowers. Management includes the use of insecticides and the use of cultural practices, such as removing overwintering sites (Stafne *et al.*, 2017).

Major Diseases and their Management

Cranberry production is also affected by several diseases, which can reduce yield and quality. One such disease is "fruit rot," which is caused by several fungal pathogens and can result in a reduction in fruit yield and quality. Management includes the use of fungicides, the removal of infected fruit, and the use of cultural practices, such as good drainage and air circulation (Stafne *et al.*, 2017). Another disease is "leaf spot," which is caused by several fungal pathogens and can result in defoliation and a reduction in fruit yield and quality. Management includes the use of fungicides and the removal of infected leaves (Stafne *et al.*, 2017).

Maturity Indices

The fruit of cranberry is harvested when it reaches the desired maturity indices. The desired maturity indices depend on the intended use of the fruit. The maturity of cranberries can be determined using several indices, including color, firmness, and sugar content. Cranberries are typically harvested when they reach their maximum color and firmness. The sugar content of cranberries is also an important indicator of maturity and can be measured using a refractometer (Fajardo *et al.*, 2019).

Harvesting and Post-harvest Management

Cranberries are typically harvested using mechanical harvesters, which shake the berries from the vines. After harvesting, the berries are cleaned and sorted, and damaged or diseased berries are removed. Cranberries can be stored for several months in cold storage, and proper post-harvest management practices can reduce losses due to decay and mechanical damage (Fajardo *et al.*, 2019). Cranberry fruit can be stored for up to 2-3 months if stored at 0-2°C and 90-95% relative humidity.

Yield

Cranberry yield is influenced by several factors, including genetics, planting density, nutrition, water management, pest and disease control, and environmental conditions. To maximize yield, it is essential to select the appropriate variety for the growing region, maintain optimal plant density, provide proper nutrition and water management, and effectively manage pests and diseases (Gardner *et al.*, 2018). The yield of cranberry varies depending on the variety, plant age, and management practices. The average yield of cranberry is about 200-250 barrels per acre.

Future Prospects

The cranberry industry has experienced significant growth over the past few decades, driven by increasing demand for cranberry products due to their perceived health benefits. The global cranberry market is projected to continue to grow in the coming years, and there are several emerging opportunities for cranberry production. One such opportunity is the production of cranberry extract, which has several potential health benefits, including reducing the risk of urinary tract infections and improving gut health (Yadav *et al.*, 2019). Another opportunity is the use of cranberry byproducts for the production of value-added products, such as cranberry juice and cranberry seed oil (Gardner *et al.*, 2018).

The genetic improvement of cranberry has been limited due to its complex genetics, low seed set, and long juvenile period. However, recent advances in genomic research and biotechnology have the potential to accelerate the development of new cultivars with improved yield, disease resistance, and quality. Additionally, the expansion of global markets for cranberry and the increasing demand for natural health products have created new opportunities for the cranberry industry.

Advances in production technology have played a significant role in the growth and success of the cranberry industry. The development of new varieties, improvements in pest and disease management, and advancements in harvesting and post-harvest management practices have all contributed to increased yield and quality. To continue to thrive and meet the growing demand for cranberry products, it will be essential to stay at the forefront of these technological advances (Gardner *et al.*, 2018).

References

Eck, P., Reiss, K., & Nennich, T. T. (2017). Cranberry production in North America. In Advances in Plant Breeding Strategies: Fruits (pp. 129-155). Springer, Cham.

Ehlenfeldt, M. K., Polashock, J. J., & Stretch, A. W. (2017). The American cranberry: Its botany, horticulture, and use. Horticultural Reviews, 45, 255-364.

Fajardo, D. A., Reekie, J. Y., & Chalker-Scott, L. (2019). Cranberry. In D. A. Fajardo & J. Y. Reekie (Eds.), Fruit Crops: Diagnosis and Management of Nutrient Constraints (pp. 191-212). CABI.

Fajardo, D. A., Schuster, W. H., & Lee, J. (2019). Cranberry Production in the United States: An Overview of Production, Harvest, and Post-harvest Management. Horticulturae, 5(1), 8. doi: 10.3390/horticulturae5010008

Fajardo, D. A., Vargas, A. M., & Guevara, J. C. (2019). Tissue culture in cranberry propagation: a review. In Vitro Cellular & Developmental Biology-Plant, 55(3), 259-269.

Food and Agriculture Organization of the United Nations. (2021). FAOSTAT database: Production quantity of cranberries. Retrieved from http://www.fao.org/faostat/en/#data/QC/visualize

Gardner, K. M., Luby, J. J., & Pavek, M. J. (2018). Cranberry. In G. A. Loebenstein & G. Thottappilly (Eds.), The Encyclopedia of Fruit and Nuts (pp. 357-361). CABI.

Larsen, R. C., Draper, A. D., & McManus, P. S. (2021). The impact of pruning on the cranberry fruit rot complex in Massachusetts. Plant Disease, 105(1), 184-189.

Lee, J., Behe, B. K., & Dennis, J. H. (2020). U.S. cranberry industry: Trends, production, and marketing. HortTechnology, 30(1), 17-25.

Liu, H., Ren, J., Yang, M., & Wu, X. (2020). Evaluation of three herbicides for weed management in cranberry fields. Journal of Environmental Science and Health, Part B, 55(3), 255-263.

Liu, J., White, J. W., & Polashock, J. (2020). Cranberry (Vaccinium macrocarpon Ait.) breeding, genetics, and genomics. In Breeding in a World of Scarcity (pp. 1-24). Springer, Cham.

Liu, Y., Xing, Y., Li, B., & Wang, L. (2020). Effect of Pre-Emergent Herbicides on Weed Control and Crop Yield in Cranberry (Vaccinium macrocarpon Ait.) Production. Agronomy, 10(3), 389. doi: 10.3390/agronomy10030389

Martin, M. P., Medeiros, M. A., & Handley, D. T. (2019). Cranberry (Vaccinium macrocarpon Ait.). In D. B. Smith & R. M. Bell (Eds.), Specialty crops for Pacific Island agroforestry (pp. 31-36). College of Tropical Agriculture and Human Resources, University of Hawaii at Manoa.

Poling, E. B., Sandhu, A. K., & Poling, J. R. (2018). Cranberry. In A. K. Dhawan & P. A. Wadhwani (Eds.), Fruit crops: Diagnosis and management of nutrient constraints (pp. 215-240). Springer.

Santos, R. A., Silva, J. F. M., & Pimenta, L. P. S. (2017). Cranberry (Vaccinium macrocarpon): An overview of the nutritional composition, phytochemistry, and medical uses. Food and Nutrition Sciences, 8(11), 1105-1115.

Savage, S. D. (2015). Cranberry production in the United States. In Horticultural Reviews (Vol. 43, pp. 293-346). Wiley.

Stafne, E. T., Ellis, M. A., & Harbut, R. E. (2017). Cranberry. In Diseases and Pests of Vegetable Crops in Canada (pp. 140-149). doi: 10.1007/978-3-319-57803-9_12

Stafne, E. T., Lehman, J. S., & Bush, E. A. (2017). Cranberry Production in the United States: A Brief History and Current Status. HortTechnology, 27(5), 541-549.

Vorsa, N. (2018). Cranberry biology and cultivation. In C. J. DeMoranville & J. J. Polashock (Eds.), Cranberry production guide (pp. 3-16). Rutgers University.

Wang, Y., Liu, Y., Liu, X., & Sun, H. (2017). Cranberry phytochemicals and their health benefits. Nutrients, 9(10), 1088.

Yadav, P., Aydarous, A., Johnson, B., & Johnson, M. (2019). Cranberry Extract and its Beneficial Effects on Health: A Comprehensive Review. Complementary Therapies in Medicine, 44, 29-38.

18

Advances Production Technology of Blueberry

Sajeel Ahamad[1]*, Amit Kumar[2], Menaka M[3], Chhail Bihari[4], Maneesh Kumar[5] and Ganesh Kumar Choupdar[6]

[1,3,6] Ph.D. Research Scholar, Division of Food Science and Postharvest Technology, ICAR- Indian Agricultural Research Institute New Delhi 110012
[2]Ph.D. Research Scholar, Department of Fruit Science, Sardar Vallabhbhai Patel University of Agriculture and Technology, Meerut, Uttar Pradesh, India
[4]Ph.D. Research Scholar, Department of Horticulture Vegetable Science Nagaland Central University, SASRD Medziphema, Nagaland, India
[5]Ph.D. Research Scholar, Department of Horticulture, G.B. Pant University of Agriculture & Technology Pant Nagar, U.S. Nagar, Uttarakhand, India
**Email: sajeelahamad04@gmail.comIntroduction*

The blueberry (*Vaccinium* spp.) is an evergreen or perennial deciduous shrub that is a member of the Ericaceae family and it is native to North America. It is oblat at maturity and forms dark blue color. Blueberry is also known as "the king of berries," have real blue fruits, which is very rare in nature (Chu *et al.*, 2018). Tropical, subtropical, and temperate climates make up the majority of the distribution of blueberries. The United States, Chile, Canada, Spain, China, Japan and Morocco are the top producer in the world (Yousuf *et al.*, 2016). Traditionally, 58% of the world fresh blueberries are consumed in North America. The lowbush and highbush varieties combine produced 850,886 tonnes of blueberries in 2020. The United States alone produced around 35% and Peru for 21%. (FAOSTAT 2022).

Fig. 1.1: Blueberries

According to the South Africa Berry Producers' Association, blueberries were originally brought to South Africa in the 1970s and are currently one of the country's fastest-growing agricultural sectors. There are many places where blueberries are grown, with the Western Cape produced around (60%) followed by Limpopo (15%), North-West (10%), Gauteng (8%), Eastern Cape (4%), Free State (2%) and Mpumalanga (1%) (Steyn *et al.,* 2022). Federick Coville of the United States Department of Agriculture began the first blueberry domestication in 1908. (USDA). High and low bushes are the most widely grown kind of blueberry, however rabbiteyes are only significant in South and North America. High and low bushes hybrids are very common in the upper Midwest of the United States. The main growing countries for north high bush are Australia, Germany, and France.

The blueberry is a member of the *Vaccinium* and *Cyanococcus* subgenera. Highbush blueberries (*V. corymbosum* L.) and rabbiteye blueberries (*V. ashei* Reade) are typically regarded as the two blueberry species with the greatest commercial importance. Three varieties of highbush blueberries exist southern highbush blueberry, half highbush blueberry, and northern highbush blueberry (SHB). Blueberries are considered a functional food because of their high bioactive chemical content, particularly flavones and other polyphenolic compounds that have powerful antioxidant activities and promote health (McAnulty *et al.,* 2011).

Chemical composition of blueberries (Ijaz *et al.*, 2022)

Nutrients	Amount per 100 g fresh weight (Fresh)	Nutrients	Amount per 100 g fresh weight (Fresh)
Energy	57 Kcal	Vitamin C	9.7 mg
Carbohydrate	14.5 g	Vitamin E	0.57 mg
Water	84%	Vitamin B1	0.04 mg
Fat	0.33 g	Vitamin B2	0.04 mg
Protein	0.74 g	Vitamin B6	0.05 mg
Dietary fiber	2.4 g	Phosphorus	12 mg
Pottassium	77 mg	Magnesium	6 mg
Fructose	4.97 g	Zinc	0.16 mg
Total beta carotene	31 µg	Manganese	0.336 mg
Calcium	6 mg	Niacin	0.418 mg
Iron	0.28 mg		

Botany

There are numerous different cultivars, and each one has unique traits for both the plant and the fruit. *Vaccinium corymbosum* L. makes up the majority of highbush blueberry cultivars, although through interspecific hybridization, several more recent cultivars have genetic components from other *Vaccinium* species (Forney *et al.*, 2022). Although highbush blueberries have self-fertile flowers, the benefits of cross-pollination outweigh the self-fertility to a greater or lesser extent depending on the cultivar. The calyxes of highbush blueberry blossoms measure 6 to 10 mm in length. Before pollination, the white petals combine to create a bell- or tubular-shaped corolla that hangs with its open end downward (Prodorutti *et al.,* 2007). Blueberry *vaccinium* genera *Vaccinium corymbosum* L. (High bush berry), *Vaccinium virgatum* or *Vaccinium ashei* (Rabbiteye blueberry), and *Vaccinium angustifolium* (Low bush blueberry) are the three most often farmed species (Retamales and Hancock, 2018). Northern highbush blueberries are the most widely grown plant species in terms of surface area (75% northern highbush, 10% southern highbush, and 15% rabbiteye) (Rodriguez Saona *et al.*, 2019). Depending on how hardy they are over the winter and how many chilling hours they need for flower development, high bush blueberry cultivars are further divided into northern, southern, and intermediate varieties.

Taxonomy

The main species that are related to the cultivated and semi-cultivated blueberries that are descended from *Cyanococcus* are *V. corymbosum* L., *V. angustifolium* Ait., *V. virgatum* Ait., *V. elliottii* Chapm., and *V. darrowii* Camp. Tetraploid *V. corymbosum* (2n=4x=48) has a relatively small genetic base. *V. amoenum* Ait., *V. constablaei* Gray, *V. tenellum* Ait., *V. hirsutum* Buckley, and other non-cultivated species of the section *Cyanococcus* have proved particularly significant in interspecific hybrids. The *Vaccinium* sections *Oxycoccus*, *Vitis-idaea*, *Myrtillus, Pyxothamnus, Bathodendron, Polycodium*, and *Hemimyrtillus* belong to a different group (Ballington 2001). Breeding *Oxycoccus* and *Cyanococcus* species together has led to some intriguing findings that could be useful for gene resources. (Lyrene and Ballington, 1986).

Some important species and their description

Vaccinium caesariense (New Jersey blueberry) comes from the Eastern United States. It belongs to the Vaccinium genus, which also includes the blueberry, cranberry, huckleberry, and bilberry species.

Vaccinium corymbosum (Northern highbush blueberry) The northern highbush blueberry is a kind of blueberry native to North America that has developed into a major food and cash crop. It is indigenous to eastern Canada and eastern and southern America. Blue huckleberry, tall huckleberry, swamp huckleberry, high blueberry, and swamp blueberry are some more frequent names for the berry.

Vaccinium darrowii (Evergreen blueberry) A species of Vaccinium of the blueberry family is known by the popular names Darrow's blueberry, evergreen blueberry, scrub blueberry, or southern highbush blueberry.

Vaccinium elliotti (Elliott blueberry) is a blueberry-related species of the *Vaccinium* genus (*Vaccinium* sect. *Cyanococcus*). Its natural range in the United States is from south eastern Virginia to Florida in the south, and west to Arkansas and Texas in the west.

Vaccinium formosum (Southern blueberry) A blueberry species that is indigenous to the South eastern United States goes by the common name highbush blueberry, southern blueberry, southern highbush blueberry, and swamp highbush blueberry.

Vaccinium fuscatum (Black highbush blueberry) is a deciduous upright shrub. Because its stems and abaxial leaf surfaces are pubescent with dingy hairs and its dark-colored fruit is devoid of a glaucous coating, it can be recognised from the similarly shaped *Vaccinium corymbosum*.

Vaccinium hirsutum (Hairy-fruited blueberry) *Vaccinium myrsinites* (Shiny blueberry) *Vaccinium myrtilloides* (Sour top, Velvet leaf, or Canadian blueberry)

Vaccinium pallidum (Dryland blueberry) *Vaccinium simulatum* (Upland highbush blueberry) *Vaccinium tenellum* (Southern blueberry) *Vaccinium virgatum* (Rabbiteye blueberry; syn. *V. ashei*) (Plunkett *et al.*, 2018)

Health benefits

The blueberry is a rich source of bioactive components, including phenolics, flavonoids, anthocyanins, and other phytochemicals, and it has biological actions including anti-oxidant, anti-inflammatory, anti-cancer, anti-obesity, and neuroprotective (Zou *et al.*, 2022). According to research by, blueberries are a top source of bioactive compounds with high antioxidant activity and potential health benefits including anti-diabetic properties, lowering blood pressure and blood cholesterol, and inhibiting the growth of cancer cells in the colon breast for these several health benefits it also referred to as "golden berries," (Milivojevi *et al.*2020). It is rich in nutrients and many antioxidant enzymes such as sugars, vitamins, amino acids, anthocyanins, catalase (CAT), superoxide dismutase (SOD), peroxidase (POD), and other enzymes, prevent and treat hypertension, clear capillaries and slow aging (Yang *et al.*, 2022, Wang *et al.*, 2019).

Flavouring compounds

In Southern highbush blueberries, GC-O analysis revealed that linalool and (E)-2-hexenal were the two most aroma-active compounds among four cultivars, while the contribution of other aroma-active compounds varied depending on the cultivar and included geraniol, eugenol, 2-heptanol, and (E, Z)-2,6-nonadienal, geraniol, and hexenal. In comparison to geraniol, (Z)-3-hexen-1-ol, 1-octen-3-one, -terpineol, and linalool in highbush fruit, the most aroma-active chemicals in lowbush fruit were ethyl 2-methylbutanoate, methyl 2-methylbutanoate, ethyl 3-methylbutanoate, and ethyl propanoate. (Forney *et al.*, 2022)

Soil and climate

Blueberries are bushes with little root hairs, insignificant primary roots, and minimal drought tolerance (Lui *et al.*, 2017). The right amount of soil acidity is necessary for the growth of blueberries. Lowbush blueberries grow naturally in North American soil that is acidic (pH 3.5–5.5) and rich in organic materials (Hancock and Draper, 1989). By using the proper nutrients, the pH of the soil can be lowered if it is higher than 5.5. Acidic soil that drains well, sufficient moisture retention, and a high organic matter content are all necessary for blueberry growth. Low bush type cultivars can withstand temperatures as low as -30°C and require at least 1000 chilling hours (less than 7°C). Less than 600 chilling hours are needed for rabbiteye cultivars, because flower buds cannot tolerate temperatures below freezing. North high bush cultivars need 800-1000 hours of chilling time and can withstand temperatures as low as -20°C in the winter. Southern high bush varieties need 550 cooling hours and are not resistant to very cold winters. Moderate high bush cultivars need to be chilled for somewhere between 400 and 800 hours (Retamales and Honcock 2018)

Commercial varieties

Northern Highbush Blueberry (NHB)- Berkely, Bluecrop, Bluegold, Bluejay, Blueray, Brigitta, Chandler, Darrow, Duke, Early Blue, Herbert, Jersey, Spartan,

Sunrise, Sunshine, Legacy, Reka, Elizabeth, Bluehaven, Liberty, Pioneer, Washington, Pacific

Southern Highbush Blueberry (SHB)- Jubile, Misty, Oneil, Ozarkblue, Bluecrisp, Farthing, Jewel, Millenia, Primadonna, Scintill, Snowchaser, Springhigh, Sweetcrisp, Windsor, Emerald, Sharpblue, Star, Arlen, Lenoir, Sampson, Collins

Half Highbush Blueberry (HHB) -Northcountry, St.Cloud, Northblue, Bluegold, Northland, Friendship, Polaris

Rabbiteye (RE)- Alapaha, Austin, Brightwell, Briteblue, Centurion, Climax, Premier, Ochlockonee, Powderblue

Due to its high fruit quality for fresh markets and tolerance to subtropical and tropical production zones, the southern highbush blueberry (SHB, Vaccinium corymbosum L. interspecific hybrids) has been crucial to this expansion.

Propagation

The preferred technique for growing rabbiteye and southern highbush blueberries is softwood propagation. Rabbiteyes and southern highbush also propagated through hardwood cuttings. In North Carolina, hardwood propagation has long been employed as a less-intensive propagation method that nonetheless produces a 60–70% stand of cuttings. Blueberry suckers are branches that grow several inches from the plant's main crown and are transferred to a permanent spot in the field or laid out in a nursery bed for one season before being moved to the field. (Bill Cline and Gerard Krewer 2003). Both lingonberry (*V. vitis-idaea*) and lowbush blueberry (*V. angustifolium*) micro propagated plants have been demonstrated to have an earlier and higher rhizome development in comparison to plants propagated through cuttings (Gustavsson & Stanys 2000). Often, cuttage is used to propagate blueberries. The propagation bed, heat treatment of the rooting solution, insertion depth, and length of cuttings in order to maximize the rooting ratio of hardwood cuttings. Cuttings rooted in a mist spray house are preferable to cuttings rooted in a straightforward tunnel culture in the instance of highbush blueberries (*Vaccinium corymbosum* L.). (Morrison *et al.,* 2000).

A straightforward tunnel propagation bench works best for rabbiteye blueberry (*Vaccinium ashei* Reade), though. Both the rabbiteye blueberry and the highbush blueberry, which are difficult to multiply by cutting in a mist spray house, are more likely to root after the rooting medium has been heated. In a mist spray home, the shallow-insertion cuttings of rabbiteye blueberry rooted noticeably better than the deep-insertion ones. Also, the common lengths of highbush blueberry cuttings, which are small (5 cm) and long (10 cm), both rooted. Nevertheless, rabbiteye blueberry cannot be used with the shortcut technique.

High Density Planting (HDP)

The standard spacing for planting southern highbush blueberries is 2.75 to 3.00 m between beds and 0.76 to 0.9 m between bushes within the row. Innovative farmers use containers for production in some circumstances, decreased the across-bed spacing to increase single row planting density. High-density plantings have the important benefit of speeding up the process of a field reaching positive cash flows. High-density plantations may improve weed control and utilization of fertilizer and water. In comparison to varieties with a spreading growth habit, upright-growing cultivars like "Abundance," "Chickadee," and "Meadowlark" may be more suitable for planting in large densities. As an alternative, reflective plastic mulches can be used to raise the amount of radiation that is photosynthetically active in the lower portions of the canopy, enhancing photosynthetic rates and increasing fruit quality (Fang *et al.*, 2020).

Irrigation and Fertilizer Management

The level of soil moisture heavily influences the appropriate fertilizer application, and higher moisture can cause nutrient components to move to the root surface and absorb more quickly. As a result, plants can grow and develop more quickly with the use of fertilizer and water in moderation. This demonstrates how crucial the water to fertilizer ratio is for the successful and exact planting of rabbit-eye blueberries. The best fertilization and irrigation regimes for growing rabbiteye blueberries were medium fertilizer doses with medium to medium-high irrigation or high fertilizer doses with medium-high irrigation (Dai *et al.,* 2019). In their natural environment, blueberries coexist in harmony with mycorrhizal fungus that help plants absorb nutrients like nitrogen (N), phosphate (P), and iron. Highbush blueberry fruits contained more soluble solids after being fertilized with boron (B). When N, P, and K were sprayed at rates of 0.8, 2.0, and 7.2 g m1, respectively, liming increased fruit anthocyanin content in both years (Albert *et al.,* 2010).

Training and pruning

Plant vigour and age should also be considered when harvesting by machine and altered as necessary. In order to enhance light penetration into the canopy and harvest effectiveness, pruning should provide an open core. To make fruiting wood accessible to harvesters, try to position it close to the edge of the bush. Growth that is too wide or outside of the trellised area might not fit inside a harvester's throat. To maintain good picking height, which is governed in part by the cultivar, pruning should be done. Maintaining a thin crown and removing low growth below the catcher plates will guarantee that the catcher plates fit snugly around the base of the plant, further optimizing pruning and training for machine harvest. The aim of trellising to position canes and branches upright so they fit better within a machine. (DeVetter *et al.*, 2022)

Pollination

The most frequent and active pollinators in a number of blueberry cultivation areas are bumblebees. fruits produced in blueberry plots using precise pollination control techniques as opposed to standard methods. By identifying the following crucial points, we hope to create the framework for the application of good pollination techniques utilizing *Apis mellifera* as the pollinator. Selection of high-quality bee hives, getting them ready to enter the field before flowering, placement and distribution of the hives within the field, bee nutrition and management within the field, and bee health so that no pesticides are sprayed during the pollination service period are the first five steps. In contrast to plots managed using conventional pollination, *V. corinbosum* plots managed using the precision pollination service would exhibit higher bee visitation rates to their flowers, resulting in a greater proportion and higher quality of the fruits developed. In contrast to distribution in medium groups (5 hives) to large (30 hives) positioned in strategic locations, each hive should be placed strategically and uniformly (Cavigliasso *et al.,* 2021).

Insect pest management

The most dangerous pest in New Jersey, Michigan, Maine (United States), and Eastern Canada was the blueberry maggot fly (*Rhagoletis mendax*), followed by the cranberry fruit worm (*Acrobasis vaccinii*), cherry fruit worm (*Grapholita packardi*), and plum curculio (*Conotrachelus nenuphar*). The significant blueberry illnesses blueberry scorch virus and blueberry shoestring virus are transmitted by aphids and a phytoplasma that causes stunt is transmitted by leafhoppers. Scales and stemborers should be managed by pruning old canes, cranberry fruit worm and leafroller populations should be controlled by handpicking afflicted fruit, and black vine weevil (*Brachyrhinus sulcatus*) populations should be reduced by eliminating alternative hosts. Due to its widespread efficacy, spinosad has been widely used in the production of organic blueberries. (Rodriguez Saona and colleagues 2019)

Disease management

The first confirmed finding of Mummy Berry Disease (*Monilia vaccinii-corymbosi*), one of the most significant plant pathogenic fungi affecting blueberries in America, was in Austria in 2002. Due to potential unfavorable climatic conditions (such low temperatures) or misunderstanding with similar fungi, the fungus, which was probably imported to Europe years ago and is prevalent in other blueberry plants, has not been recognized (e.g. *Botrytis cinerea*). The EB, which possesses some tolerance to this pathogen, has yet to be determined how this sickness will impact it. Blueberry fruit rot caused by anthracnose *Clletotrichum acutatum*, cranberry bitter rot caused by *Clletotrichum gloeosporioides*, and blueberry leaf rust caused by *Naohidemyces vaccini* (Slav *et al.*, 2018). The cause of a blighting disease of highbush blueberries is the blueberry scorch virus (BlScV), commonly known as sheep pen hill sickness. In the most susceptible cultivars, BlScV symptoms might range from total blighting of flowers and young leaves to twig dieback to no evident damage symptoms. Similar to BlScv blueberry shock virus (BlShV) causes symptoms, however after a few years of symptoms, the infected plants typically recover (Bristow *et al.*2002). These asymptomatic plants could serve as inoculum reservoirs. Since BlShV, like other *ilarviruses*, is found in pollen, wind and pollinators may be important in the disease's propagation.

The most common virus that affects highbush blueberries is called blueberry shoestring sobemovirus (BBSSV). The aphid *Illinoia pepperi* spreads it in a continuous and circular fashion (MacGillivray). Few cultivars exhibit resistance because to the lengthy latent period, making rouging of infected plants impractical for commercial production. Stunt in blueberries is brought on by a phytoplasma. The sharp-nosed leafhopper (*Scaphytopius spp.)* is the only species that is known to carry this phytoplasma.

Harvesting

The start of harvest is in the south, and it sent to north. Early April marks the start of Florida production. Early May is usually when the California harvest starts, and it lasts for roughly eight weeks. The blueberry season typically lasts from early July through mid-September. Fruit for the fresh market is typically picked by hand. This makes it possible to choose fruit that is ripe and barely bruised. Five-quart collection buckets are worn by pickers. To remove ripe fruit, mechanized harvesters cross the row, shaking the bushes with their mechanical fingers. About 20 pounds of fruit are transported in plastic field lugs after the ripe berries fall into a catch frame and are transferred there. Machine-harvested fields are typically harvested several times to collect all of the mature fruit.

Postharvest management

While harvesting a crop during the day, keep it shaded or covered to prevent heat buildup from solar radiation. Reduced heat buildup may be achieved by using shade canopies, empty flats stacked on top of filled flats, or even reflective covers. The design and operations of a packing facility properly can also enhance fruit quality. Fruit must first be cooled to remove field heat. As long as there is no dew, picking at night or early in the morning can help to preserve a more optimum berry temperature, although refrigeration or forced air chilling can also be very helpful. In addition to excellent postharvest management techniques in the field, such as avoiding harvest activities during high temperatures, forced air cooling can quickly lower internal fruit temperatures.

Despite developments in technology, mechanical harvesters still lack the selection and gentleness of a skilled hand harvest crew. Depending on the cultivar, between 30 and 80 percent of fruits gathered mechanically with a soft catch system are free of interior bruising damage, compared to more than 85 percent of fruits selected by hand (Sargent *et al.*,2021).

Storage

The blueberry fruit is highly perishable in nature harvested fruit should be stored at high relative humidity higher than 95% and the cold room temperature should be maintained at 32 to 34°F. In modified storage 10 to 12% CO2 and 1 to 3% O2, a regulated or altered environment can also be a viable choice. Depending on the cultivar, these characteristics may need to be further adjusted. (Devetter *et al.*,2022)

Yield

To determine prospective productivity, 100 lowbush blueberry (*Vaccinium angustifolium* Ait.) clones were randomly selected from a commercial field. The yield data showed a normal distribution with a mean of 7726 kg/ha and a range of 300 to 17,000 kg/ha. The potential production of native lowbush blueberry fields could be increased by commercially using chosen clones or superior cultivars through new plantings, interplanting into existing clones, or replacing low-yielding clones in native stands, as well as intensifying field management (Helper *et al.*, 1991).

Future prospects

In blueberry very scanty work done on its varietal improvements so its need to develop high yielding, better berry quality, disease and pest resistance, and cold and heat resilience, research is being done. Moreover, technological advancements are being used to better manage, harvest, and package crops.

References

Albert, T., Karp, K., Starast, M., & Paal, T. (2010). The effect of mulching and pruning on the vegetative growth and yield of the half-high blueberry. *Agronomy Research*, *8*(1), 759-769.

Ballington JR (2001) Collection, utilization, and preservation of genetic resources in Vaccinium. *HortScience* 36, 213-220

Ballington, J. R. (2001). Collection, utilization, and preservation of genetic resources in Vaccinium. *HortScience*, *36*(2), 213-220.

Blueberry production in 2020; Crops/Regions/World list/Production Quantity (pick lists)". UN Food and Agriculture Organization, Corporate Statistical Database (FAOSTAT). 2022. Retrieved May 16, 2022.

Calò, R., & Marabini, L. (2014): Protective effect of Vaccinium myrtillus extract against UVA- and UVB-induced damage in a human keratinocyte cell line (HaCaT cells). *Journal of Photochemistry and Photobiology B: Biology*, 132, 27-35.

Cavigliasso, P., Negri, P., Viel, M., Graziani, M. M., Challiol, C., Bello, F., & Saez, A. (2021). Precision management of pollination services to blueberry crops. *Scientific Reports*, *11*(1), 20453.

Chu, W., Gao, H., Chen, H., Fang, X., & Zheng, Y. (2018). Effects of cuticular wax on the postharvest quality of blueberry fruit. *Food Chemistry*, 239, 68–74. https://doi.org/10.1016/j.foodchem.2017.06.024

Dai ZG, Fei L, Huang DL, zeng J, Chen L, Cai YH (2019) Coupling effects of irrigation and nitrogen levels on yield, water and nitrogen use efficiency of surge-root irrigated jujube in a semiarid region. Agri Water Manage; 213, 146–154.

DeVetter, L. W., Chabert, S., Milbrath, M. O., Mallinger, R. E., Walters, J., Isaacs, R., ... & Eeraerts, M. (2022). Towards evidence-based decision support systems to optimize pollination and yields in highbush blueberry. *Front Sustain Food Syst*, *6*, 1006201.

Fang, Y., Nunez, G. H., Silva, M. N. D., Phillips, D. A., & Munoz, P. R. (2020). A review for southern highbush blueberry alternative production systems. *Agronomy*, *10*(10), 1531.

Forney, C. F., Qiu, S., Jordan, M. A., McCarthy, D., & Fillmore, S. (2022). Comparison of volatile compounds contributing to flavor of wild lowbush (Vaccinium augustifolium) and cultivated highbush (Vaccinium corymbosum) blueberry fruit using gas chromatography-olfactometry. *Foods*, *11*(16), 2516.

Giacalone, M.; Di Sacco, F.; Traupe, I.; Topini, R.; Forfori, F.; Giunta, F. Antioxidant and neuroprotective properties of blueberry polyphenols: A critical review. *Nutr. Neurosci.* 2011, 14, 119–125.

Gustavsson BA, Stanys V (2000). Field performance of 'Sanna' lingonberry derived by micropropagation vs. stem cuttings. *Hortscience* 35, 742-744

Han, T., Wu, W., & Li, W. (2021). Transcriptome analysis revealed the mechanism by which exogenous ABA increases anthocyanins in blueberry fruit during veraison. *Frontiers in Plant Science,* 12, Article 758215. https://doi.org/10.3389/ fpls.2021.758215

Ijaz, N., Ain, H. B. U., Bashir, S., Tufail, T., Ameer, K., Imran, S., ... & Saleem, A. (2022). Health Promoting Properties and Extraction of Specific Bioactive Compounds in Blueberries: Bioactive Compounds in Blueberries. *Pakistan BioMedical Journal*, 18-20.

Krewer, G., & Cline, B. (2003). Blueberry propagation suggestions. *Southern Region Small Fruit Consortium: Tifton, GA, USA*.

Lui HG, He XL, Li J, Li FD, Gong P, Zhang J, and Yang G. Effects of water-fertilizer coupling on root distribution and yield of Chinese Jujube trees in Xinjiang. *Int J Agri Biol Eng*. 2017; 10(6): 103–114.

Lyrene PM, Ballington JR (1986) Wide hybridization in Vaccinium. Hort science 21, 52-57

Lyrene, P. M., & Ballington, J. R. (1986). Wide hybridization in Vaccinium. *HortScience*, *21*(1), 52-57.

McAnulty, L. S., Nieman, D. C., Dumke, C. L., Shooter, L. A., Henson, D. A., Utter, A. C., ... & McAnulty, S. R. (2011). Effect of blueberry ingestion on natural killer cell counts, oxidative stress, and inflammation prior to and after 2.5 h of running. *Applied Physiology, Nutrition, and Metabolism*, *36*(6), 976-984.

Milivojević, J. M., Radivojević, D. D., Maksimović, V. M., & Dragišić-Maksimović, J. J. (2020). Variation in health promoting compounds of blueberry fruit associated with different nutrient management practices in a soilless growing system. *Journal of Agricultural Sciences (Belgrade)*, *65*(2), 175-185.

Morrison, S., Smagula, J. M., & Litten, W. (2000). Morphology, growth, and rhizome development of Vaccinium angustifolium Ait. seedlings, rooted softwood cuttings, and micropropagated plantlets. *HortScience*, *35*(4), 738-741.

Pertuzatti, P.B., Hermosín-Gutiérrez, I., & Teixeira Godoy, H. (2016). Chapter 1: Market, cultivars, chemical composition and antioxidant properties. In Malcolm Marsh (Ed.) Blueberries. Ed. (pp. 1-31). New York: Nova Science Publishers, Inc.

Prodorutti, D., Pertot, I., Giongo, L., & Gessler, C. (2007). Highbush blueberry: Cultivation, protection, breeding and biotechnology. *The European journal of plant science and biotechnology*, *1*(1), 44-56.

Retamales, J. B., & Hancock, J. F. (2018). Blueberry taxonomy and breeding. In *Blueberries* (pp. 18-60). Wallingford UK: CABI.

Rodriguez-Saona, C., Cloonan, K. R., Sanchez-Pedraza, F., Zhou, Y., Giusti, M. M., & Benrey, B. (2019). Differential susceptibility of wild and cultivated blueberries to an invasive frugivorous pest. *Journal of chemical ecology*, *45*, 286-297.

Sargent, S.A., F. Takeda, A. Berry, and J. Williamson. 2021. Mechanical Harvesting of Three Southern Highbush Blueberries with Two Over-the-Row Rotary Harvesters. *Agronomy* 11: 1412.

Slav, M., Hoza, D., & Asănică, A. (2018). Researches on the presence and agresivity of the blueberry root rot (Phytophthora cinnamomi) in a Dâmbovița county plantation. *Journal of Horticulture, Forestry and Biotechnology*, *22*(4), 7-12.

Steyn, J., Lötze, E., & Hoffman, E. W. (2023). The seasonal progression of the reproductive phenology of two southern highbush blueberry (V. corymbosum L. interspecific hybrids) cultivars in the Western Cape, South Africa. *Scientia Horticulturae*, *307*, 111493.

Wang, H., Wu, Y., Yu, R., Wu, C., Fan, G., & Li, T. (2019). Effects of postharvest application of methyl jasmonate on physicochemical characteristics and antioxidant system of the blueberry fruit. *Scientia Horticulturae*, 258, Article 108785. https://doi. org/10.1016/j. scienta.2019.108785

Yang, H., Wu, Y., Zhang, C., Wu, W., Lyu, L., & Li, W. (2022). Growth and physiological characteristics of four blueberry cultivars under different high soil pH treatments. *Environmental and Experimental Botany*, *197*, 104842.

Yousuf, B., Gul, K., Wani, A. A., & Singh, P. (2016). Health benefits of anthocyanins and their encapsulation for potential use in food systems: a review. *Critical Reviews in Food Science and Nutrition*, 56, 2223–2230. https://doi.org/10.1080/ 10408398.2013.805316.

Zou, H., Ye, H., Zhang, J., & Ren, L. (2022). Recent advances in nuclear receptors-mediated health benefits of blueberry. *Phytomedicine*, 154063.

Plunkett, B. J., Espley, R. V., Dare, A. P., Warren, B. A., Grierson, E. R., Cordiner, S., ... & Schwinn, K. E. (2018). MYBA from blueberry (Vaccinium section Cyanococcus) is a subgroup 6 type R2R3MYB transcription factor that activates anthocyanin production. *Frontiers in Plant Science*, *9*, 1300.

Hepler, P. R., & Yarborough, D. E. (1991). Natural variability in yield of lowbush blueberries. *HortScience*, *26*(3), 245-246.

19

Advances Production Technology of Underutilized Temperate Fruits

*Anu Seng Chaupoo**[1], *Kuruba Ajay Kumar*[2], *Lalrinchhani Chhangte*[3], *Alemla Imchen*[4] *and Anish Sunwar*[5]

[1]*Ph.D. Research Scholar, Department of Floriculture,School of Agricultural Sciences and Rural Development, Nagaland University, Medziphema- 797106, Nagaland, India*

[2]*Ph.D. Research Scholar, Department of Fruit Science, School of Agricultural Sciences and Rural Development, Nagaland University, Medziphema- 797106, Nagaland, India*

[3]*Ph.D. Research Scholar, Department of Genetics and Plant Breeding, School of Agricultural Sciences and Rural Development, Nagaland University Medziphema- 797106, Nagaland, India*

[4]*Ph.D. Research Scholar, Department of Fruit Science, School of Agricultural Sciences and Rural Development, Nagaland University, Medziphema- 797106, Nagaland, India*

[5]*M.Sc. (Ag.), Department of Rural Development and Planning, School of Agricultural Sciences and Rural Development, Nagaland University Medziphema- 797106, Nagaland, India*

Cape gooseberry (*Physalis peruviana* L.)

Cape gooseberry, scientifically known as *Physalis peruviana* L. (*Physalis edulis* Sims) belongs to family Solanaceae. Fruits are enclosed inside a bladder-like husk which turns papery when mature making all seeds inside visible. There are more than 70 species under the genus *Physalis*. Out of these species only few of them are of economic value. Cape gooseberry is most widely known among these species as it bears superior fruit in comparison to other species. The fruits are good in taste and a good source of minerals and vitamin such as vitamin B complex, vitamin C and pro-vitamin A, rich in pectin.

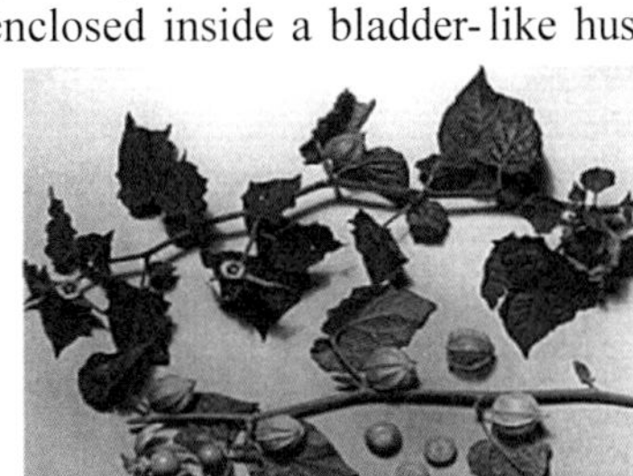

Fig. 1.1: Cape gooseberry

Species diversity

Cape goosberry is also known by different names in Latin America: *capuli, aguaymanto, tomate sylvestre,* or *uchuba,* in Bolivia; *uvilla*in, in Peru; *capuli* or *motojoboboembolsado,* in Colombia; *topotopo, chuchuva*, in Ecuador; *uchuva, vejigón* or *guchavo* in Venezuela; *capuli, amor en bolsa,* or *bolsa de amor,* in Chile; *cereza del Peru* in Mexico; *teparee, tiparee, makowi,* etc., in India. *Physalis pruinose* L. commonly known as strawberry tomato, husk tomato or ground cherry is also grown for its small yellow fruits. Inmild-temperature climates, the fruit is used to make sauce, pies, and preserves. Unripe druits aretoxic. In Australia, the plant is thought to have caused illness and death in cattle.

Uses

Canned as whole, preserved as sauce, jam, puddings, pies, ice cream, chutneys, fresh fruit salads and fruit cocktails. It also has medicinal uses; for example, in Colombia, the leaf decoction is used as a diuretic and antiasthmatic. In South Africa, the heated leaves are used as poultices on inflammations and the Zulus use the leaf infusion as an enema to treat abdominal ailments in children. A minor steroidal constituent, *physalolactone C*, was isolatedfrom the leaves by Indian chemists.

Botany description

Cape gooseberry is a perennial herbaceous or soft-wood plant that can grow to a height of 2 to 3 ft (1.6-0.9 m) or even 6 ft (1.8 m). The branches are ribbed, purplish incolor, spread out, and covered with fine hairs. The heart-shaped leaves are 6-15 cm long and 4-10 cm wide, and are nearly opposite. The leaf axils bear 2 cm wide bell-shaped, nodding flowers. These yellow flowers have a 5 dark purple-brown spots in the throat and are cupped by 5 purplish green hairy calyxes. After the flower falls, the calyx expands and forms a straw-colored husk enclosing the fruit. The fruits are globular berry with 1.25 to 2 cm width.The surface of the berry is smooth and glossy, orange-yellowish in color. There are numeroustiny seed in the juicy pulp. In contrast to the bitter and inedible husk, the fruit has a sweet andpleasant grape-like tang flavor.

Nutritional Value per 100 g of Edible Portion

Contents	Amount with Unit	Contents	Amount with Unit
Moisture	78.9 g	Iron	1.23 mg
Protein	0.054 g	Carotene	1.613 mg
Fat	0.16 g	Thiamine	0.101 mg
Fiber	4.9 g	Riboflavin	0.032 mg
Ash	1.01 g	Niacin	1.73 mg
Calcium	8.0 mg	Ascorbic Acid	43.0 mg
Phosphorus	55.3 mg		

Origin and Distribution

Cape gooseberry is known to be native to Peru and Chile. The plant is said to have grown by early settlers at the Cape of Good Hope prior to 1807, and it was later commercially cultivated in South Africa. Then it was carried to Australia where it was plantedin a large scale and acquired its common English name. it is also grown in Queensland, South Australia, Victoria, Northern Tasmania and Western Australia,

New Zealand. In England, Cape gooseberry was first reported in 1977, and mostly grown in home garden. Whereas, little attention was given about this fruit in United states even though Dr. David Fairchild reported in his well-loved book, *The World Was My Garden*. With a view to encourage growing of Capegooseberry culture in Florida, the Bahamas, and the West Indies, seeds have been repeatedly purchased from the Stanford Nursery and distributed for trial. A good growth was reported however, there was no incentive to make further plantings. Even in China, India and Malaya Cape gooseberry is not widely grown, and in India, it is not common to see it intercropping with vegetables.

Climate

In temperate climates, Cape gooseberry is an annual while in tropical climates; it is grown as a perennial plant. Its tolerance toward frost is variable asit was report to be undamaged by 3°C of frost in England while the plants was killed to the ground after a temperature drop to -0.75^{o}C in South Africa. When the crop matures, it only requires minimal care, full sun, lots of rain, and shelter from strong winds during the entire growing season.

Soil

Any well-drained soil will support the growth of the Cape gooseberry, loam with sand to gravel. A highly fertile alluvial soil encourages more vegetative growth and inhibits the development of fruit color. Plantings should be on gentle slopes where there is a drainage issue, or the rows should be molded. In times of drought, plants go dormant.

Propagation

Main propagating material is seed at a rate of28g/acre or 70g/ha. Seed germination rateis low. There are 5,000 to 8,000 seeds in 28g of Cape gooseberry. For uniform sowing, the seeds are mixed with wood ash or pulverized soil in India. The other method of propagation is by using 1-year-old stem cuttings, which has a success rate of 37.7 per cent. Plants grown from cuttings flower and produce well, but they are less vigorous than seedlings. Air-layering is also effective but rarely used.

Planting

Worldwide, planting times vary; in India, seeds are broadcast from March to May. In India, plants of 15-20 cm high should maintain 45 cm row to row and 0.9 m plant to plant distance. On sandy loam, it is advised to apply 90-180 kg of complete fertilizer per acre. Fruitquality is improved by applying 1 per cent potassium chloride to the leaves just before and after blooming. By pruning after each harvest, some growers keep their plant production foras long as 4 years, but is has been discovered that these plants are more prone to pests and diseases.

Flowering and fruiting

The creamy yellow flowers are solitary and somewhat bell-shaped with five fused petals, each with a brown or purple spot at the base. Like the tomatillo, to which it is related, the plant is noted for the inflated baglike calyx (fused sepals) that encloses a fleshy orange berry. Pollination, in England, growers shakes the flowers gently in summer to improve distribution of the pollen, or they will give the plants a very light spraying with water.

Fruits ripen in some areas of India in February, but the main crop is grown in the southfrom January to May. From the beginning of April to the end of June, the crop is harvested in Central and Southern Africa. In England, plants from spring-planted seeds start bearing fruit inAugust and continue doing so until a hard frost.

Harvesting and Yield

In India, the fruits ripen in February, fruits are picked only when the plants are dry, and picking is done by hand every 3 to 3 weeks. One plant may produce 300 fruits. 1800-1500 seedlings per acre could produce 3000 lbs. Before being delivered to markets or processors, the fruits are typically dehusked. Only 10 to 12 lbs. (4.5-5.5 kg) of husked fruits can be produced by manual laborers per hour. Therefore, the University of Hawaii has developed a mechanical husker that is 4-5 times more effective.

Storage

Cape gooseberries have good storage potential. The fresh fruits can be kept for severalmonths in a sealed container that is kept in a dry environment. While for long transportation, thehusk should remain for protection.

Pests and Diseases

Different pests are reported on Cape gooseberry from different countries. Red spiders after in mature plants; cutworms in seedbeds; and the potato tuber moth if planted close to potato fields. Hares harm young plants, birds (francolins) eat the fruits if they are not scared away, mites may cause defoliation, and it has been said that the leaves are covered in what appear to be flea beetles from the family Chrysomelidae. Controlling whitefly and flea beetle infestations on flowering plants and young plants, respectively, is required.

Powdery mildew, soft brown scale, root rots are important diseases of Cape gooseberry. Farmer prefers biennial crops because viruses thrive in poorly-drained soil and can be carried over to the following year. In Queensland, bacterial leaf spot (*Xanthomonas* spp.) has also been reported. A tobacco mosaic strain has been reported in India.

Mulberry (*Morus* spp.)

Species diversity

A total of 69 species have been identified for mulberry; *Morus* spp. under moraceae family. Majority of them are white, black and red. White mulberry is known as *Morus alba*, black mulberry as *Morus nigra* and red mulberry as *Morus rubra*. India and China are known to be major country cultivating mulberry including Japan, America, Australia, Korea and Nepal.

Origin and Distribution

White mulberry is indigenous to northern, eastern, and central Asia, while black mulberry is thought to have originated in south west Asia and red mulberry in eastern North America.

Mulberries of these three varieties are common in southern Europe, the Middle East, northern Africa, and the Indian subcontinent. In the 17th century, black mulberry was imported into Britain in the hope of being used in silkworm cultivation.

Botany

Mulberry is a deciduous shrub or a tree with a deep-rooted system and perennial in nature, it is fast growing shrub that can attain even 10 to 15 meters, while the growth of mulberry eventually declines once attain certain maturity. Mulberry has a simple lobed and serrated leaves that arrange in alternate manner, Lobes are mostly seen on young plant than matured trees. Number of lobes varies from 1-5. Mulberry fruit is a multiple fruit that is about2-3 cm long. Its colour changes on the process of ripening, from green, white or pink to pale yellow. On complete ripening the fruit turns into dark purple or black with a sweet flavour when fully ripe. The stem has a grayish brown rough bark. The presence of idioblasts, an enlarged epidermal cell in the leaf, is a distinguishing feature of members of the Moraceae family, particularly *Morus* spp.

Soil and Climate

One should avoid saline soil. Slightly acidic soils (6.2 to 6.8 pH) free from injurious salts are ideal for good growth of mulberry plant.

Mulberry can grow up to 800 MSL. A range of atmospheric temperature between 13°C to 37.7°C is required for optimum growth and good sprouting of mulberry. However, a temperature range between 24°C and 28°C with 65 to 80 per cent relative humidity and sun light duration of 5 to 12 hours per day is ideal for cultivation of mulberry. If low rainfall occurred, a supplemental irrigation of 50 mm once in 10 days is required for mulberry.

Varieties

S-36: Leaves are heart shaped with light green colour. The average yield of leaves is15,000- 18,000 kg/acre.

V-1: Released in 1997. Oval and broad shaped leaves, which are dark green in color.

The average leaf yield is 20,000-24,000 kg/acre.

Some other important varieties of mulberry are as follows:

Saharanpur: They are very early blooming variety and resistant to popcorn disease.

Black Beauty: very late blooming variety and susceptible to leaf spot.

Trader, Tehama: They are very sweet and do not contain any acid.

Sweet lavender is susceptible to leaf spot; **Northrop** has small sized fruit and **Beautiful day** is a mid late variety with white skin and very sweet flavor.

Fig. 1.2: *Morus alba*

Fig. 1.3: *Morus nigra*

Fig. 1.4: *Morus rubra*

Propagation of mulberry

Mulberry is mostly propagated by semi-hardwood cuttings from 8 to 12 months well- established garden. A full grown 10-12 mm diameter thick stem 15-20 cm long with 3-4 active buds slanted at an angle of 45°, free of pest and diseases are ideal for cuttings. Cuttings can be planted directly in the field or raised in a nursery, and the rooted saplingscan be planted in the main field at a depth of 15-20 cm after 90-120 days. Transfer of rooted cuttings is more preferred as it is easier to establish in the main field. One should avoid splitting of the bark at the ends of cutting. For quick cutting of propagation material, a manual/power operated mulberry cutter (stem cutting machine) is now available.

Pre-Treatment of cuttings

Azospirillum culture is used for pre-treatment of cuttings, dipping the cuttings in a solution of 1 kg of the culture in 40L of water for 30 minutes before planting can induce early rooting.

Preparation of nursery bed

An 800 m^2 area of nursery is ideal for 1ha planting, preferably the soil should be red loamy mix with 1600 kg FYM @ 20 t/ha. The location, if possible, should be nearby water source. Raise nursery beds of 4m x 1.5m size, however, length may be of more convenient size depending upon the irrigation source, slope, etc. A good drainage channel should be present, and a shady area should be avoided.

Nursery planting

Before planting the cuttings, the nursery should be irrigated and applied VAM @ 100 g/m^2. Before planting the cuttings VAM @ 100 g/m2 of nursery area should be applied. The cuttings should be planted at 15 cm x 7 cm spacing slanting at 45^0 angles. One active bud in each cutting should be exposed.

Nursery management

Nursery should be irrigated once in three days, dusting of either malathion 5D or quinalphos 1.5D to help avoid termite attack. In order to avoid root rot and collar rot, one should drench the soil with carbendazim 50 WP @ 2g/l or apply *Trichoderma viride* @ 0.5 g/m^2 using rose can. Application of urea @100g/m^2 after weeding between 55 and 60 daysafter planting to promote growth.

Time of planting

The ideal time for planting is during rainy season, winter and summer months should be avoided.

Planting methods

Paired row system: In 1ha area, 27,780 cuttings can be accommodated at 60 x 60 cm spacingand 12,345 cuttings can be maintained at 90 x 90 cm spacing. The ideal spacing for cuttings/saplings is 75 / 105 cm x 90 cm. Raise intercrops in the wider inter row spaces (amenable for mechanization also).

Planting method	**Spacing (cm)**	
	Irrigated	**Rainfed**
Ridges and furrows	60 x 60 / 90x90	90 x 90
Pit system	90 x 90	90 x 90

Nutrient management

Management of nutrients for mulberry is as follows:

a. Rainfed (Kg/ha)

	N	**P**	**K**
Recommendation	100	50	50
First dose	50	50	50
Second dose	50	-	-

b. Irrigated / semi irrigated (kg/ha)

	Row system			**Pit system**		
	N	**P**	**K**	**N**	**P**	**K**
Recommendation	300	120	120	280	120	120
Split doses						
First crop	60	60	60	60	60	60
Second crop	60	-	-	40	-	-
Third crop	60	60	60	40	-	-
Fourth crop	60	-	-	60	60	60
Fifth crop	60	-	-	40	-	-
Sixth crop	-	-	-	40	-	-

The first and second doses should be administered to coincide with South West and North East monsoons respectively.

Bio-fertilizers

Biofertilizers such as *Azospirillum* @ 20 kg/ha or Phosphobacterium @ 10 kg/ha mixed with 50 kg FYM can be applied in five split doses and two split respectively. Immediate irrigation should be given and mixing of biofertilizers with inorganic fertilizers should be avoided.

Irrigation

Ridges and furrows method: This method uses less water than other methods; the furrows act as a drainage channel during heavy rainfall.

Flat-bed method: The initial cost of this irrigation system is high, and it requires more land and labors, but it provides better crop growth. It is appropriate for undulating terrain, and fertilizers can be applied in addition to irrigation.

Weed management

Mulberry plantation is both infested by grassy and broad leaves weeds such as nut grass,bermuda grass, carrot grass, pig weed, hog weed etc. These weeds can be managed through cultural method and mechanical method with chemical method can be applied for control of weed. Post-emergence application of weedicide such as paraquat (grammaxone) @ 2-3 L/ha orglycel @ 7.5 ml with 10 gram of ammonium sulphate/ litre of water is common. Theapplication method is flooding, deflector, or fan type nozzle spraying. Weedicide should be applied immediately after pruning or within 2-3 days.

Intercropping

Intercropping with short duration pulse crops like black gram, green gram or cowpeaenriches the soil, generates extra income, and inhibits the growth of weeds.

Mulching

Mulching with pruned mulberry twigs and other materials such as straw and dried leaves aids in control of weeds, moisture conservation, increased water infiltration and soil temperature reduction.

Pruning Methods

Bottom pruning: This pruning is done once in a year. The plants are cut at ground levelleaving 10-15 cm stump above the ground.

Middle pruning: After bottom pruning, subsequent cuts are made at 45-50 cm height. The branches are cut at 40-60 cm above the ground level.

Kolar or strip system: This type of pruning is done in closely planted area. The branchesare cut at a ground level. This pruning is done five times in a year. Since this type of pruning is severe the plant requires heavy fertilization and irrigation.

Flowering

Yellowish green color bloom will arise in spring. The most common reason for bloom/fruitfailure in mulberry tree is lack of soil nutrients and late spring frost that kills the flower buds.

Harvesting

The leaves are harvested mainly in the morning hours by leaf picking or branch cuttingor whole shoot harvest. The leaves are picked with or without the petiole in the leaf picking method. Leaf picking begins 10 weeks after bottom pruning and is repeated every 7-8 weeks. In the branch cutting method, entire branches are cut and fed to the worms. However, toppingis done before harvesting to ensure uniform maturity of the lower leaves. The branches in the whole shoot harvest system are cut at ground level by bottom pruning, and shoots areharvested every 10-12 weeks. This allows for 5 to 6 harvests per year. To keep the harvested leaves cool and fresh, store them in a chamber or wet gunny bags.

Persimmon (*Diospyros kaki* L.)

Persimmon (*Diospyros kaki* L.) is a sub-tropical climacteric fruit that has attractive appearance, sweet flavor with little acidity. Persimmon belongs to the Ebanaceae family which includes 6 genera and 400 species, and the genus *Diospyros* includes 200 species of which only four species are cultivated, and is widely consumed in Asia. Persimmon are not recommended to cultivate under tropical conditions, it required low chilling temperature, and have a higher tolerance to heat, water stress and logging than other deciduous fruit crops. Persimmon fruits are reddish orange and orangish red color with orange color flesh whichis fibrous pulpy in texture. The shape is flat-globose, conical globose, with orange, reddish orange and orangish red color. Fruit weights range from 50 to 300 g.

Origin and distribution

Persimmon is known to be primarily originated from central China and secondarily from Japan, major Persimmon producing countries are China, Korea, Japan, Brazil, Italy, Israel, Australia and India. In India Persimmon is mainly grown in Himachal Pradesh, Jammu and Kashmir, Tamil Nadu, Uttarakhandetc.

Soil and Climate

Loamy soil is considered best for Persimmon, light, sandy soil should be avoided. Heavy clay can also be used to grow persimmons with good drainage. This fruit can be grown best in soil with a pH between 5.5 and 6.5. Both warm and sub-tropical region are suitable for cultivation of persimmon. It is successfully cultivated in Nepal at elevations ranging from 1000 to 1800 meters; the agro-climatic conditions surrounding the Kathmandu valley are considered suitable for persimmon cultivation. A temperature range of 14-22°C is suitable whereas for sweet persimmon production temperature range of 14.6-15.7°C is suitable.

Propagation

Grafting is the most common method of propagation of persimmon while fresh mature seeds are used for raising rootstocks. Stratification of seed is required for improved germination. Seeds can be stratified at 45°C for 60-90 days. Once the rootstock become dormant, grafting should be done with ¼ to ⅓ inch diameter of 3 to 5 inches long scion having 2 to 4 buds. During the winter dormancy period, the scion should be taken from a healthy and well-lignified parent plant.

Varieties

The Persimmon cultivars are divided into two categories, namely Astringent and Non- astringent persimmons.

Non-astringent: Tannin content is greatly reduced the moment it turns from green to orange. Non-astringent cultivars have lost their astringency by maturity and can be eaten crisp like an apple or at various stages of softness. Examples: Fuyu, Jiro, Gosho and Suruga.

Astringent: This taste is caused by astringent substances such as tannins. The astringent tasteis dominant in unripe persimmons. Examples: Hiratanenashi, Hachiya, Aizumishirazu, Yotsumizo and Yokono.

Fig. 1.5: Varieties of Persimmon

Planting

The grafted persimmons are transplanted in the main field during July or early August in a pit of 1 m^3. The spacing required depends on the cultivar and soil type. For dwarf, semi-dwarf and vigorous cultivars spacing should be 5.0 ×2.5 m, 5.0 ×3.0 m and 6.0× 4.5 m respectively.

Training and Pruning

To provide structural strength, better light penetration and easier intercultural operationçentre leader system of training is mainly done in dwarf cultivar. Semi-dwarf cultivars, on the other hand, are well suited to the palmette system. Pruning is an important practice as well. Healthy branches are also trimmed to 1/3 to a bud. Summer pruning is also done to improve the size and color of the fruit. Pruning should be kept to a minimum.

Flowering and fruiting

Persimmon trees are typically dioecious, meaning male and female flowers are produced on separate trees. Some trees have both male and female flowers and in rare cases may bear a perfect flower. Male flowers are pink and appear in groups of 3. They have 24 stamens in 2 rows. Female flowers are creamy-white and appear singly. They have a large calyx, yellow corolla, 8 undeveloped stamens, and a rounded ovary bearing the style and stigma.

Persimmon fruit matures late in the fall and can stay on the tree until winter. In color,the ripe fruit of the cultivated strains range from glossy light yellow-orange to dark red- orange depending on the species and variety. The calyx generally remains attached to the fruitafter harvesting, but becomes easy to remove once the fruit is ripe.

Harvesting and Yield

Persimmons should be harvested when the fruit is mature, still firm and the color nearly fully developed. Fruits are mostly mature in September to October; the fruit is harvestedby clipping r breaking the stems, leaving the calyx lobes attached to the fruit. The full-grown tree yields around 150-200 kg fruits in a year.

Hazelnut (*Corylus avellana*)

Fig. 1.6: Hazelnut

The common called Hazel nut (*Corylus avellana* L.) is a European hazel nut also known as Cobnut or Filbert. It is a fruit of Hazel tree that belongs to the family Betulaceae. Nuts of any species that belong to genus *Corylus* are known as hazel nut. This nut tree is a typical temperate zone nut crop that mostly grows in Turkey, Italy, Spain, Germany, France and England. In UK a distinction is made between cobnut and filbert. Filbert is known to have thehusk longer than the nut, and in cobnut the husks are shorter than the nut. Turkey and Italy arethe two largest producers of hazel nut.

Species

There are two types of European hazel nut; *C. avellana* (European hazel) that has a wide distribution and *C. colurna* (Turkish hazel).

Uses

Hazel nuts are used in baking, desserts making, confectionary and in praline. It is also used for extracting cooking oil, and used in making chocolate bars, Nutella, Frangelico liqueur and chocolate truffles in combination with chocolate. The food industry uses 90% of the world's production as kernels, and the remaining 10% is sold in-shell for fresh consumption.

Origin and Distribution

Hazel nut is indigenous to Europe and Western Asia. European hazel has wider distribution whereas Turkish hazel is restricted to Balkans, Romania and northern Turkey. The European hazel is found from the Mediterranean coast of North Africa northward to British Islands, Scandinavian Peninsula, Ural Mountains of Russia, Caucasus Mountains, Iranand Lebanon. It provides important cultivars to Europe and Turkey. Turkey is the leading producer and exporter of hazel nuts accounting for 71% of global production, followed by Italy (13%), the United States (4.1%) and Spain (2.8%). In India, hazel is growing wild in Pangi region of Chamba district of Himachal Pradesh and locally known as Thangi.

Soil and climate

Moderately fertile soil with pH ranging from 4.5 to 8.5 or 7 is considered best for growing hazel nut. Heavy clay should be avoided since it cannot tolerate wet soils, on the other hand moisture retention is necessary since the tree cannot tolerate excessive dry summer heat and hot winds. Hazel tree is shallow rooted than most fruit and nut trees.

Even though hazel nut tree is quite hardy it required moderate climate to produce quality fruit. Temperature lower than 10°C can be critical especially if accompanied by wind since the tree has both pistillate and staminate flowers. The chilling requirement of hazel nut is about the same as most commercial cultivars of apple so; it can be cultivated in apple growing regionssuch as Jammu and Kashmir, Himachal Pradesh, Uttarakhand and North-eastern Himalayan regions of India.

Varieties

Tonda, Romana, Barcelona, Negret, Tonda giffoni, Tond Gentile delle Langhe, Pauetet and Tombul are important varieties of hazel nut. There are more than 200 cultivars of hazel nutaround the world.

Planting and Planting Density

Planting is usually done in the months of winter, depending on soil's fertility, rainfall and choice of variety; tree spacing can be variable in different countries. A recommended planting density is 860 trees/ha with 1-5 m row to row and 2-3m within row spacing. Thiswill allow machinery access. Planting at least 10% different varieties evenly in the field will ensure adequate pollination.

Manures and Fertilizers

Both organic and inorganic fertilizers should be added from prior to planting and up to bearing age, 30 ton of FYM per ha if the soil organic matter is below 2% is recommended as organic fertilizers. Leaf and soil analysis should be done to determine inorganic fertilizers dose. Mostly, the recommended dose of NPK is 120-150:60-70:100 kg/ha.

Training and Pruning

Multi stem brush training system are used in major production areas such as Turkey, Italy and Spain whereas one stem bush training systems are used in new orchards in the United States, France, Italy, and Spain. Since hazel nut bears fruits in lateral and terminal wood of previous season's growth pruning should be done in such that it stimulates a moderate new growth each year.

Weed Control

Since weed affects the yield and quality of nut as it competes with moisture and nutrition. Both manual weeding and used of herbicides is recommended. Simazine, Diuron, Napromide, Oxifluorphen, Propyzamide, Trifluraline as pre-emergence and paraquat, 2, 4-D, Aminotrizol, Glyphosate, Ammonium gluphosinate, etc. as post-emergence are recommendedherbicides.

Flowering

Hazelnuts (*Corylus avellana*) are monoecious, wind-pollinated, and self-incompatible, which guarantees cross-pollination and genetic diversity. Also, they are dichogamous, and in many locations male flowers open earlier than female flowers. While female flowers became receptive from early February in some years, the shed of pollen spanned around 2-3 weeks, usually in March. A severe frost during February and/or March can damage the catkins, interruptpollination, and potentially decrease nut yield.

Harvesting and Yield

Harvesting is usually done in mid-autumn when the nuts change its color from green to brown. On average five years old tree can produce 2-2.5kg.

Black Currants (*Ribes nigrum*)

Black currant (*Ribes nigrum* L.), also known as black currant or cassis, is a woody perennial deciduous shrub native to temperate parts of central and northern Europe and northern Asia. It is a member of the Grossulariaceae family. It is cultivated for its edible berries; raw berries are a good source of polyphenols and vitamin C. Black currants can be consumed raw, but they are more frequently cooked in sweet or savoury dishes. The domesticand commercial cultivation of it are both widespread. The size of the crop may be affected bycold weather during the spring flowering period even though black currant is a crop that can withstand the winter. Small glossy black fruits are developed in bunch which can be harvestedalong with the stems in summer by hand or machine. The fruits are used for making jams, preserves, and even grown commercially for juice market. Alcoholic beverages and dyes are also prepared from the fruit.

Soil and Climate

Soil rich with compost at pH range 6 to 6.5 is preferred for growing black currant. Flowering time is critical stage and therefore cold weather during flowering may reduce the size of the fruit.A location which receives morning sun and afternoon shade is best locations for growing blackcurrants.

Varieties

Baldwin: It is an English currant variety that is very productive with medium quality fruit and highly susceptible to white pine blister rust.

Ben Lomond: Have excellent fruit quality and slightly susceptible to white pine blister rust.

Ben Nevis: It has excellent fruit quality and high production. It is slightly susceptible to white pine blister rust.

Titania: It has excellent yields and large fruit and is immune to white pine blister rust and resistance to powdery mildew.

Propagation

Black currant can be propagated through hardwood cutting from a healthy disease-freeplant from mid-autumn to winter. 12-inch-long cuttings from a year-old stem containing a healthy bud is taken during spring when pruning is done. The cutting does not require much care as compared to other crop.

Manuring

Besides artificial fertilizers such as balanced fertilizer pr granular fertilizers, a composted manure over the root zones of the plants at 2.5 cm thick is recommended during late fall. Slow growing plants can be fed again in early summer.

Spacing

Plant to plant at 90 cm apart with 90 cm row to row is recommended spacing.

Sowing and Planting

Spring is the ideal season for planting black currant since the plants emerge from dormancy. The plant should be transplanting in a pit slightly deeper than their nursery pots. Mulching is ideal as it keeps soil moist and cool.

Flowering and fruiting

It is winter hardy, but cold weather at flowering time during the spring may reduce the flowering and fruiting percentage. Fruit bunches are small, glossy black develop along the stems in the summer.

Harvesting

Once the fruits are dark and ripe, they are supposed to be harvested and give chilling treatment for longer self-life. Black currants are known to freeze very well.

Quince (*Cydonia oblonga*)

Quince (*Cydonia oblonga* Mill.), also known as common quince, is the only species of the Rosaceae family's genus *Cydonia*. It is known by several common names known in different parts of the world. In the Indian Ayurveda (sanskrit) system, common names include Amritphala, Imbitkaa, and pataalaa. Whereas, Behi (Urdu/Hindi), Wen po in Chinese, Ajva in Russian, Avya in Turkey, Al Sefarjal in Arabic, Quitte/Quittenbaum in German, Cognaisseror Coig in French, Marmaelo in Portuguese, Marumero in Japanese, Bum choonth in Kashmiri etc. are the several names of Quince. Cooler subtropical areas to cold temperate regions are the usual place where quince thrives as it has culture similar to that of pears and apples.

Species

Four species previously included in *Cydonia* are now treated in separate genera. These are the Chinese Quince *Pseudocydonia sinensis*, native to China, and remaining three flowering quinces are native to East Asia. The Chinese flowering quince (*Chaenomeles cathayensis*) reaches 3m height. It produces white to pink flowers and bears the largest fruit of the genus, 15 cm (5.9 inches) long. The Japanese quince (*C. japonica*) is popularly grown in bonsai and has provided several horticultural varieties with red, pink, or white flowers. The common flowering quince (*C.speciosa*), frequently used in informal hedges, bears red, pink, or white flowers and grows to about 2 meters (6.6 feet).

Origin and Distribution

Transcaucasia region is known to be the centre of origin for Quince; this region includescountries like Armenia, Azerbaijan, Iran, Turkmenistan and southwest Russia. Quince is a native plant in of Western Asia's rocky foothills and woodlands. From these centre of origin and native place, it spread to the countries in the region of Himalayan Mountains in theeastern side extending to China, Europe and in the Mediterranean region.

Uses

Quince can be eaten fresh, stewed, canned, or made into jams and jellies. Jellies are a vibrant shade of pink, and the majority of them cool to a lovely shade of pink that can range from light to even yellow.

Soil and Climate

Most soils are suitable except light and shallow that is likely to dry out. Well-grown fruits are often found on trees that grow in wild near creek banks where heavy and moist soil are often found. Well drainage system is still necessary even in heavy and moist soil. This fruitcrop prefers slightly acidic soil as it is susceptible to iron deficiency under alkaline conditions.Quince adapted well to cooler parts of subtropical and temperate regions. When choosing a location for a quince orchard, four key considerations must be made.

a) **Aspect:** Successful plantings are found in hillsides exposed to south but these sites often experience cold southerly airstreams that can make the environment too cold for pollinating insects to move around during the blossoming season.
b) **Slope:** The slope on which plants can be planted is determined by the safe and efficient use of machinery for easy of cultivation.
c) **Frost:** Since lower temperature needed to be avoid just before and after the blossoming period, location such as valley floors or lower parts should are not ideal unless they have enough slope and dense timber for obstruction.
d) **Wind:** Wind can be very damaging to crops and can impede tree training and development. Windbreaks are required in the orchard due to the location and general topography; artificial windbreaks can also be considered.

Propagation

Quince is typically propagated form hardwood cuttings of desirable varieties. The cutting should be about 25 cm long and taken during late autumn and early winter.

Cutting is the most convenient method of propagation, but the resulting trees produce suckers that must be removed on a regular basis. Whereas, budding method do not produce suckers and bear fruit on about the fifth year. Trees from cutting method of propagation bear fruit earlier than budding method.

Planting

Planting is done during late June to early July, when the trees are completely dormant.Planting distance is maintained at 5 x 2.5-3 meters spacing, this allows 667-800 trees per hectare.

When seedling rootstock is used in a deep, fertile soil, it can be anticipated that the trees will grow larger, whereas planting distance should be maintained at 6 x 4.5 meters, this spacing can accommodate 370 trees per hectare.

Fertilizers

Mixed fertilizers are considered best, a total of 350 g N for each tree of full bearing capacity annually or a heavier dose of N at 450g per tree during spring followed by applicationof 50-100 kg P/ha during autumn. Application of single super phosphate will directly promote the growth of clovers, trefoils and other species, ultimately improving the organic soil matte status of the soil.

Flowering

Blooming in late winter or early spring, the flowers have five petals and can be up to 4.5 cm (1.8 inches) in diameter.

Yield

A mature planting of around 250 tree per hectare will yield about 25-35 tonnes perhectare. A single tree may yield 100-150 kg per tree.

Chilgoza (*Pinus gerardiana*)

Fig. 1.7: Chilgoza

Chilgoza (*Pinus gerardiana* Wall.), also known as neja is a pine belonging to the family Pinaceae and is native to the northwestern Himalayas in Afghanistan, northern Pakistan, northwestern India and Waziristan. This pine is generally found growing at elevations 1800 and 3350 m. The trees are tall (10-25 m in height), with deep, wide open crowns and long, erect branches. They have needle-like leaves, 6-10 cm in length which are spread stiffly and have glossy green outer surface and blue-green stomatal lines on the inner surface. Chilgoza isconsidered to be the secondary host for the Himalayan dwarf mistletoe.

Due to overcutting and overgrazing, this plant is now listed as a near threatened speciesand therefore, The Himachal Pradesh State Forest Department has come up with the idea of artificially regenerating chilgoza pines in various places. Although, this has come to very littlesuccess as the seedlings performed very poor.

Uses

The pine nuts of Chilgoza are known to be edible and rich in carbohydrates and proteins. It is one of the most important cash crops in Kinnaur and Pangi valley in the HimachalPradesh district of Chamba. The indigenous people of this areas sell the seeds of the pine which fetch a very high price, approximately around 2000 to 3600 rupees per kg inthe market.

Medlar (*Mespilus germanica*)

Fig. 1.8: Medlar

Mespilus germanica L., commonly known as medlar or common medlar, of family Rosaceae, is a large decidous shrub that can reach a height of 8 meters. The wild form of the plant is usually thorny while the thorns are either absent or reduced in thecultivated forms.

Uses

It is usually eaten raw and can also be cooked along with different dishes. The fruits are available in winter and eaten when bletted.

Propagation

Medlar is usually propagated by grafting on different rootstocks like *Crataegus spp.* (hawthorn), mountain ash, quince or pear. It takes almost 6-7 years for the grafted plant tocome to full bearing stage and fruiting can continue til 20-25 years. The grafts take slower growth if it is grafted on medlar seedlings.

Soil and Climate

The plant can grow well in a variety of soil types, but it thrives in deep, fertile, well- drained soil. It should be high in organic matter and have a high water retention capacity. It can be grown in both full sun as well as partial shade.

Harvesting

It can be harvested in the fall before the fruits are fully ripe, or leave until they ripen if frost is nota threat. Fruit must be 'bletted' before it can be eaten raw. Leave them in a cool place until they turn a dark red-brown and soft.

Yield

Yield can vary on variety and age. Average fruit production can be between 30-70 kg/tree/year.

Bilberry (*Vaccinium myrtillus*)

Fig. 1.9: Bilberry

Bilberry (*Vaccinium myrtillus* L.), also known as European blueberry belongs to the familyEricaceae, is a decidous shrub having blue edible fruit, commonlycalled as whortleberry, winberry, huckleberry, blaeberry. It is native to Europe, Greenland, Northern Asia and Western United States.

Uses

Bilberries are used to treat scurvy, kidney stones and also act as a remedy for retinal problems. The leaves are used as a treatment for diabetes. The berries are rich in flavanoids which have antioxidants properties.

Soil and Climate

Bilberry requires a relatively low pH soil and can be grown in forests having medium fertility. Since the roots of the bilberry plants are shallow, a good amount of humus must be available for the better growth and development of the plant.

Bilberries grow wild in the temperate zone and hence, care should be taken while cultivation so as to avoid frost during flowering, drought and extreme wind conditions. If the temperature goes below -3°C, it may be hazardous to the developing flowers. Windbreaks canbe established in order to protect the plant from severe frost injury.

Propagation

Bilberry can be propagated by seeds as well as vegetatively by rhizomes and cutting. Seeds should be stratified in the presence of light at a temperature of 15.5°C for 12 weeks or more. The seeds starts to germinate when the temperature ranges between 15-24°C. Bilberry has an extensive branched network of rhizomes which measures about 3-6 mm in diameter andthis can be successfully use as propagating materials when seeds are not available. Cutting is also done in Bilberry but for good survival, the cuttings should first be treated with rooting hormone and the planted in peat or sand.

Varieties

Allamanda, Castelletto, Chiutin, Grill, Pra Martino, Bluecrop, patriot, Gretha, Spartan,Bluetta, Ama, 13/16 A.

Planting

Planting of bilberry plants is done in July. Generally, one year old plants obtained from cuttings are used for planting in the field.

Water Management

Excess irrigation can be deleterious to the plant. When the plant is in non-bearing stage,irrigating the field can increase the bud formation which in turn increases the yield of the crop.

Harvesting and Yield

Fruits can be harvested manually with the help of rakes or mechanically using tractors mounted or walk behind picking heads. The yield of the plant can vary from year to year and also depends on the location, varieties, climate and soil condition. The average yield of bilberryis about 100 kg fruits per hectare and can exceed 500 million kg per year.

Bird Cherry (*Prunus napaulensis*)

Fig. 1.10: Bird Cherry

Bird Cherry *(Prunus napaulensis* Serr.) is belongto family Rosaceae, native to the eastern foothills of the Himalayas, including Nepal, Myanmar and China. Itcan reach up to a height of 27m and preferably grows at 1800-3000m above sea level. In India, it is mostly cultivated in Ukhrul District and Assam, Manipur for its fruits and also wood, which is used for making furniture. Fruits are astringent in taste and are used for making fruit juice known as Un Soh-Iong in Khasi tribe of Meghalaya. The leaves of the plant are diuretic are used to treat dropsy. The plant flowers in May and the fruits areripened by August. Bird Cherry is a hermaphrodite plant with both the male and female organs in the same plants and it is usually pollinated by insects.

Soil

Bird cherry prefers a moist, well-drained loamy soil. It can grow in a variety of soil types,from light sandy to heavy clay. The plant can become chlorotic when excess lime is present in the soil. The plant is shallow rooted and will produce sucker if roots are damaged. Care should be taken to avoid infection as the plant is very susceptible to honey fungus.

Propagation

Seeds and cuttings can be used to propagate the plant. Seed should be cold stratified for 2-3 months and should be sown in a cold frame for best result. The seeds may germinate very slow, sometimes even taking 18 months. The seedlings should be prick out into individual pots when they are large enough to be handled. Softwood cuttings from strong growing plantscan be taken in spring to early summer. Layering can also be done in spring.

Paniala (*Flacourtia jangomas*)

Fig. 1.11: Paniala

Paniala *(Flacourtia jangomas*, Lour.) also known as Indian coffee plum or scramberry, is a lowland and mountain rain forest tree in the Salicaceae family. It is widely grown in Southeast and East Asia. Although its wild origin is unknown, it is thought to be in tropical Asia, preferably India. It is very common in Southern India, particularly Kerala, where it is known as lubikkaor lololikka. The plant is edible as well as medicinal, and it is sometimes harvested for its lumber. The plant is one of the primary hosts of the Queensland fruit fly (*Bactrocera tryoni*).

It is a diminutive, decidous tree or shrub with a 6-10 m height. The tree is cover with blunt woody thorn when young but becomes thorn less when they are older. Flowers are small white to whitish green in color and are fragrant while the fruits are round, pink to dark red in color, containing 7-12 seeds and are very juicy. The fruits and leaves of the scramberry are used to treat diarrhea and the dried leaves for bronchitis. Roots are considered to be effective against toothache. In the tribal areas of Western Ghats, the bark paste is used to cure many ailments as it has various antifungal and antibacterial properties.

Jalpai (*Elaeocarpus floribundus*)

Fig. 1.12: Jalpai

Indian Olive, *Elaeocarpus floribundus* is an evergreen moderate sizedtree which belongs to the family Elaeocarpaceae that grows well inthe evergreen forests of India, Indo-china, Thailand, Peninsular, Malaysia, Sumatra etc. In India, it grows well in the Northeast region, up to 1500 m elevation. The fruits of the plant are sour in taste, light green in color and botanically drupe. It can be eaten raw but people usually preferred it in pickles. The fruit also have some medicinalqualities like curing ulcers, constipation and also used as a tonic.

Bignay (*Antidesma bunius*)

Fig. 1.13: Bignay

Bignay (*Antidesma bunius* L.), commonly known as wild cherry, currant tree and mamao dong is a medium growing tree belonging to family Phyllanthaceae which produces small berries that are sour and looks like cranberry when immature but becomes tart and sweet when ripe. It is found in the rainforests and semi-evergreen tropical forests of Southeast Asia and northern Australia. They thrive in both sunny and shady locations and can be grown in a variety of soil types.

The flowers are small and without petals, and they are dioecious, with male andfemale flowers borne on separate trees. Fruits can be eaten raw when ripe and can also be usedin making jam, jelly, wine and sauce. Ripe fruits are also used medicinally for treating gastric intestinal problems and leaf extracts have antioxidant, anti-diabetic and liver-protective properties. Matured leaves are also used to treat snake bite. Bignay leaf tea is also consumed in Thailand, Singapore and the Philippines.

Bignay tree starts fruiting from April-September and produce small grape-like clusters of 20-30 fruits, each containing a single seed. When immature, berries are greenish white in color which change to deep purplish red when ripen.

The plant should be fertilized 2-3 times a year with a general purpose fertilizer during the warm months. The plants should be well protected from frosts and drying winds.

Bay berry (*Myrica* Hook.)

Myrica nagi

Fig. 1.14: *Myrica nagi*

It belongs to the Myricaceae family. The plant's fruit are eaten raw and also used to make pickles.The powdered stem bark is used to treat dysentery.

Myrica farquhariana

Fig. 1.15: *Myrica farquhariana*

It belongs to Myricaceae family. Ripe fruits are edible and can be used to make refreshing drink. The bark is used externally as rheumatism stimulant and in theproduction ofyellow dye.

Myrica rubra

Fig. 1.16: *Myrica rubra*

It belongs to the family Myricaceae. The fruits are consumed fresh as well as dried, canned and fermented alcoholic beverages.

References

Chadha, K.L. (2019). Hand Book of Horticulture. Second Edition, Volume 1, Indian Council of Agricultural Research, New Delhi.

Joshi, P. S. (2014). Temperate fruit cultivation in India. *Jaya Publishing House*, Delhi. pp- 122-128.

Ghosh, S. N. and Kumar, A. (2019). Cultivate Minor Temperate Fruits Scientifically. *Jaya Publishing House*, New Delhi. Volume 1, pp: 1-26.

Ghosh, S. N. and Kumar, A. (2019). Cultivate Minor Temperate Fruits Scientifically. Jaya Publishing House, New Delhi. Volume 2, pp: 489-575.

Ghosh, S. N. (2015). Breeding of Underutilized Fruit Crops. *Jaya Publishing House*, New Delhi. Volume 1, pp: 31-35.

Farjon, A. (2013). "*Pinus gerardiana*". IUCN Red List of Threatened Species. 2013: e.T34189A2850009.

Hawksworth, F. G. and Wiens, D. (1998). dwarf Mistletoes. DIANE Publishing, U.S.A. pp: 264-265.

Nayik, G. A. and Gull, A. (2020). Antioxidants in Fruits: Properties and Health Benefits. pp 397-416.

Boccacci, P. and Botta, R. (2009). Investigating the origin of hazelnut (*Corylus avellana* L.) cultivars using chloroplast microsatellites. *Genetic Resources of Crop Evolution*, 56: 851–859.

Thompson, M. M., Lagerstedt, H. B. and Mehlenbacher, S. A. (1996). Hazelnuts. In: Janick J, Moore JN (eds) Fruit breeding: nuts. Wiley, New York. Volume 3, pp: 125–184.